I0055349

Nanomaterials for Alcohol Fuel Cells

Edited by

**Inamuddin[1,2,3], Tauseef Ahmad Rangreez[4], Fatih Şen[5]
and Abdullah M. Asiri[1,2]**

[1]Centre of Excellence for Advanced Materials Research, King Abdulaziz University,
Jeddah 21589, Saudi Arabia

[2]Chemistry Department, Faculty of Science, King Abdulaziz University,
Jeddah 21589, Saudi Arabia

[3]Department of Applied Chemistry, Faculty of Engineering and Technology,
Aligarh Muslim University, Aligarh-202 002, India

[4]Department of Chemistry, National Institute of Technology, Srinagar, Jammu and Kashmir-
190006, India

[5]Department of Biochemistry, Dumlupinar University, Kutahya-43266, Turkey

Copyright © 2019 by the authors

Published by **Materials Research Forum LLC**
Millersville, PA 17551, USA

All rights reserved. No part of the contents of this book may be reproduced or transmitted in any form or by any means without the written permission of the publisher.

Published as part of the book series
Materials Research Foundations
Volume 49 (2019)
ISSN 2471-8890 (Print)
ISSN 2471-8904 (Online)

Print ISBN 978-1-64490-018-5
eBook ISBN 978-1-64490-019-2

This book contains information obtained from authentic and highly regarded sources. Reasonable efforts have been made to publish reliable data and information, but the author and publisher cannot assume responsibility for the validity of all materials or the consequences of their use. The authors and publishers have attempted to trace the copyright holders of all material reproduced in this publication and apologize to copyright holders if permission to publish in this form has not been obtained. If any copyright material has not been acknowledged please write and let us know so we may rectify this in any future reprints.

Distributed worldwide by

Materials Research Forum LLC
105 Springdale Lane
Millersville, PA 17551
USA
http://www.mrforum.com

Manufactured in the United States of America
10 9 8 7 6 5 4 3 2 1

Table of Contents

Preface

Today, one of the most important problems for countries for economic and social development is to provide a cheap, clean, reliable and easily obtainable energy source. For this purpose, intensive research is being carried out to develop new and renewable energy sources and efficient and inexpensive energy production/use technologies. Especially the increasing population, urbanization and industrialization have caused the need for energy to be met with fossil fuels for many years. As a result of the intense burning of fossil fuels, greenhouse gases in the atmosphere, especially CO_2, and consequently the warming of our world, are defined as global warming due to the greenhouse effect. Due to increasing CO_2 concentration in the last century, the average temperature of our world has been increased. The increase in the temperature of the earth will cause warming in places close to the earth, and cooling in the upper parts of the air sphere. In this case, it is estimated that high-pressure systems will be affected and consequently extreme climate conditions will be observed. By making some changes in the production and use of energy, it is possible to reduce the amount of these gases that cause the greenhouse effect. For this reason, some of the alternative energy devices should be used as fuel cells. Fuel cells are the systems that convert the chemical energy of fuels into electricity. The fuel cell consists of an ionic conductivity electrolyte and anode and cathode where different reactions take place. Fuel cells are electrochemical cells, such as batteries; however, there are some differences from the batteries. The most important difference between fuel cells from batteries is the need to continuously supply fuel from the outside. In other words, if the fuel is supplied from outside, the electrochemical cell that can operate continuously is called fuel cell. Given the energy systems of the future, fuel cells are a very important alternative to commercialization today. Fuel cells have many advantages, especially when used as portable energy systems. These energy systems are generally considered as an alternative to batteries and internal combustion engines. Fuel cell systems have higher energy densities than batteries, but they do not have problems such as charging or running because they can operate continuously if fuel is supplied. The biggest obstacle to the use of fuel cells in daily life is the synthesis of an active catalyst. Recent intensive research efforts have focused on the development of more efficient, stable and less costly nanomaterials as catalysts. For this purpose, this book aims to develop alcohol fuel cells and nanomaterials required for these fuel cells. We hope that the study will be useful for those interested in energy use and management. This book on "Nanomaterials for Alcohol Fuel Cells" discusses various types of catalysts, membranes and supports used in alcohol fuel cells.

We are thankful to the authors and their co-authors and publishers, copyright holders. We may need to offer our sincere regrets to any copyright holder if unexpectedly their advantage is being encroached.

Inamuddin[1,2,3], Tauseef Ahmad Rangreez[4], Fatih Şen[5] and Abdullah M. Asiri[1,2]

[1]Chemistry Department, Faculty of Science, King Abdulaziz University, Jeddah 21589, Saudi Arabia
[2]Centre of Excellence for Advanced Materials Research, King Abdulaziz University, Jeddah 21589, Saudi Arabia
[3]Department of Applied Chemistry, Faculty of Engineering and Technology, Aligarh Muslim University, Aligarh-202 002, India
[4]Department of Chemistry, National Institute of Technology, Srinagar, Jammu and Kashmir-190006, India
[5]Department of Biochemistry, Dumlupinar University, Kutahya-43266, Turkey

Nanomaterials for Alcohol Fuel Cells   Materials Research Forum LLC
Materials Research Foundations **49** (2019) 1-78   doi: https://doi.org/10.21741/9781644900192-1

Chapter 1

Carbon-Based Nanomaterials for Alcohol Oxidation

M.Selim Çögenli[1], A.Bayrakçeken Yurtcan[1,2]*

[1]Graduate School of Science, Department of Nanoscience and Nanoengineering, Ataturk University, Erzurum 25240, Turkey

[2]Faculty of Engineering, Department of Chemical Engineering, Ataturk University, Erzurum 25240, Turkey

ayse.bayrakceken@gmail.com

Abstract

Carbon is a unique material for utilization in direct alcohol fuel cells as catalyst support material due to its outstanding properties including high surface area and high electrical conductivity, which are the most crucial properties for fuel cell applications. This chapter focused on the carbon-based nanomaterials used as catalyst support material in most studied directs of alcohol fuel cells especially for direct ethanol, methanol fuel cells, and other alcohol fuel cells. The targeted carbon-based nanomaterials include carbon black, carbon nanotubes, carbon gels, graphene, mesoporous carbon, and heteroatom-doped carbon nanomaterials.

Keywords

Carbon, Direct Alcohol Fuel Cells, Ethanol Oxidation, Methanol Oxidation, Nanomaterials

Contents

1.   Introduction

Carbon is a crucial element for over 95% of all known chemical compounds [1] and also very crucial for our lives. Carbon at different forms has a very wide range of application fields from energy to rubber industries. What makes carbon so unique in material science? Carbon in its ground state has an electronic configuration of $1s^2 2s^2 2p^2$ which enables it to form sp^3, sp^2 and sp hybrid bonds due to hybridization and various chemical bonds which brings carbon unique properties. In its ground state, carbon has two unpaired electrons in the outer shell which means that it can form 2 bonds, but in reality, this is not the case. Binding of molecular orbitals occurs with the binding of the same kind of orbitals, e.g. two s or two p orbitals bind together in an antibonding and bonding way. In carbon, due to the hybridization, a binding ability for four electrons are observed. In hybridization, atomic orbitals are mixed in a way that the direction of the chemical bonds is changed and the total energy of the molecule is decreased. This phenomenon resulted in new hybrid orbitals, which affect the bonding properties. In carbon, hybrid mixtures of s and p orbitals are formed by overlapping. These new orbitals showed different characteristics depending on the combination of one s orbital with different numbers of p orbitals[2].

In sp^3 hybridization, one s orbital is combined with three p orbitals (p_x, p_y, p_z) resulted in a tetrahedral arrangement with an angle of $109.5°$. The energy and shape of these hybrid orbitals are identical. In this hybridization, carbon is bonded to the other four atoms with no unpaired electrons. Covalent bonds in diamond, known as one of the hardest materials, formed due to sp^3 hybridization[2].

In sp^2 hybridization, one s orbital is combined with two p orbitals (p_x and p_y) resulted in a flat trigonal or triangular arrangement with an angle of $120°$ and forming σ bond between hybrid orbitals. Other orbital p_z forms a π bond, which is perpendicular to sp^2 hybrid orbitals. In this hybridization, carbon is bonded to the other three atoms with two single and one double bond. A typical example for sp^2 hybridization is graphite, which is comprised of parallel carbon layers. sp^2 hybrid orbitals have strong bonds whereas π bonds between the layers are weak Van der Waals forces which makes graphite as one of the softest material[2].

In sp hybridization, one s orbital is combined with one p orbital (p_x) resulted in a linear arrangement with an angle of $180°$ and carbon is bound to two carbon atoms resulting either two double bonds or one single plus one triple bond. Carbynes have sp hybridization. Carbon can be classified according to its hybridization type and its crystallinity. Diamond, graphite, carbines, amorphous and nano forms are some commonly used classifications of carbon[2].

There are many types of fuel cells. Proton exchange membrane fuel cell (PEMFC) that utilizes direct feeding of hydrogen is the most commonly used type of fuel cells in order to obtain electrical energy. Instead of using direct hydrogen, different types of liquid fuels such as methanol, ethanol, propanol and ethylene glycol can be used for liquid-feed fuel cells. Alcohol fuel cells are very interesting fuels due to a lot of advantages such as relatively high reactivity, ease of storage and supply, low toxicity (except for methanol). They can be directly electro-oxidized in a direct alcohol fuel cell (DAFC) [3]. A typical alcohol fuel cell comprises an anode and a cathode electrode which are physically separated by a membrane. While the alcohol is fed to the anode, oxidant (air or O_2) is fed to the cathode. The two electrochemical reactions occurring at the anode and cathode electrode side are combined to form an overall reaction. Direct alcohol fuel cells produce electricity directly from the electro-oxidation of alcohols. The oxidation reaction occurring through many successive and parallel paths involving many adsorbed intermediates and by-products. The oxidation reaction needs a catalyst to increase the reaction rate and to modify the reaction pathway in order to reach more rapidly the final step. The main ability of the catalyst is to obtain hydrogen and CO_2 by oxidizing the fuel as a result of reaction in acidic media. The use of catalyst support materials are essential for nanostructured catalysts to achieve enhanced catalytic performance and stability with significantly reduced catalyst loadings [4]. Carbon-based materials such as carbon blacks (CBs), graphene, carbon nanotubes, carbon aerogels, carbon-carbon and carbon-metal oxide hybrid structures are used as catalyst supports. Carbon-based nanomaterials are very crucial especially for proton exchange membrane (PEM) and direct alcohol fuel cells (DAFCs) which operate at low-temperature conditions. In low-temperature fuel cells, the most critical component is the membrane electrode assembly (MEA) which is comprised of the electrodes that the half cell reactions occur and also the electrolyte that allows the ion transfer. Mostly in these types of fuel cells, the electrode materials are comprised of carbon supported metal nanoparticles. There are lots of different supported catalyst preparation techniques all of which are targeted to disperse the metal nanoparticles over the support material homogeneously and provide high electrical conductivity. Support materials used have to provide some properties in order to get promising fuel cell performances. These properties can be summarized as follows [5]:

- High surface area
- High electrical conductivity
- Good support and nanoparticle interaction
- Proper porosity
- Corrosion resistance
- Inertness and pureness
- Promising thermal resistance and conductivity
- Able to operate with either anion or cation exchange membranes
- Able to adsorb gas or liquid

The electrochemical reactions occur in the interface of the electrodes and the electrolyte. The electrochemical activity of the catalysts is directly proportional to the availability of the active sites in the catalyst and the activity of these active sites. Since the support material affects the dispersion of the catalysts, the activity of the synthesized catalyst also varies with the support material. Depending on the full range of change in the properties of carbon-based nanomaterials, it is clear that carbon nanomaterials are promising support materials for fuel cell catalyst applications. In this chapter, it is aimed to summarize the commonly used carbon-based nanomaterials used for alcohol-based fuel cells including the relevant fuel cell literature.

The electro-oxidation of alcohols is a complex reaction involving multi-electron transfers and several intermediate steps. The reaction mechanisms for alcohols are involving several different adsorbed intermediates and numerous products and by-products. It is more difficult to elucidate exactly the mechanism of alcohol electro-oxidation because of the fact that alcohols are electrochemically oxidized through different pathways on different catalytic surfaces or in different media. In a DAFC the total electro-oxidation to CO_2 of an aliphatic oxygenated compound C_xH_yO containing one oxygen atom (mono-alcohols, aldehydes, ketones, ethers, etc.) involves the participation of water (H_2O) or of its adsorbed residue (OH_{ads}) [3]. In alcohols, there are three kinds of bonds: carbon bonded to hydrogen (C-H bond) which breaks most easily, carbon bonded to carbon (C-C bond) which breaks more difficult, and carbon bonded to oxygen (C-O bond) which do not break, even on a catalyst like platinum [6]. The most basic alcohol to understand the oxidation of alcohols is methanol. Methanol, being the simplest alcohol with no C−C bond, has been studied widely and has been shown to be able to completely oxidize to CO_2 giving $6e^-$, but can also undergo multistep electron transfer reactions resulting in several reaction products and intermediates such as formaldehyde and formic acid [7]. Higher alcohols which contain the C−C bonds are more difficult to oxidize as they form various adsorbed and intermediate species instead of completely oxidizing to CO_2.

The oxidation of small organic molecules, such as methanol and ethanol, in DAFC, is a very active field [8, 9]. Nevertheless, other liquid fuels such as ethylene glycol, glycerol, and to a less extent 2-propanol and allyl alcohol have emerged as feasible alternatives for fuel cells [10-15]. DAFCs are mostly used acidic proton exchange membrane as the electrolyte. CO_2 generated in the anodic reaction can be easily removed with strong acidic electrolyte membrane [16]. A significant contribution to the relatively low DAFC performance is from kinetic constraints in the alcohol oxidation reaction in acidic media. In addition, the ionic current in alkali fuel cells is due to conduction of hydroxide ions and the reverse direction to that in proton conducting systems [17]. Because of the complex oxidation of alcohols and slow reaction kinetics, the strongly binding reaction intermediates play a vital role [18]. The most common technique for investigating the catalytic activity in both alkaline and acidic media of catalysts for alcohol oxidation is cyclic voltammetry (CV). In the CV measurements, the alkaline experiments are usually carried out with NaOH or KOH, whereas in acidic experiments H_2SO_4 or $HClO_4$ are used as the electrolytes. The development of new anode and cathode catalyst systems is more likely in alkaline media because of the full range of options for the materials support and catalyst, as compared to acidic media [19]. Furthermore, using alkali electrolytes allows for a greater possibility for application of non-noble and less expensive metal catalysts [16]. Some studies can provide a discussion of the literature concerning the extensive

Materials Research Forum LLC
doi: https://doi.org/10.21741/9781644900192-1

research carried out in acidic and alkaline electrolytes for alcohol oxidation [20, 9, 18] Theoretical energy potentials and types of electrolyte for DAFCs are listed in Table 1 [21]. Table 2 shows the different carbon support structures used in alcohol oxidation which are considered in this book section. Poor reaction rates and the difficulty in breaking the C-C bond of alcohols leads to the need for high over-potentials, which decreases the DAFC efficiency [22]. CV measurements are taken to determine the oxidation of alcohols. There are two well-defined peaks at the forward and backward scans were obtained for the catalysts (except some metals), I_f (the right peak) and I_b (the left peak), which were produced by the forward and backward sweeps, respectively. The graphs obtained from CV measurements represent the electro-oxidation of alcohols. Depending on the catalysts used, different intermediates are formed, and different reaction mechanisms are realized [23-27]. The higher current value ratio of (I_f/I_b) shows a richer catalytic performance. This rate can be used to define the catalyst tolerance to carbonaceous species (like CO) adsorbed to the catalyst [28] which means a better activity for alcohol oxidation [29, 30]. Also, both the onset potential of the oxidation peak shifts to lower values, reinforcing the role of water in the reaction rate and current of I_f which could present the performance of the catalysts [31, 32]. Higher currents and more negative onset potentials (i.e., higher activities) are observed in the alkaline medium than in sulfuric acid electrolyte [33, 34]. In addition to the use of different electrolytes, the CV operating temperature is a significant parameter that affects the alcohol oxidation. Carbon-supported catalysts were also used to investigate the temperature effect on alcohol oxidation [35].

In addition to CV measurements, chronoamperometric (CA) and electrochemical impedance spectroscopy (EIS) measurements are taken to obtain information about the catalyst poisonings and resistance. For all of the catalysts, the CA measurements applied at a fixed potential value were used to estimate the stability of the catalysts, the current densities decayed in the initial stage and then reached to relatively stable values, because of the catalyst poisoning by CO-like intermediates [28, 36, 37]. The diameter of the primary semicircle taken from EIS measurements is always used to measure the charge transfer resistance of the catalyst [38].

Table 1. Theoretical energy potentials and types of electrolyte for DAFCs.

Fuel	Formula	Type	E^0 (V)	Number of Electrons
Hydrogen (gas)	H_2		1.23	2
Methanol	CH_3OH		1.21	6
Ethanol	CH_3CH_2OH	Alcohol	1.15	12
Ethylene Glycol	$(CH_2OH)_2$		1.22	10
1-Propanol	$CH_3CH_2CH_2OH$		1.13	18
2-Propanol	$CH_3CHOHCH_3$		1.12	18

Table 2. The different carbon structures for alcohol oxidation.

Carbon Based Materials	Type
Carbon Blacks	Vulcan XC-72
	Black Pearls
	Ketjenblack
	Acetylene black
Carbon Nanotubes (CNTs)	Single-wall carbon nanotubes (SWCNTs)
	Multi-wall carbon nanotubes (MWCNTs)
	Functionalized CNTs
Carbon Gels	Carbon xerogel
	Carbon aerogel
	Graphene hydrogel
	Graphene aerogel
Graphene	Graphene oxide(GO)
	Reduce graphene oxide(rGO)
	Graphene nanoribbons
Mesoporous Carbons	Ordered mesoporous carbons
	Mesoporous graphitic carbons
Heteroatom Doped Carbon	N doped carbon
	N doped mesoporous carbon
	N-doped graphene aerogel
	N-doped carbon nanotubes
	Boron and nitrogen doped graphene aerogel
	Phosphorus doped carbon
Hybrid Supports	Carbon based-Carbon based hybrids
	Carbon based-metal oxide hybrids

2.   Carbon blacks

Carbon blacks are the most commonly used support materials for direct alcohol fuel cells, which provide high surface area, and appropriate electrical conductivity for fuel cell catalysts. Carbon black is comprised of spheroidal particles with a pronounced ordering of the carbon layers (graphene layers) wrapped around a very disordered nucleus with a preferential orientation parallel to the particle surface [39]. Carbon black global manufacturing is made by using two common processes: oil-furnace process [40] or thermal process. The dominant process is an oil-furnace process, which includes the partial combustion via cracking and polymerization of hydrocarbon molecules followed by the dehydrogenation of the polymer species to spherical fine carbon particles [41, 42]. Denka Black is acetylene black, a carbon black, obtained from thermal decomposition of acetylene. The characteristics of commonly used carbon blacks are reported in the literature [5, 43]. All these types of carbon blacks provide a homogeneous dispersion of the metallic nanoparticles. Fig. 1 schematically shows the structure of carbon black and metal nanoparticle doped carbon black catalyst.

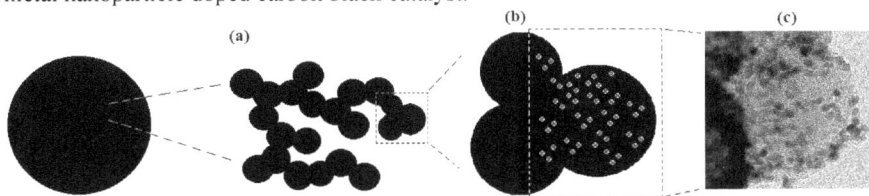

Fig. 1. Structures of carbon black (a) metal-doped carbon black (b) and TEM image of metal-doped carbon black (c).

Different carbon materials with high electrical conductivity have been studied as electrocatalyst support material for alcohol oxidation reactions [44]. The various applications such as alcohol oxidation, durability, impedance, and cell performance of Pt-based and Pd-based catalysts have various mixture, alloy, and core-shell structures with specific supports in DAFCs are discussed. The particle sizes and distributions significantly depend on the properties of the support materials before the preparation method and the components of the catalysts. Different carbon black support materials for the noble metal catalysts, such as Vulcan XC-72, Black Pearls, Ketjenblack and Acetylene black have been investigated for alcohol oxidation reactions. Each carbon black supported catalysts with different metals and different metal loadings compared for alcohol oxidation activities and corresponding electrochemically active surface areas (ECSAs) are summarized in Table 3.

2.1. Carbon black based nanomaterials for methanol oxidation

2.1.1 Vulcan XC72 carbon black

Carbon black like Vulcan XC-72 is the most widely used group of carbon materials as supports for methanol oxidation [28, 45-48, 44]. Carbon blacks (CBs) have the sp^2-hybridized graphite carbon atoms, which facilitate the interaction between the metallic NPs and the surface of CB [49]. Calvillo et al. [44] used different carbons such as carbon nanofibers (CNFs), carbon nanocoils (CNCs), ordered mesoporous carbons (gCMK-3) as support materials, and the results were compared with carbon black (Vulcan XC-72R, 218 m^2 g^{-1}). Support material's characteristics are showed to have significant influence on both the crystallite size, dispersion, structure of Pt nanoparticle and electrochemical properties toward alcohol oxidation of the electrocatalysts. Yang et al synthesized highly durable and carbon monoxide (CO) tolerant polymer-coated carbon black catalysts and they explained that polymer-coated carbon black catalyst (PVPA) weakened the binding energy between the Pt and CO species [50].

In bulk metals, two kinds of metal elements often provide an alloy structure. Alloy nanoparticles show different structural and physical properties than bulk samples. Alloy and core-shell types of nanoparticles supported on carbon represent one group of catalysts that have been demonstrated to exhibit synergistic properties for the MOR. Core-shell structured nanoparticles over carbon black materials are also used in order to improve the methanol oxidation. Au@Pt core-shell nanoparticles were successfully synthesized on Vulcan XC-72 carbon surface whose average diameter is 30 nm and surface area is 250 m^2/g, and the peak current density of Au@Pt/C (1:2) catalyst is about 2.5 times as large as that of Pt/C catalyst without Au core [51]. Bimetallic metal nanoparticles can also be decorated over the carbon black support material. CV measurements showed that PtSn/Vulcan XC72 showed better electrochemical catalytic activity for methanol oxidation, as it is possessed a higher electroactive surface area than Pt/Vulcan XC72 [52, 28]. PtRu nanoparticles on Vulcan XC72 prepared as catalysts, and it received more attention due to its high CO tolerance [53, 28]. In another study, PtRu/C catalysts were prepared with the reduction treatment at different temperatures [54]. The synthesized PtRu on Vulcan XC72 at 300 ^{0}C catalyst showed the highest activity for adsorbed CO oxidation and methanol electro-oxidation due to the increase of alloy degree and electrochemical surface area, as well as the formation of grain boundary among the interconnect particles. In another study, Pd/CB exhibited a much higher mass current density toward methanol oxidation, which is almost 2.5 times as much as that of commercial Pd/C [49].

2.1.2 Black Pearl carbon black

Carbon blacks have a wide range of diversities regarding their structural properties including surface area and pore sizes. In some studies, a high surface area carbon black Black Pearls 2000 (BP-2000) and Vulcan XC-72 (XC) carbon blacks were used as carbon supported materials either for Pt or PtRu based catalysts for methanol oxidation [55, 56]. Pt particle size changes concerning different carbon black material were investigated. The average Pt particle sizes increase with increasing Pt loadings on XC support material (245 $S_{BET}/m^2 g^{-1}$); however, the average Pt particle sizes are unexpectedly independent of Pt loadings on BP-2000 support material (1562 $S_{BET}/m^2 g^{-1}$) [55]. The results offer evidence that the surface chemistry of the carbon supports plays an important role in Pt particle sizes, which affect the methanol oxidation. In another study, high activity was expected with the BP-2000 support, because of its high surface area (1485 $S_{BET}/m^2 g^{-1}$). However, the results showed that PtRu/BP-2000 have lower activity than other catalysts due to many agglomerates and no uniform particle size distribution for the PtRu/BP-2000, in addition, PtRu/XC catalyst showed a very homogenous and uniform particle distribution and PtRu/XCR also have a less homogeneous distribution [56].

2.1.3 Ketjen black

Ketjenblack is also a unique electro-conductive type of carbon blacks. Yang et al. [57] compared two electrocatalysts prepared over different carbon support materials (Vulcan XC-72R and Ketjenblack), and the Ketjenblack (KB) support material showed higher CO tolerance and long-term durability for methanol oxidation. In addition, the BET results showed that the specific surface areas of the KB and XC72R were determined to be 1232 and 235 m^2/g, respectively. In another study, Pt-Ru nanoparticles were deposited on KB (800m^2/g) in order to obtain the catalyst, and PtRu/KB catalyst showed higher DMFC performance, but lower catalytic activity toward methanol oxidation than mesoporous carbon (1693 m^2/g) supported PtRu catalyst [58]. A composite catalyst consisting of N-doped graphitic carbon and KB supported Pt nanoparticles were synthesized, and the composite support material exhibited excellent electrocatalytic properties than single KB for methanol oxidation [59]. In addition to, though the specific surface area of composite support material (N-doped graphitic carbon/KB) was lower than XC-72, it was still high enough and showed a nice electrocatalytic performance.

2.1.4 Acetylene black

Zhang et al. [60] suggested the addition of acetylene black as a support material makes it possible to prepare electrocatalysts with high electrocatalytic performance and stability for methanol oxidation by using ND material as support with relatively low or non-electrical conductivity. Typical sp^2-bonded carbon materials provide transferring electrons during the reaction process, sp^3-bonded nanodiamond (ND) possesses high oxidation resistance with low conductivity [60]. The Acetylene black (90 m^2 g^{-1}) supported RuNi catalyst showed some positive specific features in the process of electrocatalytic oxidation of methanol and other low-molecular-weight alcohols [61]. Also, Galiyev et al. [62] used to check the effect of support materials with various types of carbon blacks (more oxidized Vulcan XC72R (250 m^2 g^{-1}) and more hydrophobic acetylene black AC-1 (150 m^2 g^{-1}) for specific steady-state activity towards methanol oxidation. AC-1 supported Pt catalyst showed better steady-state polarization curve for methanol oxidation than XC72 supported Pt catalyst.

2.1.5 Carbon black and metal oxide hybrids

In recent years, it is found that certain metal oxides, such as CeO_2 [63], RuO_2[64, 32], SnO_2 [65, 66], Fe_3O_4[67] and TiO_2[68] can enhance the catalytic activity for methanol electro-oxidation through synergetic interaction with metals such as Pt, Pd or Ru. The conductivity of the support material, which can be provided by carbon hybrid structures, is important for alcohol oxidation. The increased catalytic activity can be ascribed to the increased conductivity of the support material [67]. The PtRu/C-CeO$_2$ electrocatalyst exhibited higher performance for methanol oxidation at room temperature in comparison with PtRu/C-Er$_2$O$_3$, PtRu/C-CeO$_2$, PtRu/C-La$_2$O$_3$, and PtRu/C-Nd$_2$O$_3$, this situation could be explained by its remarkable oxygen storage capability and catalytic properties for the removal of adsorbed CO [69]. In another study, Yousaf et al. [70] have developed a highly active and durable electrocatalyst for methanol oxidation comprising Pt$_3$Pd$_1$−CeO$_2$/C, synthesized through a facile surfactant-free method. The alloying of PtSn$_2$ and existence of SnO$_2$/Vulcan XC72 improved the electrical properties of methanol oxidation and enhanced the resistance ability to CO oxidation [71]. SnO$_2$-modified PtRu/C catalyst prepared by the deposition of PtRu before SnO$_2$ gave better catalytic activity for CO$_{ads}$ and methanol electro-oxidation compared to PtRu/C [72]. This situation was explained with CO does not adsorb on the surface of SnO$_2$ particles and OH$_{ads}$ was formed on SnO$_2$ sites, so the removal of CO$_{ads}$ on Pt or Ru sites proceeded via its reaction with OH$_{ads}$ on SnO$_2$ sites.

2.2. Carbon black based nanomaterials for ethanol oxidation

2.2.1 Vulcan XC72 carbon black

Chu et al. [73] have used a mixture of carbon black which has a high surface area and nanographite which has a high electrical conductivity as support material for Pt catalyst, and they showed that the mixed support enhanced the electrocatalytic performance of the Pt catalyst for ethanol oxidation reaction (EOR). Because of the superior properties of the palladium (Pd), it is another noble metal used for alcohol oxidation [74]. Cyclic voltammogram studies indicated that ethanol oxidation was favored on carbon-supported Pd particles [45, 75]. The as-prepared highly dispersed Pd nanoparticles (NPs) with an average size of 2.7 nm on commercial carbon black (CB, Vulcan XC72) catalyst showed significantly high catalytic activity than the commercial Pd/C catalyst toward electro-oxidation of alcohols in alkaline medium [49]. In another study, the results suggested that Nb modified Pd nanoparticles were supported on the XC72 as an electrocatalyst to obtain a reaction path with high selectivity, only a single determining step and low production of the intermediates for ethanol oxidation reaction [76]. In addition, synthesized electrocatalysts became more hydrophobic than Vulcan XC72.

Bimetallic PtSn on carbon catalysts are known to be active catalysts for EOR in acidic media; However, Jiang et al. [34] showed that the EOR currents in the alkaline solutions were much higher than those in the acid solutions. CVs for ethanol oxidation showed that the mass activity on PtSn/Vulcan XC72 electrode reaches 259.0 mA mg^{-1} Pt in acidic media [77]. CVs provide insights about the CO poisoning tolerance of Pt-Sn supported on carbon nanofibers (CNFs), Vulcan and oxidized Vulcan catalysts, which is an important point to be considered for their use in direct alcohol fuel cell anodes, as CO is the main poisoning intermediate formed during alcohol oxidation [33]. The results showed that the mesoporous area of CNFs is much higher than that of Vulcan, facilitating the mass transfer of the ethanol molecule from the active sites of the catalysts. In addition, oxidized Vulcan showed higher activity than non-oxidized Vulcan. Binary PtIr, PtSn and ternary PtSnIr electrocatalysts on Vulcan XC72 were also used for EOR because of the presence of oxygenated species (OH_{ads}) in large amounts is necessary for the complete oxidation of poisoning intermediates such as CO [78]. This behavior could be explained by a better dispersion of metals on carbon and/or the formation of ultrafine particles.

2.2.2 Acetylene black

The oxidation current, which is observed at a potential of 0.3 V, is equal to 5.48A per gram of the 15 wt % RuNi/acetylene black (68:32 at. %) catalyst for ethanol [61]. The activity of this catalyst at the same conditions is lower than the activity of the platinum onto carbon black XC72.

2.2.3 Ketjenblack

The Pt/SnO$_2$(1:1)/ Ketjen Black (KB) electrode, both had less positive onset potential also had higher specific activity than the Pt/KB electrode for ethanol oxidation due to the synergistic effect of Pt and SnO$_2$ on KB [79].

2.2.4 Carbon black and metal oxide hybrids

The effects of combining noble or non-noble metal particles acting as catalysts and oxides as support materials were also investigated [80]. Pt nanoparticles (20 wt%) on the hybrid supports were prepared by mixing suspensions of carbon powder (Vulcan XC-72, Cabot Corp.) with 80 wt% and commercial nanopowders of transition metal oxides (ZrO$_2$, SnO$_2$, MoO$_3$, WO$_3$, CeO$_2$ and TiO$_2$) with a weight percentage of 20wt.%[81]. These materials promote changes in the electronic and catalytic properties that produced the lowest and highest EOR currents (Pt/C-CeO$_2$ and Pt/C-TiO$_2$). PtSn(50:50)/CeO$_2$–C electrocatalyst showed a significant increase of performance for ethanol oxidation compared to PtSn/C catalyst [82]. The forward peak current, I$_f$, which could present the performance of the catalysts, for Pt@PdSn–SnO$_2$/C catalyst (336 mA mg^{-1}) is significantly higher than those for Pt@PdSn/C (298 mA mg^{-1}) catalyst [83].

2.3   Carbon black based nanomaterials for other alcohols oxidation

2.3.1 Ethylene glycol oxidation

Ethylene glycol has a low cost because it can be produced by heterogeneous hydrogenation of cellulose [84]. Among different types of alcohols for DAFCs, ethylene glycol has attracted significant attention because of its high energy density and low toxicity. Ethylene glycol which is higher alcohol has the problem of breaking the C–C bond which is being the determinant factor for the yield of CO$_2$[85]. Pt- and Pd-based nanocatalysts are commonly regarded as the most effective catalysts toward ethylene glycol oxidation [86], however difficult to oxidize on platinum or platinum alloys [8], in usually alkaline medium [87, 88], rather than in acidic medium [89]. Pd/CB exhibited a much higher mass current density toward EG oxidation, which is almost 2.2 times as much as that of commercial Pd/C [49]. There are some disadvantages still needed to be improved for Pd based catalysts, such as the comparatively poor durability for alcohol oxidation. Maumau et al. [45] show that the incorporation of Pd on Au using Vulcan XC72 enhances the onset potential for the oxidation of ethylene glycol and the highest stability is observed for ethylene glycol, followed by glycerol and then ethanol oxidation. The voltammetric results show that current–potential curves are shifted by about 0.2 V towards more negative potentials through the promoting effect of Ru to Pt on Vulcan XC-72 [85].

Tarasevich et al. [61] investigated the RuNi catalysts on acetylene black, which are designed for the electrooxidation of methanol and other low-molecular-weight alcohols in alkaline solutions. The results showed that there is a considerable growth of the rate of the reaction of ethylene glycol molecule has two hydrogens which can participate in hydrogen bonding oxidation as compared with methanol. Li et al. [90] showed that Pd and Bi nanoparticles are uniformly dispersed on the surface of Vulcan XC72-doped TiO_2 hollow spheres, the Bi addition into Pd/TiO_2HS-C catalyst can remarkably enhance the electrocatalytic activity for EG oxidation.

2.3.2 1- or 2- Propanol Oxidation

1-propanol and 2-propanol were investigated as the fuels for DAFCs because they have the smallest secondary alcohol and less toxicity than methanol [91]. Many years ago, the electrochemical behavior of n-propanol on Pt in acid solutions was investigated by Pastor et al. [92], and an increase in the quantity of adsorbed CO was observed. Therefore, there is more interest in alkaline media for n-propanol oxidation on Pd based catalysts [93]. Liu et al. [91] showed that Pt has a lower activity than Pd for 1-propanol and 2-propanol electro-oxidation in alkali medium.

Pd and PdAu catalysts supported on carbon black (Vulcan XC-72R)were synthesized as electrocatalysts for 2-propanol oxidation in alkaline media [94]. This study shows that synthesized PdAu (4:1)/C catalyst has higher catalytic activity in alkaline media for the 2-propanol electrooxidation than both Pd/C and commercial Pt/C (ETEK) electrocatalysts. In another study, Pd/CB is much more active than commercial Pd/C catalyst for electro-oxidation of n-propanol and isopropanol [49]. In comparison with the n-propanol, Pd/CB possesses a much higher mass current value for electro-oxidation of isopropanol in alkaline medium.

3. Carbon nanotubes

There are two different classes of CNTs: single-wall carbon nanotubes (SWCNTs) and multi-wall carbon nanotubes (MWCNTs). The work functions of SWCNTs are very close to those of graphene in the armchair conformation [95]. Some works present a detailed comparison between MWCNTs and SWCNTs in an effort to understand which can be the better candidate of a future supporting carbon material for electrocatalyst in direct alcohol fuel cells [96, 97]. The physicochemical properties of CNTs were greatly changed after oxidation by different acidic solutions. These modifications include the increase in surface functional groups and surface basic sites, which enhance the adsorption capacity of alcohols. Carbon nanotubes (CNTs) are suitable for many different applications from energy to electronics, catalysis, and reinforcement of the composites due to their superior properties. A CNT is a tubular form of graphene layer wrapped into a seamless tube having diameter of nanometer order but length in micrometers in which the number of tubes defines whether it is a single walled or multi-walled CNT [98]. The length and direction of the rolling vector determine the chiral vector which defines different arrangements depending on the carrier network [99]. These arrangements are an armchair, zigzag or chiral lattice. Fig 2 shows the schematic structures of graphene sheet (a) rolled to form three chiralities of nanotubes (b) armchair, (c) zigzag and (d) chiral lattice nanotubes and (e) multi-walled carbon nanotube.

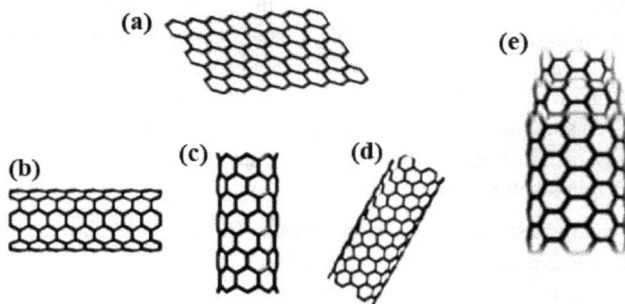

Fig. 2. Schematic structure: (a) graphene, (b) Armchair, (c) Zigzag, (d) Chiral lattice arrangements and (e) multi-walled carbon nanotube.

In comparison with the more widely used Vulcan XC-72R carbon black support, CNTs have significantly higher electronic conductivities and present higher mesoporous volumes for comparable or higher surface area materials [5]. The electrical conductivity of CNT (149.2 S cm^{-1}) is better than that of commercial carbon support XC-72 (29.2 S cm^{-1}) [100]. Lehman et al. [101] studied the characteristics of multiwall carbon nanotubes. MWCNTS are functionalized before loading the metal nanoparticles. Main functionalization methods of MWCNTs include a covalent modification which is achieved by chemical binding of molecules on the MWCNTs or non-covalent modification which is carried out by physical adsorption of molecules on MWCNTs [102]. Electrocatalytic activity measurements indicated that the supported oxidized multi-walled carbon nanotubes were the most active supports for methanol>ethanol>ethylene glycol oxidation [103]. The effects of carbon nanotube supported catalysts on electrocatalytic activities of different alcohol oxidation reactions are summarized in Table 4.

3.1 Carbon nanotubes based nanomaterials for methanol oxidation

MWCNTs were investigated because of their high mechanical strength, high electrical conductivity and used as an ideal support material for loading noble and non-noble metal nanoparticles toward methanol oxidation [104, 105]. CNTs were demonstrated to be more efficient support, which showed much higher specific activity than Vulcan XC72 due to CNTs having a large surface area and high conductivity [106]. The fabricated doubly polymer-coated carbon nanotubes supported electrocatalyst showed 1.5 times higher CO tolerance compared to the commercial carbon black supported catalyst due to the phosphorus in PVPA weakened the binding energy between the Pt and CO species [107].

Functionalization of carbon nanotubes strongly affects the metal-support interactions of the catalysts. Multi-walled carbon nanotubes (MWNTs) were functionalized with -COOH groups [104] and -OH groups [108] enhanced the metal loading with good uniform distribution which is used as support materials for methanol oxidation. Raman measurements confirmed that the nanotubes have still tubular graphitic structure after chemical functionalization without showing any significant damage [109]. The effect of oxidation on the structural integrity of CNTs through nitric acid, hydrochloric acid, sulfuric acid/hydrogen peroxide and ammonium hydroxide/hydrogen peroxide agents were studied, and nitric acid (HNO_3) treatment supports the destruction of better graphitic integrity and the subsequent formation of small graphitic fragments [110]. Chitosan and Nafion were used as dispersing agents for the preparation of catalytic inks. Thus, Pd–MWCNT catalysts prepared with chitosan as a dispersion medium increased the electrocatalytic activity toward oxidation due to suppressing methanol crossover [104]. Yang et al. [108] deposited the Pt nanoparticles on the MWNTs after wrapping by the dihydroxy-polybenzimidazole (2OH-PBI). They showed that the -OH groups were accelerated the formation of the $Pt(OH)_{ads}$, which enhances the CO tolerance of the catalyst.

To overcome the degradation of catalysts due to carbon corrosion from the catalyst supports, metal oxide and CNT composite supports are important in catalyst preparation. The presence of metal oxide in the hybrid support for catalysts decreases the onset potential of alcohol electrooxidation and increases the rate of the oxidation because the OH_{ad} forms on some metal oxide surfaces and helps to remove the CO intermediate product on the catalyst. This is called a bifunctional mechanism, as in alloy metals. Carbon nanotubes (CNTs) and CeO_2 [111], RuO_2 [112], SnO_2 [113], carbon [100] or rGO [114] composite support materials have been used toward methanol oxidation through the bifunctional mechanism. Peng et al. [115] showed that the forward anodic peak current (I_f) in cyclic voltammogram of methanol oxidation by as prepared PtRu/CNTs catalyst to have a little higher value than that of the commercial PtRu/Vulcan catalyst, in addition Pt/RuO2/CNTs catalyst has much larger forward anodic peak current than the other catalysts. Another study was aimed to investigate the interaction between CeO_2-CNTs hybrid support and Pt nanoparticles for methanol oxidation. The results showed better activity and durability than Pt/CNTs because of the stronger Van Der Waals interaction between Pt and CeO_2 and the more contact surface between Pt sites and CeO_2 available for the removal of the intermediary CO on the Pt surface [116]. Kakati et al. prepared a CNT/SnO_2 composite supported PtRu catalyst that exhibited the presence of SnO_2 layer over CNT can further improve the electrocatalytic activity of PtRu alloy nanoparticles for methanol oxidation [117].

Tsai et al. [118] prepared highly dispersed Pt and PtRu nanoparticles that were successfully deposited on the surface of CNTs directly grown on carbon cloth by pulsed potentiostatic electrodeposition and reported a high catalytic activity for methanol oxidation.

3.2   Carbon nanotubes based nanomaterials for ethanol oxidation

Researchers treated their CNTs with acid treatment for carboxylic groups functionalized multi-walled nanotubes (MWNTs-COOH) as support materials. The electrochemical performance on EOR was hindered by the long term acid-treated CNT–Pd due to sidewall damage of CNT [38].

The electrochemical performance toward EOR demonstrated that the Pd/MWNTs-2:1 exhibited the highest specific Pd activity and tolerance stability [119]. Wen et al. observed that the current density of forwarding anodic peak (i_f) of Pd/CNTs (2824 mA mg^{-1}) is 1.52 times that of the commercial Pd/C (1858 mA mg^{-1}) in alkaline media. In addition to the current density of catalysts decreases with an increase in the number of cycles, and the activity decay of Pd/CNTs catalyst is steadier than that of Pd/C [120]. They also showed that uniform-sized Pd nanoparticles distributed on the surface of the CNTs, without any obvious aggregation occurring compared to commercial Pd/C.

Nanostructured Pt and Pt–Ru catalysts were successfully electrodeposited on oxidized multi-walled carbon nanotubes (MWCNTs), and electrocatalytic activity measurements indicated that the bimetallic catalyst was the most active electrode for alcohol oxidation [103]. They also explained that the –OH groups in MWCNTs could also be formed on

Materials Research Forum LLC
doi: https://doi.org/10.21741/9781644900192-1

atoms of Pt, improving the activity of the catalyst in the alcohols oxidation reaction. Pt and Pt-Sn nanostructured materials of different compositions were prepared on oxidized carbon nanotubes, and cyclic voltammetry measurements at room temperature showed that the bimetallic catalyst with 40 at. % Sn exhibited the highest activity for ethanol oxidation [121]. This behavior could be explained considering the synergistic effect between the facilitation of alcohol oxidation via oxygen-containing species adsorbed on Sn atoms, the presence of Sn weakening CO adsorption on the Pt surface, thus reducing catalyst poisoning. Trimetallic nanocatalysts were supported on MWCNTs and the Pt–Sn–Ir/MWCNTs exhibited both a better catalytic activity for ethanol oxidation and higher ECSA from among all the catalysts prepared [122]. In another study, relatively high dispersive PdSnNi nanoparticles loaded on the surface of functionalized MWCNTs which have a diameter of about 20 nm [123]. These catalysts have reduced the cost compared to pure platinum and palladium based electrodes.

Some results demonstrated that the addition of oxide structures such as ZrO_2 and TiO_2 to CNTs significantly increased the catalytic activity of Pt for ethanol oxidation [124, 125]. Pang et al. [126] explained that the adsorption and decomposition of ethanol and its intermediate reaction products happen on Pt active sites, while the decomposition of water occurs over SnO_2 sites to form oxygen-containing species (OH_{ads}), which can conveniently react with the CO-like species produced from ethanol oxidation to free Pt active sites. Also, they explained that the current density value increased when CNTs are used due to the unique electrical properties of CNTs. In another study, the synthesized Pd/MWCNTs catalyst showed largely enhanced electrocatalytic activity and durability toward ethanol oxidation in alkaline media as compared to the commercial Pd/C catalyst. Also, graphene nanoribbons (GNRs) which synthesized by step-oxidation unzipping of MWCNTs showed the highest activity and durability towards EOR [127].

3.3 Carbon nanotubes based nanomaterials for other alcohols oxidation

3.3.1 Ethylene Glycol Oxidation

Platinum (Pt) and platinum-ruthenium (Pt–Ru) nanoparticles are deposited on surface-oxidized multi-walled carbon nanotubes (MWCNTs), and excellent activity is observed for oxidation of ethylene glycol in acidic solution because of the efficacy of carbon nanotubes acting as a good catalyst support and uniform dispersion of nanoparticles on CNT surfaces [128]. In addition to, they explained that the carbon nanotube materials were considered to have more advantages than conventional support materials such as higher conductivity, little impurities which poisons to catalysts, reactant flow is easier due to the structure also chemically stable and resistant to thermal decomposition. When compared to ethanol, ethylene glycol and methanol oxidation, the cleavage of the C–C bond is the main obstacle in ethanol and ethylene glycol electro-oxidation unlike methanol oxidation [103]. In another study, MWCNTs were acid-functionalized by acid treatment for electro-decoration of Pt/Ru nanoparticles towards oxidation of ethylene glycol (EG) [129].

The optimum Pt: Sn composition on carbon nanotubes which decorated with carboxylic acid and other oxygen-containing groups synthesized as catalyst exhibited high catalytic activity for ethylene glycol oxidation [121]. Sun et al. [130] studied to prevent agglomeration of Pd nanoparticles on sulfonated MWCNTs surfaces, and they found that functionalized sulfonated MWCNTs/Pd catalysts exhibited better long-term stability and electrocatalytic activity than the unsulfonated support material toward EG oxidation. In another study, bimetallic nanoparticles loaded on sulfonated MWCNTs which enhanced EG oxidation than single metallic catalyst [131].

3.3.2   1 or 2 Propanol oxidation

The metals on multiwalled carbon nanotubes (MWCNTs) and their electrocatalytic activity toward oxidation of 1 or 2 propanols did not report too much in the literature. MWCNTs which functionalized with sulfuric acid and nitric acid were used as support material toward 1 and 2 propanol oxidation [123]. In addition to, Pd based nanoparticles with small sizes of 3.5–3.8 nm are uniformly dispersed on MWCNTs as a catalyst, and the cyclic voltammograms of the catalysts showed that electrooxidation anodic peak current density (69.4 mA cm^2) of 1-propanol on the prepared catalysts is much larger than that of 2-propanol.

PdNi nanoparticles were deposited on MWCNT modified by physical adsorption with β-Cyclodextrin (β-CD) which played an essential role in the dispersing of nanoparticles and these catalysts used toward ethanol and propanol oxidation reactions [102]. Compared to the ethanol oxidation, the propanol oxidation presented lower current density and higher electrochemical resistance. Results showed that propanol oxidation is a more complicated process than ethanol oxidation due to two strong C–C bonds in propanol molecule; it is difficult to be broken down because of its high stability. In another study, high surface area, electrical conductivity and a decrease in the amount of impurities in the carbon of the functionalized MWCNTs lead to better dispersion of the nanoparticles which used as a catalyst toward isopropanol oxidation in acidic solution [132].

Zhang et al. [133] showed that the binary PdAg/MWCNT catalyst gave higher current density for the oxidation of ethanol and propanols than the Pd/MWCNT catalyst, suggesting that the addition of Ag can enhance the activity of Pd-based catalyst for the electro-oxidation of alcohols. They also showed that the anodic current density of the n-propanol oxidation on the forward-going scan is much higher than that of the isopropanol oxidation due to the difference of molecular conformation of these alcohols.

4.   Carbon gels

In order to obtain a highly active supported catalyst, it is necessary to control the particle size of the catalyst which is significantly affected by the utilized support material. Carbon gels including xerogels, aerogels, and cryogels have recently attracted attention as a new form of interconnected mesoporous carbon having controlled porous textures which can be substituted for the supporting materials [134]. Sol-gel processing is used in order to synthesize carbon gels through the polymerization of hydroxybenzenes and aldehydes [135]. The final carbon gel product is obtained by the carbonization of organic gels coming from a sol-gel process. The steps followed for carbon gels synthesis can be summarized as follows [135]:

a) Polymerization of resorcinol and formaldehyde in aqueous solution with a base catalyst

b) Gelation and aging for aggregation and crosslinking

c) Drying and

d) Carbonization in an inert atmosphere

The drying process can be achieved by evaporation [135], supercritical drying [136] and freeze drying [137]. The textural properties of carbon gels such as surface area, pore volume, and pore size distributions can be controlled easily. The effects of carbon gels supported catalysts on electrocatalytic activities of different alcohol oxidation reactions are summarized in Table 5.

4.1   Carbon gels based nanomaterials for methanol oxidation

Carbon-based gels can be obtained by using different polymer sources via carbonization. Carbon xerogel was obtained by carbonization of a polymer xerogel previously synthesized via sol-gel condensation of resorcinol and formaldehyde [138]. The electrocatalytic tests showed better performance of the catalysts when impregnated with this carbon xerogel having a high BET surface area ($724\,m^2/g$), and nearly 75% of mesoporosity. Carbon aerogel has a mesoporous structure which is preferred for the mass transfer and proton diffusion within the electrode [139]. Carbon aerogels have different pore structures (ranging from 28nm to 40nm) used as support materials showed higher electrocatalytic activity than commercial Vulcan support toward methanol oxidation [140]. Pt supported on carbon aerogels (Pt/CAs) electrocatalysts exhibited about 3.9 times higher anodic peak current density and more negative onset potential toward methanol oxidation than Pt/Vulcan XC72 electrocatalyst due to the improved mass transfer for the electrochemical reactions with mesoporous structure and high surface area (616 m^{2}g^{-1}) of CAs [141].

Pt nanoparticles loaded on the graphene hydrogel (G-Gel) and deposited in the micropores of nickel foam (NF) are more easily accessible to methanol molecules with the interconnected porous structure of G-Gel [142]. Pt/G-Gel/NF catalyst showed nearly 2.6 times higher electrochemical activity toward methanol oxidation when compared to the conventional Pt/rGO composite catalyst. In addition, XPS and Raman's results indicated that the oxidized areas of the GO sheets were partially removed, forming small conjugated domains. Graphene oxide aerogel (GOA) was used as support material to disperse Pt particles, and the prepared Pt/GOA showed approximately 2 times greater current density than commercial Pt/C toward methanol oxidation [143]. In addition to, The ECSA of Pt/GOA is approximately 5 times higher than Pt/rGO due to the GOA could effectively prevent the stacking of graphene nanosheets. Zhoa et al. [144] studied on three-dimensional (3D) structured Pt/graphene aerogel (GA) catalysts for methanol oxidation. They showed that the forward peak current density on Pt/GA lost about 20% of its initial value after 1000 cycles, while nearly 30% for Pt/graphene, although initial I$_f$/I$_b$ value for Pt/Graphene was higher than Pt/GA.

A metal oxide such as CeO$_2$ modified carbon aerogels (860 m^2/g) showed 3 times higher activity toward methanol oxidation compared to unmodified PtRu/C catalyst because of the contributions from Ce 4f orbitals located mainly within the valence band [145]. GA and Vulcan XC-72 carbon black hybrids designed as a support material and CV results whichwere taken before and after the stability tests showed that Pt/C catalyst lost nearly 40% of its mass activity at 1000 cycles, compared with only 16% for Pt/C/GA [146].

4.2 Carbon gels based nanomaterials for ethanol oxidation

Palladium loaded graphene aerogel (GA) deposited on nickel foam (Pd/GA/NF) for electro-oxidation of ethanol in order to prevent the NPs from agglomerating during the alcohol oxidation [147]. XPS results showed that the intensity of oxide functional groups was strongly reduced after the synthesis when compared to the C1s spectrum of GO. Compared to the amount and the type of surface oxygen functional groups, structure of the supports played an essential role in determining the catalytic performance of the prepared supported catalysts [148].

The peak current density value of the non-noble nickel/GA catalyst was much higher than Ni/MWCNTcatalyst due to the highly porous ultrafine network structure of the aerogel and to the high electrical conductivity of graphene [149].

4.3   Carbon gels based nanomaterials for other alcohols oxidation

4.3.1   Ethylene glycol oxidation

Carbon aerogel supported molybdenum and tungsten carbide nanoparticles were prepared as a support material toward EG oxidation [150]. Carbon aerogel on carbides showed higher activity due to the synergistic effect between binary carbide with aerogel and Pd when compared to XC-72R supported catalyst.

Pd nanoparticles were loaded on both 3D reduced graphene oxide (rGO) aerogel and 2D rGO nanosheets [151]. The Pd/3D rGO aerogel exhibited higher catalytic activity and stability toward ethylene glycol oxidation in alkaline media than Pd/2D rGO due to ultra-high porosity of the 3D rGO support leading to high active ECSA of the as-synthesized Pd catalyst.

5.   Graphene

Graphene is a 2D allotrope of carbon made out of carbon atoms arranged on a honeycomb structure of the hexagons [152]. The superior properties of graphene including high Young's modulus (~1,100 GPa), fracture strength (125 GPa), thermal conductivity (~5,000W $m^{-1}K^{-1}$), mobility of charge carriers (200,000 cm^2 V^{-1} s^{-1}) and specific surface area (calculated value, 2,630 m^2 g^{-1}), plus fascinating transport phenomena such as the quantum Hall effect make it a very special material for different application fields [153]. In 3D space, graphene has some defects other than intrinsic corrugations including topological defects (e.g., pentagons, heptagons, or their combination), vacancies, adatoms, edges/cracks, adsorbed impurities, and so on [154]. Graphene can be synthesized by using following methods: a) chemical vapor deposition (CVD) and epitaxial growth b) micromechanical exfoliation from patterned graphite c) epitaxial growth on electrically insulating surfaces such as SiC d) creation of colloidal suspensions[153]. Fig. 3 shows the structure of graphite, multilayer graphene, graphene oxide, reduced graphene oxide, and graphene. The effects of graphene supported catalysts on electrocatalytic activities for oxidation of alcohols are given in Table 6.

5.1   Graphene-based nanomaterials for methanol oxidation

In recent years graphene and graphene-based materials are commonly used in methanol oxidation reactions. Gao et al. [155] attempted to add carbon black (Vulcan) to the 20 wt.% Pt/Graphene during the catalyst preparation and the as-prepared catalyst exhibited high catalytic performance and stability toward methanol oxidation. They showed that the ECSA of Pt/G-C is higher than both Pt/C and Pt/G catalysts and the forward peak current density which is 1.4 and 1.7 times larger than Pt/G and Pt/C catalysts, respectively. Fig 4 shows the TEM images of Pt/G, Pt/C, and Pt/G-C catalysts.

Fig 4. TEM images of Pt/G (A), Pt/C (B) and Pt/G-C (C), taken from ref. [155]

Woo et al. [156] designed a well-arranged structure of graphene-Vulcan carbon composite to prepare highly dispersed 40 wt.% PtRu electrocatalysts. Electrocatalytic activity of the PtRu catalyst for methanol oxidation was enhanced because of adding the carbon as a nano spacer to the graphene support material. The reduced graphene oxide and carbon nanotubes with various mass ratios were used as support materials for Pt nanocomposites and the as-prepared Pt/GO: CNTs (1:2) having $114 \, m^2 \, g^{-1}$ surface area exhibited high catalytic activity and stability for MOR owing to the synergic effect of rGO and CNTs [157]. Excess GO addition to the sample usually adversely affects electron transport because plenty of oxygen-containing groups cannot be reduced completely. PtCo nanoparticles onto the sandwich-structured graphene/carbon dots/graphene (GCG) composite were synthesized, and CV results indicated that PtCo/GCG catalyst has higher ECSA, much better electrocatalytic activity and stability toward methanol oxidation when compared with PtCo/graphene [158].

In Pt-based alloy catalysts, the electrocatalytic performance is highly related to the Pt d-band center, whose position can be modulated by the synergetic interaction of electronic and geometric effects [159]. Xie et al. [160] showed that Pt/rGO electrocatalysts have better catalytic activity, stability and resistance to CO poisoning when compared to Pt/carbon black catalyst. In addition to single Pt metal-containing catalysts, Pt–Fe [161] and Pt-Cu [162] bimetallic nanoparticles supported on reduced graphene oxide powder exhibit high electrocatalytic activity, tolerance for the CO poisoning and durability for electro-oxidation of methanol compared to the Pt/rGO and commercial Pt/C catalyst. In another study, the role and use of Ni@Pt nanodisks with low Pt content supported on reduced graphene oxide catalysts were considered for the development of efficient electrocatalysts for MOR [163]. GO sheets are reduced to rGO without losing their morphology, which played an essential role in the growth of the Ni nanodisks. The results showed that the catalyst rich in Ni presents good catalytic activity toward the methanol oxidation because of the presence of Ni. Similar results showed that the Ni-Pd is favorable due to Ni mediated alloyed formation for the overall alcohol oxidation reaction [164]. PtPd bimetallic catalysts deposited on graphene are active catalysts for methanol oxidation [165-167]. The addition of Pd element to Pt element is an effective way to improve the tolerance to CO poisoning [166].

Raman spectrum is used to distinguish sp^2 and sp^3 hybridized forms and analyze the defects of the surface structure of the obtained graphene [168]. Yang et al. [165] explained that the D band corresponds to defects in the graphene nanosheets, while the G band is related to the stretching mode of crystal graphite and when PtPd metal was involved in, the value of I_D/I_G shows a distinct increase. The increase in this ratio also suggests a decrease in the average size of the sp^2 [169]. Yan et al. [167] showed that the obtained Carboxylic sodium-functionalized graphene-supported Pt-Pd bimetallic nanoparticles have much high catalytic activity and long-term stability than Pt or Pd black towards methanol oxidation. The results demonstrated that the carbon radical reaction is an effective method to modify the graphene with ignorable destructiveness. In another study, the results indicated that the GO supported PtAuPd catalyst so-called tri-metallic catalyst could improve the catalytic activities toward the methanol oxidation compared to bi-metallic catalysts [170]. Reduced graphene oxide (rGO) is used as efficient support to accommodate highly dispersed Pd based Ru, Sn or Ir nanoparticle catalysts, which were prepared using the $NaBH_4$ reduction method combined with the freeze-drying procedure [171]. CO-stripping tests revealed that the onset and peak potentials for the CO oxidation appear to decrease by the addition of Ru to Pd/rGO, indicating that the electro-oxidation of CO can take place more efficiently on the PdRu/rGO catalyst. Compared to commercial Pd/C, the as-prepared PdCu/rGO which have a molar ratio of 1:1 showed 2.49 times higher mass activity and much more stable electroactivity [172]. In addition to, BET specific surface areas of the rGO nanosheets supported nanoparticles were calculated in the range of 515.8 to 589.7 m^2 g^{-1}, which are larger than the surface area of commercial Pd/C catalyst (357.62 m^2 g^{-1}).

Combination of various metal oxides such as SiO_2 [173, 174], TiO_2 [175-177] and MnO_2[178] with the excellent electronic conductivity of different graphene structures was provided as support materials toward methanol oxidation. Yang et al. [173] showed that the rGOcoated SiO_2 spheres form an rGO-encapsulated structure. Zhang et al.

showed a schematic illustrating the synergistic effect between TiO_2 and Pd on rGO [175]. Binary nano-composites of Pd, a metal (Fe or Cu) and Mn_2O_4 on graphene nanosheets (GNS) improved the catalytic activity of the Pd/GNS catalyst towards methanol oxidation [179]. The presence of the oxide material significantly improves the formation of adsorbed OH ions/OH radicals at lower potentials, and this situation increases the CO_2 conversion from CO.

5.2   Graphene-based nanomaterials for ethanol oxidation

The activities of monodisperse Pt nanoparticles with low crystalline particle size on graphene oxide are about 16.8 times higher than commercial Pt catalyst (E-TEK) for ethanol oxidation reactions [180]. Pt-Cu bimetallic nanoparticles loaded on reduced graphene oxide (Pt-Cu/rGO) exhibited better catalytic activity for the electrocatalytic oxidation of ethanol, compared to Pt/rGO and Pt/C catalysts[162]. Electronic properties would be changed due to the d-band center shifts during the formation of Pd-Cu alloy, and the synergistic effect of the composition supported on graphene increased the electrocatalytic activity towards ethanoloxidation [181, 182]. High metal loaded (60 wt%) binary PdM (M = Ru, Sn, Ir) catalysts were synthesized on reduced graphene oxide (rGO) using the borohydride reduction method [171]. Among the prepared catalysts, the addition of Ru to Pd (PdRu/rGO) improved the electrocatalytic activity regarding larger forward peak current density and lower onset potential for ethanol. Pd nanocrystals used with Au nanoparticles on graphene showed an excellent electrocatalytic activity toward ethanol oxidation because of the Au atoms can remove intermediates such as CO of EOR [183, 184].

The prepared 3D Pd/rGO nanocomposite [185] and 3D Pd-Cd/rGO nanonetwork [186] catalysts exhibited better electrocatalytic performance than commercial Pd/C catalyst because of the synergistic effect between Pd, Cd, and rGO. The high catalytic activity, high chemical intermediate tolerance, and stability testing results demonstrated that the 3D-PtPd@Graphene catalyst obtained using copper foam could be an effective catalyst for ethanol oxidation [187]. PtPd nanocubes (NCs) were uniformly deposited on the reduced graphene oxides (rGOs) and the catalysts exhibited a high CO-tolerance and durability is also remarkable with only 8.9% loss of their ECSA after 10 000 cycles [188]. Despite the noble metals such as Pd and Pt are generally used as electrocatalysts for ethanol oxidation, Cu monolayer/electrochemically reduced graphene oxide (ML/ErGO) layered nanocomposites are grown on Au substrates have the highest electrocatalytic activity toward electrochemical oxidation of ethanol in alkaline solution [189].

The low electrical conductivity of SnO_2 materials limits their practical applications as catalyst supports, but SnO_2/graphene hybrid support for the Pt catalyst may improve the electrocatalytic activity and stability for the EOR [190, 191]. In a study, Cu, Co, and Ni were individually reduced on SnO_2/Gr electrocatalyst, followed by depositing palladium metal and Cu@Pd/SnO_2–Gr electrocatalyst showed an increased quasi-steady state oxidation current by 10.77 times about Pd/Gr catalyst [192]. Ceria (CeO_2) itself does not show any catalytic activity towards oxidation of ethanol [193], but CeO_2 promoted with graphene which is used as support material for ethanol electro-oxidation [194]. Graphene nanoribbons (GNRs), synthesized by step-oxidation unzipping of MWCNTs[195], coated with MnO_2 were used as a unique supporting material for loading and dispersing Pd nanoparticles and the synthesized catalyst displayed most negative onset potential of CO oxidation, largely enhanced electrocatalytic activity and durability toward ethanol oxidation in alkaline media as compared to the Pd-based catalysts on rGO and MWCNT [127]. Since the oxidation of CO_{ad} needs the reactant pair OH_{ad}, the expected formation of OH_{ad} on Pd at lower potentials in the presence of MnO_2/GNRs may account for the negatively shifted onset oxidation potential.

In another study, NiO nanoparticles (NPs) were deposited on functionalized MWCNTs and Pd nanoparticles were formed on NiO-MWCNTs; finally the Pd-NiO/MWCNTs nanocomposite is then deposited on rGO [196]. Results showed that the enhanced catalytic activity of Pd-NiO/MWCNTs/rGO electrocatalyst for electrooxidation of ethanol couldbe attributed to the synergistic effect between Pd & NiO active sites on MWCNTs and rGO composite support.

Pt-Pd nanoparticles loaded on PEDOT/graphene nanocomposites [197] were used as a high-performance catalyst in ethanol oxidation reaction due to the poly(3,4-ethylenedioxythiophene) (PEDOT) having a low price, good electrical conductivity, high compatibility with materials and high stability.

5.3   Graphene-based nanomaterials for other alcohols oxidation

5.3.1   Ethylene glycol oxidation

The oxidation of EG in alkaline media is much easier than acidic solutions. EG shows higher reactivity in alkaline solutions and much less catalyst poisoning by the adsorbed CO like intermediates in alkaline solutions [198]. The electrocatalytic performances for hydrogen evolution reaction (HER) and ethylene glycol oxidation of the as-prepared AuPt@Pt NCs/rGO catalyst were examined in acidic and alkaline electrolytes [199]. A facile one-step aqueous method was developed to construct core-shell AuPt@Pt nanocrystals supported on reduced graphene oxide (AuPt@Pt NCs/rGO) with poly(1-

vinyl-3-ethylimidazolium bromide) (poly(ViEtlmBr)). Results showed that AuPt@Pt NCs/rGO has about 3 times better catalytic activity towards EG in the alkaline electrolyte than in the acidic electrolyte (Table 6).

In another study, the as-prepared reduced graphene oxide (rGO) supported hollow Ag@Pt core-shell nanospheres served as a promising potential electrocatalyst in direct alcohol fuel cells due to the improved electrocatalytic activity and durability toward ethylene glycol oxidation [200]. For FT-IR spectra of the GO sample, a broad and intense absorption peak is detected at 3420 cm^{-1}, which originates from the O–H stretching vibration, however, most of the bands from the oxygen-containing groups decrease after the formation of rGO.

Alcohol oxidation dramatically decreases in electrochemical activity of the catalyst due to some poisonous intermediates, such as CO, generated and strongly adsorbed on catalyst surface [201]. Oxygen-containing species at alcohol oxidation could easily form on the surface of the CeO_2 and this species at lower potential can transform CO-like poisoning species on Pt to CO_2[202]. Platinum/graphene nanosheet (GNS) and platinum-CeO_2/graphene nanosheet were prepared as electrocatalysts for the oxidation of ethylene glycol in alkaline solutions [203, 204]. Both the GNS and addition of CeO_2 supported catalysts showed higher current densities and good stability, and the GNSs provided a highway for electron transport and a large surface area to disperse Pt and CeO_2.

5.3.2   1 or 2 propanol oxidation

The PdRu/rGO catalyst performed well with significantly improved specific activity of 38% for the electro-oxidation of 1-propanol in alkaline media, as compared to Pd/rGO at forward (I_f) peak current density value [171]. Monodisperse platinum nanoparticles supported on graphene oxide (GO) used as a catalyst for C_1 to C_3 (methanol, ethanol,and propanol) alcohol oxidation reactions and prepared catalyst showed 7.61 times higher catalytic activities than the commercially available Pt (ETEK) catalyst toward 2 propanol oxidation [180]. Furthermore, it is evident that the current density of 2- propanol oxidation is higher than from methanol and ethanol oxidation that is most probably because of the lower reaction rate of the intermediate formation for 2-propanol than that of other alcohols.

6.   Mesoporous carbons

Porous carbon black particles are the most commonly used support materials, but there are different alternatives such as mesoporous carbon (MC) which demonstrated better properties than conventional carbon blacks [205]. Conventional small pore sizes (below

10 nm) of carbon supports would not be suitable in practical fuel cell environment because the bulky ionomer could not be easily distributed in small pores of carbon support which is caused diffusion resistance [206]. For the above reason, several research groups have studied mesoporous support materials with large pores and surface area toward alcohol oxidation reactions [207, 208]. Large mesopores are beneficial to improve the electrocatalytic activity and durability toward alcohol oxidation, whereas carbon with small mesoporosity exhibited good electrocatalytic selectivity toward alcohol oxidation [209]. However, mesoporous carbon contains a small amount of oxygen surface groups, which can be disadvantageous for many applications [210]. Pore sizes of MC materials are in the range of 2-50 nm which provide high surface area and conductivity. These types of carbon materials can be classified as a) ordered mesoporous carbon (OMCs) and b) disordered mesoporous carbon (DOMC) [211]. OMCs are synthesized by using either ordered mesoporous silica templates or by templating triblock copolymer structures [212, 213]. The effects of mesoporous carbon supported catalysts on the electrocatalytic activities for oxidation of alcohols are summarized in Table 7.

6.1   Mesoporous carbon-based nanomaterials for methanol oxidation

The performance of the Pt/CMK-3 catalyst, prepared using different reduction methods exhibited higher activity and better stability than commercial platinum supported carbon catalysts [214]. The enhanced activity of the platinum supported mesoporous carbon is attributed to better dispersion of the platinum, which is primarily based on large pore volume and higher surface area of the mesoporous carbon materials synthesized by hard silica template (SBA-15). In another study, CMK-3 carbons with controlled structure parameter, changeable electronic conductivity, and desirable surface functional groups are prepared from three different SBA-15 templates [215]. Pt catalysts with the CMK-3 support of large pore size and surface area possess larger ECSA and higher activity toward methanol oxidation in acidic media.

Surface oxygen groups were created by incipient wetness impregnation of SBA-15 silica, and the ordered structure of the original CMK-3 carbon was maintained [210]. TheCMK-3 carbon supported Pt–Ru catalysts were compared with Vulcan XC-72 supported electrocatalysts and commercial catalyst (E-TEK). The activity of the Pt–Ru/CMK-3 catalyst towards methanol oxidation was higher than that of the commercial Pt–Ru/C (E-TEK) catalyst. The onset potential of CO oxidation is shifted negatively for the Pt–Ru based catalysts due to the forms of Ru–OH$_{ads}$ species at lower potentials which helps to oxidize the CO$_{ads}$. There are two well-defined peaks at the forward and backward scans which exhibit the irreversible nature of the methanol electro-oxidation, however, due to the direct oxidation of methanol, these peaks do not appear in catalysts containing Ru [210, 28].

Sun et al. [216] showed that the ordered mesoporous carbons (OMC) supported Pt nanoparticles exhibited higher current density, higher I_f/I_b value, and better stability for methanol electro-oxidation than commercial Pt/C (E-TEK) catalyst. They used a triblock copolymer F127 as a soft template in the preparation of OMCs, which have a large specific surface area, enhanced mesoporosity (lower percentage of micropores), and large total pore volume. In another study, silica–polymer composites directly used for the preparation of the mesocellular carbon foam (MCF) and MCF with large pore size supported Pt catalysts have smaller Pt nanoparticles and higher catalytic performance than the commercial Pt/C catalyst toward methanol oxidation [206].

The high mesoporous surface area of carbon can significantly increase the metal dispersion and affect the particle size, which favored the progress of the electrochemical processes occurring during methanol oxidation [23]. Furthermore, the mesoporosity of the surface could assist in the more facile decomposition of CO clusters from the surface, in addition to providing facile diffusion paths to the reactants/products during the electro-oxidation of methanol [138]. Liu et al. [217] showed that the pore diameter of mesoporous carbon support played an essential role in the electrochemical activity of PtRu towards methanol electrooxidation reaction. Suitable pore size provides the easier mass transfer and higher electrochemical surface area which leads to higher activity towards alcohol electrooxidation. In another study, Li et al. [218] demonstrated that uniform Pt:Ru ratio on mesoporous carbon support can significantly affect the effectiveness of methanol oxidation.

6.2   Mesoporous carbon-based nanomaterials for ethanol oxidation

Compared to the Pt/XC-72 and Pt/CMK-3 catalysts, mesoporous graphitic carbons supported Pt catalysts (Pt/GMC) displayed better electrocatalytic activity and stability in acidic media for ethanol oxidation which may be attributed to the unique electronic properties, large pore size and the crystallographic orientation of the graphite mesopore structure [219]. In addition to, XPS results showed that there are more sp^2 bonds in GMC than in XC-72, which proves that GMC has higher graphitization degree and GMC contains more oxygen-containing functional groups which are helpful for Pt dispersion. In comparison with the commercial Pt/C catalysts (40 wt %, Johnson Matthey), the Pt supported on mesoporous carbon(MPC) displays a high activity toward ethanol oxidation and this catalyst also exhibits high durability after a long-time potential scan [220]. The surface area of MPC is much larger than other common OMC and CMK materials (Table 7).

In another study, Yan et al. showed that the current densities of ethanol oxidation on Pd nanoparticles supported on highly mesoporous hollow carbon hemispheres (HCH), having an ultrahigh surface area of 1095.59 $m^2 g^{-1}$, electrocatalyst are three times as much as Pd/C catalyst at the same Pd loadings [221]. They explained that the larger pores made molecules transport into and out of the hollow spheres much easier.

The catalytic activities of the Pt/CMK-3 and Pt/SnO$_2$-CMK-3 materials for the EOR are compared to commercial Pt/C, and Pt/SnO$_2$-CMK-3 showed an enhanced activity [222]. SnO$_2$-containing electrocatalysts can effectively split the C-C bond in ethanol, facilitating the EOR at low potentials to form CO$_2$ products as SnO$_2$ modifies the electronic structure of active sites. Pt-Sn binary and Pt-Ru-Sn ternary alloy nanoparticles (NPs) dispersed on mesoporous carbon CMK-3, and results show that Sn can potentially be used for replacing Ru in Pt-based catalysts, in order to improve the EOR activity [223].

6.3   Mesoporous carbon-based nanomaterials for other alcohols oxidation

6.3.1   Ethylene glycol oxidation

Different carbon supports (ordered mesoporous carbon (OMC), multiwall carbon nanotubes (MWCNT) and Vulcan XC72were used as support materials for Pt deposition [207]. OMC was synthesized via self-assembly in aqueous phase using a mixture of resorcinol/formaldehyde, and Pluronic F127 triblock copolymer was used as the template. The BET total specific surface area (S$_{BETsupport}$) showed that OMC has 17 times higher value than MWCNT. Cyclic voltammetry results demonstrated that OMC which promoted a more significant support catalysts interaction has a lower onset potential, delivering a higher anodic peak current density for the EGOR.

Lo et al. [223] explained the role of Sn in promoting C-C bond cleavage and improving catalyst tolerance against poisoning, and they showed that Pt$_{20}$Ru$_{10}$Sn$_{15}$ ternary alloy catalyst exhibited better activity for ethylene glycol oxidation than Pt$_{20}$Ru$_{10}$ when supported on the mesoporous CMK-3 carbon material. In this study, CMK-3 was synthesized via a replication/carbonization method using the prepared SBA-15 templates and XRD patterns showed clearly that both SBA-15 and CMK-3 exhibited well-ordered mesoporous structures, with distinct characteristic (100), (110), and (200) peaks present at $2\theta < 2.0°$.

6.3.2   1 or 2 propanol oxidation

Nan et al.[209]observed that the oxidation of n-propanol was much better as the pore size increased because the transport of alcohols is remarkably affected by the pore size. Although it is known that the smaller molecular size for alcohols leads to an easier mass transfer, the increased pore size does not have a significant contributing effect on mass transfer for isopropanol oxidation [208].

7. Heteroatom doped carbon-based materials

Along with the surface texture, the surface chemistry of carbon is also critical for the decoration of the metal nanoparticles and improvement of carbon properties. Electron density of neighboring carbon atoms can be changed with the addition of various heteroatoms to the carbon surface which causes improvement in electrical conductivity, surface wettability and binding with a target ion [224]. Typical heteroatoms mostly used in fuel cell applications are nitrogen, phosphorous and bromine. The effects of heteroatom-doped carbon supported catalysts on the electrocatalytic activities for oxidation of alcohols summarized in Table 8.

7.1 Heteroatom doped carbon-based nanomaterials for methanol oxidation

Pt-catalyst supported on nitrogen (N) doped carbon possessed electrocatalytic activity for MOR and had better stability than Pt/XC-72 because of the enhanced bond strength between Pt and N-doped support [225, 37]. Liu and coworkers[226] found that Nitrogen-Doped Porous Carbon (PCN) have 1128 m^2 g^{-1} surface area, about 5-fold higher than that of Vulcan. They showed that PtRu nanoparticles supported on PCN showed higher catalytic activity and more stable sustained current toward methanol oxidation than on carbon black XC-72, because of the higher surface area of the PCN for the catalyst dispersion. Electrochemical studies revealed that the prepared Pt/nitrogen-doped porous hollow carbon spheres (HNPHCS) catalyst possesses notably higher catalytic activity and CO-tolerance, and better stability toward methanol electro-oxidation in comparison with Pt/nitrogen-doped porous carbon and the commercial Pt/C catalysts [36]. It is likely that enhanced catalytic activity of the Pt/HNPHCS could be due to the 3.5 times higher surface area than Vulcan XC72, high dispersion of small Pt nanoparticles and the presence of nitrogen species. The electrical conductivity is enhanced when the contents of pyridinic N, pyrrolic N, and graphitic N are achieved at optimal proportions and further improving electrocatalytic activity and stability of the catalysts for methanol oxidation[227]. The results of the CVs and EIS indicate that methanol electrooxidation activity of N doped mesoporous supported catalyst is better than that of the undoped mesoporous supported catalyst.

Nan et al.[209] focused on the effect of carbon pore structure for different alcohols electrooxidation, and Fig 5 shows the schematic illustration of alcohol transport through the windows between mesopores of N doped mesoporous carbon (NCM). They showed that both the electrocatalytic activity and durability increased toward methanol oxidation with an increasing pore size of NMC. In another study, Pt nanoparticles were deposited on Nitrogen/sulfur dual-doped ordered mesoporous carbon as the catalyst for MOR, showing that have excellent MOR activity and more tolerance to the intermediate products in MOR from a high ratio of I_f/I_b, outperforming Pt deposited on commercial Vulcan XC72 carbon [228]. Synthesized ordered mesoporous carbons with mesopores ranging from 2 to 7 nm showed BET surface areas varying from 408 to 1011 $m^2 g^{-1}$.

Pt-decorated 3D nitrogen-doped graphene nanoribbons catalysts showed exceptional electrocatalytic activity, strong poisoning tolerance and superior long-term stability toward methanol oxidation due to smaller sizes of Pt NPs on porous network, larger surface area, uniform nitrogen distribution, and good electrical conductivity when compared to Pt/Vulcan XC-72, Pt/carbon nanotube and Pt/undoped graphene nanoribbons catalysts [229]. N-doped porous carbon nanospheres possessed a microporous structure with a surface area of 1010 m^2 /g and a particle size of less than 100 nm, and when compared to carbon nanospheres, N doped support material showed enhanced performance for MOR [230].

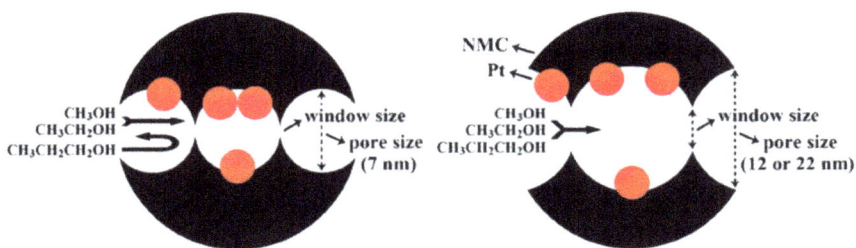

Fig 5. Schematic illustration of alcohol transport through the windows between mesopores of N doped mesoporous carbon (NCM), taken from ref. [209].

Nitrogen-doped graphene (N-G) was used as conductive support for Pt nanoparticles and compared to a catalyst of Pt loaded on undoped graphene. The Pt/N-G catalyst showed higher electrochemical activity towards methanol oxidation [231]. Nitrogen-doped graphene aerogel is highly efficient electrocatalyst supports for methanol oxidation due to providing both a high loading volume with hierarchical 3D porous architectures and synergetic effects between the doped N atoms and the Pt nanoparticles [232]. In another study, TEM results showed that Pt nanoparticles for commercial Pt/C displayed a much aggregation than Pt/NGA after stability test, so prepared N-doped graphene aerogel is quite appropriate as support for Pt-based catalyst toward methanol oxidation [233]. It was observed from accelerated potential cycling test (APCT) that the ECSAs of Pt loaded graphene, 3D-GA and NGA supported catalysts decrease gradually with the number of cycles, but the ECSA value of Pt/3D-NGA is much larger than both Pt/3D-GA and Pt/G due to high N-doped level and uniform dispersion of Pt [234].

Several research groups have reported that the doped CNTs with heteroatoms such as nitrogen [235], boron [236], phosphorus [237] and sulfur [238] are effective support materials for methanol oxidation. N-doped carbon nanotubes are suitable to support materials for anchoring the platinum nanoparticles [239]. In another study, nitrogen-doped CNTs, directly grown on the carbon cloth and platinum nanoparticles uniformly dispersed on the support material and provided rapid electron-transfer, which is a good effect for methanol oxidation [240].

In a study, XRD peaks showed that Pd/NG have broader half peak width than Pd/G, caused by the smaller size of Pd in N-doped graphene than in graphene support, which may be favorable for catalyst activities toward methanol based on more active sites [201]. In another study, boron atoms were successfully doped into graphene (Pt/BG), and the specific activity of Pt/BG catalyst towards the methanol oxidation reaction was higher than that of Pt/G and Pt/C catalysts [241]. After 1000s, the current density for Pt/BG catalyst was 38.6 mA mg_{pt}^{-1}, which was 4.2 times and 22.7 times higher than that for Pt/G catalyst and Pt/C catalyst, respectively. 3D boron and nitrogen together co-doped graphene aerogel supported Pt nanoparticles were used for methanol oxidation reaction, and the results showed that Pt/BN-GA (369.2 m^2g^{-1}) catalyst have significantly better activity compared to the conventional Pt/carbon black, Pt/graphene and Pt/graphene aerogel catalysts[242]. The D/G intensity ratio (I_D/I_G) of Pt/BN-GA (1.19) is higher than those of GO (0.84) and Pt/GA (1.00), attesting to its higher defective nature because of the B and N co-doping. The D band is associated with the disordered carbon structure in the lattice of sp^3-hybridized carbon atoms, while the G band represents sp^2 hybridization, indicating the degree of graphitization [243].

Core-shell AuPd@Pd nanocrystals supported on three-dimensional (3D) porous N-doped reduced graphene oxide hydrogels displayed enhanced catalytic activity and stability for MOR in comparison with commercial Pd-C catalysts, along with the synergistic effects between support material and nanocrystals [244].

7.2   Heteroatom doped carbon-based nanomaterials for ethanol oxidation

The superior catalytic performance of nitrogen-doped carbon nanotubes is attributed to the introduction of nitrogen which enhances the adsorption of ethanol and promotes the activation of oxygen, both of which improve the catalytic performance, especially at low temperature [245].

Palladium nanoparticles were uniformly anchored on nitrogen-doped carbon nanotubes with a three-dimensional network structure (Pd/3DNCNTs) for the EOR [246]. Pd nanoparticles tend to aggregate on the CNTs (BET surface area 101 m^2g^{-1}), while the homogeneous Pd NPsdistributed with equal size on the 3DNCNTs (BET surface area 195 m^2g^{-1}). The maximum forward current (I_f) and the minimum onset potential (E_{onset}) showed that the Pd/3DNCNTs possess an electrocatalytic activity and a significant enhancement of the kinetics for EOR higher than those of Pd/C and Pd/CNTs. In another study, Fe/Fe$_3$C (iron/iron carbide) particles embedded within nitrogen-doped nanotube support offer strong and effective interactions between metal nanoparticles and the support material [247]. In the CV, Fe-C showed the larger capacitive current which provides a larger electrochemically accessible area than CNTs.

The enhancement of the catalytic activity of Palladium supported on phosphorus-doped carbon (Pd/P-C) catalyst might be related to the presence of P increases the amount of oxygen on the electrocatalyst surface, which could contribute to the oxidation of intermediates formed during ethanol electro-oxidation process [248]. Thus, the peak current density for Pd/P-C was 50% higher than Pd/Vulcan XC72 toward ethanol oxidation in alkaline media.

Nitrogen atoms into graphene increased the catalytic activity of catalysts toward ethanol oxidation. Yu et al. [249] prepared nitrogen-doped graphene which was prepared by deriving from pyrolysis of graphite oxide (GO) with urea at different temperatures and further deposited PtSn nanoparticles and the sample at 600 °C exhibited high electrocatalytic activity and stability which indicate the promise for DEFCs applications. Pyridinic N may be beneficial for the formation of Pt^0 species that exhibits better catalytic properties for oxidations of methanol and ethanol. Pt-Pd nanoparticles loaded on N-doped graphene oxide/polypyrrole framework [250] were used as high-performance catalysts in ethanol oxidation reaction because the polypyrrole (PPy) has a low price, good electrical conductivity, high compatibility with materials and high stability.

Wang et al.[77] used iron nanoparticles wrapped inside nitrogen-doped carbon (INC) as support material for ethanol oxidation and XRD results showed that the (002) peak of carbon black was sharper than that of (INC) because carbon black had a better graphitic structure. It was also observed that both the current of double layer capacitance region is visibly much broader and catalytic activity is high toward ethanol oxidation for PtSn/INC than PtSn/C because of the INC possesses better electrical conductivity and a larger electrochemically accessible area than carbon black.

7.3 Heteroatom doped carbon-based nanomaterials for other alcohols oxidation

7.3.1 Ethylene glycol oxidation

The binary PdAu nanoflowers loaded on N-doped graphene (Pd$_1$Au$_1$-NF/NG) were fabricated, and the higher graphitization degree resulting from nitrogen doping would induce higher electron conductivity and the electronic interactions of Au into Pd enhanced their resistance to poisoning species and catalytic activity toward EG oxidation [251].

Palladium-based catalysts of metal alloys (Sn, Pb) and/or (N-doped) graphene support were synthesized, and PdSn/NG catalyst showed that the metal alloy and N-doped graphene support improved the activity and anti-poisoning effect during ethylene glycol oxidation [201]. In other words, Pd-based catalysts had much more catalytically active and better poisoning-tolerance than singly doped Pd-based catalysts because of the uniform nanoparticle dispersion on NG, a unique electronic structure and high Pd(0)/Pd(II) ratio of Pd-based catalysts.

Phenanthroline as a nitrogen source was used for traditional carbon XC72 support (PMC), and uniform Pd nanoparticles were highly dispersed on this support [252]. The results indicated that Pd/PMC has good electrocatalytic activity and stability for ethylene glycol oxidation due to both the conjugated π bond in phenanthroline provide easier electron transfer on the carbon material and metallic Pd^0 nanoparticles easily anchored in the N doped defect sites.

7.3.2 1 or 2 propanol oxidation

Nan et al.[209] focused on the different pore size effects of N doped carbon supports on the electrocatalytic performance of Pt catalysts toward oxidation of n-propanol. Enlarging the pore size of NMCs showed the better electrocatalytic performance of Pt@NMCs toward n-propanol oxidation because nitrogen doping enhanced the electrical conductivity of carbon support and the dispersivity of Pt catalysts. The surface areas of Pt loaded on NMC with 7, 12 and 22 nm pore sizes calculated from BET results are 566, 507, and 447 m^2 g^{-1}, respectively. In addition, it is believed that the cavities, especially at the edge of NMCs, are connected by small windows throughout NMCs, which offers a three-dimensional pathway for mass transport.

8.   Concluding remarks

The electrocatalytic oxidation of many alcohols has been investigated on different support materials. It has been demonstrated that the alcohol oxidation depends both on the nature and structure of the support materials and on the molecular structure of the alcohol. Different carbon supports have both advantages and disadvantages when compared with each other. Carbon blacks are the most commonly used support materials for direct alcohol fuel cell catalysts. They have a wide range of diversities regarding their structural properties including surface area and pore sizes. However, they have a considerable content of micropores which provides difficult accessibility of the reactants and used metal nanoparticles could be sunk into the micropores. CNTs have significantly higher electronic conductivities and present higher mesoporous volumes. They are chemically inert due to the lack of sufficient binding sites for anchoring nanoparticles, which leads to the aggregation of nanoparticles. So, to introduce more surface anchoring groups like –COOH and -OH, chemical oxidative processes have been employed.

Carbon gels including xerogels, aerogels, and cryogels have recently attracted attention as a new form of interconnected mesoporous carbon. Porous carbon can provide a large surface area to highly disperse catalysts and pore channels to facilitate mass diffusion. Graphene has high interest for direct alcohol fuel cell due to its high surface area for high metal dispersion, high electrical conductivity and good thermal properties. Hetero-atom doped carbons can enhance the electrical conductivity and surface area of the carbon support. This can increase the dispersion of the catalyst nanoparticles. The metal oxides are semiconductors, and the complete replacement of the carbon support don't suitable because of their lower electrical conductivity. Thus, it can be expected that the introduction of a metal oxide in the carbon support will have a positive impact on alcohol oxidation.

Table 3. Comparison of electrocatalytic activities of carbon black support materials for alcohol oxidation.

Catalyst	Support BET Surface area ($m^2\,g^{-1}$)	ECSA (Electrochemical surface area) ($m^2\,g^{-1}$)	Alcohol	Media	Forward peak (current density or mass current density)	I_f/I_b	Explanation	Ref
Pt/C: Vulcan XC72	-	55.0	Methanol	0.5 M H_2SO_4 and 0.5 M CH_3OH	12.74 mA cm^{-2}	1.03	XC72 supported Pt, PtRu, PtPd and PtSn single and bimetallic nanoparticles used as a catalyst for electrooxidation of methanol	[28]
Pt/XC-72	-	29.9	Methanol	0.5 M $HClO_4$ and 1.0 M CH_3OH	293.00 mA mg^{-1} Pt	1.01	Hierarchical porous nitrogen–doped carbon (HPNC) as the carbon support material synthesized for the methanol oxidation reaction (MOR) and compared with Vulcan XC-72.	[37]
40 wt% Pt/BP	1562	59.0	Methanol	0.5 M $HClO_4$ + 1.0 M CH_3OH	620.59 mA mg^{-1} Pt	-	Pt/C electrocatalysts with different Pt loadings, two kinds of carbon supports– Black Pearls 2000, and Vulcan XC was prepared for methanol oxidation in acidic media.	[55]
40wt% Pt/XC	245	53.0	Methanol	1.0 M CH_3OH	594.32 mA mg^{-1} Pt	-		
Pt/ Ketjenblack	1232	60.0	Methanol	0.1 M $HClO_4$ and 0.5M methanol	-	2.45	Two different Pt/C electrocatalysts using Vulcan XC-72R (VC) and Ketjenblack (KB) as the carbon support material synthesized for the methanol oxidation reaction (MOR).	[57]
Pt/ VulcanXC-72R	235	41.6	Methanol	0.5M methanol	-	1.56		
Pt$_1$Ru$_1$/C (Vulcan XC72)	230	-	Methanol	2.0 M CH_3OH and 1.0 M H_2SO_4	333.00 mA mg^{-1}	-	PtRu nanoparticles supported on both nitrogen–doped porous carbon and commercial Vulcan XC72 for electrooxidation of methanol.	[226]
Pt/Vulcan XC-72	-	22.0	Methanol	1.0 MH_2SO_4 and 2.0 M methanol	4.40 mA cm^{-2}	1.04	Nanosized Pt deposited onto 3D nitrogen-doped graphene nanoribbons and Vulcan XC-72 towards efficient methanol electrooxidation.	[229]
Pt–C Vulcan XC72	-	6.7	Ethanol	0.5 M H_2SO_4 and 0.5 M ethanol	17.50 mA mg^{-1}	1.12	Platinum deposited on titanium nitride surface and compared with the conventional support Vulcan for the electrochemical oxidation of ethanol in acidic medium.	[29]
Pt/C: Vulcan	-	66.7	Ethanol	0.5 M H_2SO_4 and	1790.00 mA mg^{-1}	0.50	Pt-RuCo nanoparticles with low noble metal content supported on Vulcan XC-72 carbon are	[31]

					(at 300rpm)			
XC72				0.5 M ethanol			successfully synthesized and compared with Pt/C catalyst.	
Pd-Au/C C: Vulcan XC 72	-	110.0	Ethanol	0.5M KOH 1 M Ethanol	45.00 mA cm^{-2}	0.69	Mono and binary Pd-Au based catalysts were supported on Vulcan XC 72 carbon black towards the oxidation of alcohols (methanol, ethanol, ethylene glycol,and glycerol) in alkaline media.	[45]
Pt/C C: Vulcan XC 72	-	32.6	Ethanol	0.5 M H$_2$SO$_4$, and 1.0M ethanol	149.90 mA cm^{-2}	-	The effect of the mixed support of carbon black and nanographite (CG) on the electrocatalytic activity and stability for ethanol oxidation.	[73]
PtSn/C C: Vulcan XC 72	-	-	Ethanol	0.5 M H$_2$SO$_4$, and 0.5 M ethanol	259.00 mA mg^{-1}$_{Pt}$	0.69	Iron nanoparticles wrapped in N-doped carbon (INC) and Vulcan XC72 carbon used as support materials.	[77]
Pd-Au/C C: Vulcan XC 72	-	110.0	EG	0.5M KOH 1.0M EG	54.40 mA cm^{-2}	0.81	Mono and binary Pd-Au based catalysts were supported on Vulcan XC 72 carbon black towards the oxidation of alcohols (methanol, ethanol, ethylene glycol,and glycerol) in alkaline media.	[45]
Pd/CB	-	126.2	EG	1.0 M NaOH and 1.0 M EG	3870.00 mA cm^{-2}	-	Palladium nanoparticles dispersed on commercial carbon black (Vulcan XC72) for methanol, ethanol, isopropanol,and n-propanol oxidation	[49]
Pd/CB	-	126.2	n-propanol	1.0 M NaOH and 1.0 M n-propanol	4542.00 mA mg^{-1}	-	Palladium nanoparticles dispersed on commercial carbon black (Vulcan XC72) for methanol, ethanol, isopropanol,and n-propanol oxidation	[49]
Pd/CB	-	126.2	Isopropanol	1.0 M NaOH and 1.0 M isopropanol	1310.00 mA mg^{-1}	-		

Table 4. Comparison of electrocatalytic activities of carbon nanotube supports for alcohol oxidation.

Catalyst	ECSA (Electrochemical surface area)	Alcohol	Media	Forward peak (current density or mass current density)	If/Ib	Explanation	Ref
Pd-Ni-P/ MWCNT	-	Methanol	0.5 M KOH 1.0M methanol	0.91 mA/cm²$_{Pd}$	1.9	Commercial MWCNTs (60-100 nm diam., 1-2 um length) were used as a support material.	[106]
Pd–MWCNT–CS	138.9 m² g⁻¹ $_{Pd}$	Methanol	0.5 M NaOH + 1 M CH$_3$OH	2541 A g⁻¹ $_{Pd}$	1.1	Commercial MWCNTs were functionalized with –COOH groups by using HNO$_3$. In addition to, catalyst ink prepared with chitosan (CS) and Nafion (Nf) dispersants.	[104]
MWNT/2OH-PBI/Pt	47.2 m² g$_{Pt}$⁻¹	Methanol	0.1 M HClO$_4$ with 1 M CH$_3$OH	About 1000 A g⁻¹ $_{Pt}$	1.3	Commercial MWCNTs were functionalized with –OH groups by using the dihydroxy-polybenzimidazole (2OH-PBI).	[108]
Pt/CeO$_2$-CNTs(4:4)	85.2 m² g⁻¹	Methanol	0.5 M CH3OH and 0.5 M H$_2$SO$_4$	124.40 A g⁻¹	-	The CNTs were boiled in HNO$_3$ for oxidation, and CeO$_2$ was deposited onto the oxidized CNTs with both different weight ratios and calcination temperature.	[111]
MWCNTs@SnO$_2$@Pt	97.3 m² g⁻¹	Methanol	0.5 M H$_2$SO$_4$ + 0.5 M CH$_3$OH	1.75 mA cm⁻² $_{Pt}$ and 1701.6 A g⁻¹ $_{Pt}$	0.74	The surface of MWCNTs was functionalized by refluxing them in concentrated nitric acid and MWCNTs hybrids synthesized with SnCl$_2$·2H$_2$O and urea.	[113]
Commercial Pt/C	71.7 m² g⁻¹	Methanol		0.38 mA cm⁻² $_{Pt}$ and 274.3 A g⁻¹ $_{Pt}$	0.67	Commercial Pt/C (Johnson Matthey, 20 wt % Pt)	
Pt-Ru/CNTs	-	Methanol	0.5 M H$_2$SO$_4$ and 1 M CH$_3$OH	1038.25 A g⁻¹ $_{Pt}$	2.94	Carbon nanotubes (CNTs, 30-50 nm in diameter, 6-8 µm in length) were grown via a thermal chemical vapor deposition process on titanium treated carbon cloths.	[118]
CNT/Pd	89 m² g⁻¹ $_{Pd}$	Ethanol	1 M KOH and 1 M ethanol	82.3 mA/cm²	0.9	CNT was refluxed in a mixture of concentrated H$_2$SO$_4$ and HNO$_3$ in the ratio of 3:1 at 90 °C for different times.	[38]

Catalyst	Surface area	Fuel	Electrolyte	Activity	Ratio	Description	Ref.
Pd/MWNTs-2:1	83.2 m² g⁻¹ Pd	Ethanol	1 M KOH and 1 M ethanol	1169 A g⁻¹ Pd and 27.55 mA/cm²	-	Carboxylic group functionalized multi-walled nanotubes (MWNTsCOOH) have outside diameter: 50–80 nm, inside diameter: 5–15 nm, length: 10–20 μm.	[119]
Pt–Sn–Ir/MWCNTs	128 m² g⁻¹ Pt	Ethanol	0.5 M H2SO4 and 1.0 M ethanol	128 mA/cm²	1.25	The commercial MWCNTs have 6–13 nm ranges of diameter, 20 μm length and 220 m² g⁻¹ BET surface area.	[122]
Pt–ZrO₂/CNTs (1:1)	538 cm² mg⁻¹ Pt	Ethanol	1 M C₂H₅OH + 1 M HClO₄	668 mA/mg	-	Commercial MWCNTs were pretreated with 5 M HCl and 3 M H₂SO₄ + 3 M HNO₃ solutions for oxidation.	[124]
Pd/MWCNT	92.5 m² g⁻¹	Ethanol	1.0 M NaOH + 1.0 M C2H5OH	1200 mA/mg	-	Commercial MWCNTs (length: 10–30 μm and OD: 20–30 nm) were used as support material.	[127]
Pt/CNT	Higher than Pt/C	EG	0.5 M H₂SO₄ and 0.5 M EG	7.45 mA/cm²	-	Commercial multi-walled carbon nanotubes have been functionalized with appropriate functional groups (–COOH) via HNO₃ solution, for 48 h under refluxing conditions, which will favor the adherence of metal nanoparticles on its surface.	[128]
PtRu/CNT				12.62 mA/cm²			
S-MWNTS/Pd	-	EG	0.2 M KOH and 0.5 M EG	177.2 mA mg⁻¹ Pd	1.94	MWNTS (20–40 nm in diameter) were produced by a chemical vapor deposition method and sulfonated-MWNTS derived from concentrated sulfuric acid and acetic anhydride through a facile synthesis.	[130]
SF–MWCNT–PdSn_mix	63.6 m² g⁻¹	EG	0.5 M Ethylene Glycol in 0.5 M KOH	51.9 mA/cm²	2.1	Commercial MWCNTs have 0.5–200 μm length, 7–15 nm O.D and 3–6 nm I.D. Mixture of H₂SO₄ and acetic anhydride were used for sulfonated multi-walled carbon nanotubes.	[131]
Pd₄Ag₁/MWCNT	-	n-propanol	1 M NaOH and 0.5 M n-propanol	165.5 mA/cm²	-	Commercial Multi-walled carbon nanotubes (MWCNTs) treated in the mixed acid of sulfuric acid and nitric acid to obtain the functionalized MWCNT.	[133]

Table 5. Comparison of electrocatalytic activities of carbon gels supports for alcohol oxidation.

Catalyst	BET Surface area (m2/g)	ECSA (Electrochemical surface area) (m2g-1)	Alcohol	Media	Forward peak (current density or mass current density)	If/Ib	Explanation	Ref
Pt-Ni/CX	724	-	Methanol	0.5 M CH$_3$OH and 0.5 M KOH	26.0 A g^{-1} (Pt) At 0.6V	-	The polymer precursor was synthesized by condensation of resorcinol and formaldehyde using sodium carbonate. The gel was pyrolyzed at 800 °C under inert atmosphere.	[138]
Pt/CA-125	618	87.40	Methanol	0.5 M H$_2$SO$_4$ + 2.0 M CH$_3$OH	94.4 mA cm^{-2}	-	Carbon aerogels (CAs) with different pore sizeswere prepared by condensation of formaldehyde and resorcinol/CTAB with different molar ratios.	[140]
Pt/CAs	616	67.00	Methanol	1.0 M NaOH + 1.0 M CH$_3$OH	66.0 mA cm^{-2}	7.3	Carbon aerogels (CAs) were prepared by sol-gel polycondensation of resorcinol and formaldehyde.	[141]
Pt/G-Gel/NF-4	-	150.30	Methanol	1 M CH$_4$OH + 0.5 M KOH	105.8 mA cm^2	18.2	Commercial nickel foam (NF) were immersed in GO/H$_2$PtCl$_6$ suspension. It is transferred to plastic tubes containing vitamin C and kept undisturbed for 1 h and subsequently heated at 60 C for 2 h.	[142]
Pt/GOA	-	95.50	Methanol	0.5 M H$_2$SO$_4$ + 0.5 M CH$_3$OH	876.0 mA mg$^{-1}_{Pt}$	-	Synthesized crosslinked hydrogels after several different processes were supercritically dried with CO$_2$.	[143]
Pt/GA	-	41.20	Methanol	0.5 M H$_2$SO$_4$ and 0.5 M CH$_3$OH	576.2 mA mg^{-1} and 1.40 mA cm^{-2}	-	Pt/graphene aerogel was prepared by one-pot solvothermal reduction with Teflon-lined autoclave, followed by freeze-drying.	[144]
Pt/C/GA	-	70.40	Methanol	0.5 M H$_2$SO$_4$ and 0.5 M CH$_3$OH	405.3 mA mg^{-1}	-	3D Pt/C/GA was synthesized by hydrothermal assembly into a Teflon-lined autoclave of GO and Pt/C, subsequently with freeze-drying.	[146]

Catalyst	ECSA (Electrochemical surface area)	Alcohol	Media	Forward peak (current density or mass current density)	I_f/I_b	Explanation	Ref
Pd/GA/NF	-	Ethanol	1.0 M EtOH and 1.0 M KOH	300.0 (A g^{-1} Pd) at 1000th cycle	1.95	The GO was prepared through the modified Hummer's method. Freeze-dried GO and metal precursor stirred in DI water. Then porous nickel foam was immersed into the dispersion.	[147]
Pd/WC–Mo$_2$C	-	EG	0.5 M EG and 0.5 M KOH	600.0 mA mg^{-1} Pd	-	Carbon aerogel has been synthesized by polycondensation of resorcinol and formaldehyde.	[150]
Pd/3D rGO aerogel	96.26	EG	1.0 M EG in 2.0 M KOH	163.2 mA cm^{-2}	-	3D rGO aerogel was synthesized by a hydrothermal reduction method in a Teflon-lined autoclave. rGO powder was further dried by freezing drying for 3D rGO aerogel.	[151]

Table 6. Comparison of the electrocatalytic activities of graphene supports for alcohol oxidation.

Catalyst	ECSA (Electrochemical surface area)	Alcohol	Media	Forward peak (current density or mass current density)	I_f/I_b	Explanation	Ref
Pt/Gr-C	50.5 m^2 g^{-1}	Methanol	0.5 M H$_2$SO$_4$ + 0.5 M CH$_3$OH	59.8 mA cm^{-2}	0.80	GO was synthesized from graphite powder using a modified Hummers method and then Vulcan XC-72 carbon was inserted into graphene through ultrasonication.	[155]
Pt/GO:CNTs (1:2)	117.3 m^2 g^{-1}	Methanol	1 M CH$_3$OH + 0.5 M H$_2$SO$_4$	691.1 mA/mg	1.03	GO was prepared by chemical oxidation of natural graphite powder according to the improved Hummers' method and a certain amount of GO and CNTs were sonicated and stirred mechanically.	[157]
PtCo/GCG	88.2 m^2 g^{-1} Pt	Methanol	0.5 M H$_2$SO$_4$ and 0.5 M methanol	437.0 mA mg^{-1} Pt	1.57	Carbon dots were synthesized by hydrothermal method. Graphene is produced from natural graphite by using a high-intensity cavitation field in a pressurized ultrasound reactor. Then the carbon dots were deposited on the surface of graphene through ultrasonic treatment.	[158]

Catalyst	Surface area	Fuel	Electrolyte	Activity	Ratio	Preparation	Ref.
PtAg/graphene	20.8 $m^2 g^{-1}$	Methanol	0.1 M HClO$_4$ + 1.0 M CH$_3$OH	1.48 mA cm^{-2} and 580 mA mg^{-1} Pt	1.20	Commercial graphene was used in this study	[159]
Pt/RGO	-	Methanol	0.5 M H$_2$SO$_4$ + 1 M CH$_3$OH	18.5 mA/cm^2	6.51	Hummers method was used for the preparation of graphene oxide from a solution of graphite	[160]
Pt-Fe/rGO	18.0 $m^2 g^{-1}$	Methanol	0.5 M H$_2$SO$_4$ + 1 M CH$_3$OH	4.7 mA/cm^2	3.00	The graphene oxide (GO) was prepared from natural graphite by Hummers method,and as-synthesized GO was then purified. The reduction of graphene oxide is achieved with NaBH$_4$ as a reducing agent.	[161]
Pt-Cu/RGO	65.9 $m^2 g^{-1}$ $_{Pt}$	Methanol	0.5 M H$_2$SO$_4$ 0.5 M CH$_3$OH	1.6 mA cm^{-2}	1.67	GO was prepared from purified natural graphite according to a modified Hummer's method.	[162]
Pt-Pd/carboxylic sodium-functionalized graphene (CS-G)	-	Methanol	1 M H$_2$SO$_4$ + 1 M CH$_3$OH	15.7 mA/cm^2	2.27	GO nanosheets were synthesized by a modified Hummers' method. The rGOwas prepared by using hydrazine hydrate. Then, firstly the graphene–cyano was produced,and it was dispersed in sodium hydroxide and methanol to obtain CS-G with refluxing	[167]
Pt$_1$Pd$_1$/GNS	46.1 $m^2 g^{-1}$ $_{Pt}$	Methanol	0.1 M H$_2$SO$_4$ + 0.5 M CH$_3$OH	0.4 mA/cm^2 404 mA mg^{-1} Pt	0.81	Graphene oxide (GO) was synthesized from natural graphite by a modified Hummers' method. Triblock pluronic copolymer P123 was used as reducer, stabilizer,and structure-director.	[165]
Pt$_3$Pd$_1$/Graphene	82.8 $m^2 g^{-1}$	Methanol	1 M CH$_3$OH and 0.5 M H$_2$SO$_4$	198.0 mA mg^{-1} Pt	1.29	The one-step synthesis of few-layer graphene supported Pt-Pd NPs were prepared by DC arc discharge evaporation	[166]
PdRu/RGO	52.0 $m^2 g^{-1}$	Methanol	1.0 M NaOH + 1.0 M methanol	0.2 mA m^{-2}	-	Pre-oxidized commercial graphite was further oxidized to graphite oxide according to Hummer's method.	[171]
PdCu/rGO	-	Methanol	1.0 M KOH and 1.0 M CH$_3$OH	916.0 mA mg^{-1} Pd	-	Modified Hummers' method was employed to produce graphene oxide nanosheets from natural graphite powder.	[172]

Catalyst	Surface area	Fuel	Electrolyte	Activity	Value	Synthesis method	Ref.
Pd/SiO$_2$@RGO	41.4 m^2 g^{-1}	Methanol	1M KOH with 1 M methanol	1533.0 mA mg^{-1}	3.90	GO was synthesized by the oxidation of graphite via an improved Hummer's method. Mono-dispersed SiO$_2$ nanospheres around 200 nm were prepared according to the Stöber process. Hybrid catalysts were synthesized by hydrothermal treatmentinan autoclave.	[173]
Pd-TiO$_2$/RGO	105.0 m^2 g^{-1}	Methanol	0.5 M NaOH and 1.0 M CH$_3$OH	764.1 A/g$_{Pd}$	1.64	Modified Hummers' method was utilized to prepare graphene oxide (GO) from natural flake graphite. Mixed GO and TiO$_2$ suspension were transferred to a Teflon-lined autoclave for hydrothermal synthesis.	[175]
Pd/MnO$_2$:RGO(1:7)	-	Methanol	0.5 M KOH + 1 M CH$_3$OH	20.4 mA cm^{-2}	4.30	GO was prepared from natural graphite flakes by a modified Hummers method. MnO$_2$-GO was prepared with GO and KMnO$_4$ dispersed uniformly in deionized water, and then ethylene glycol was added dropwise in suspension.	[178]
Pd-8%Fe Mn$_2$O$_4$/GNS	302.0 cm^2 mg^{-1}	Methanol	1 M KOH + 1 M CH$_3$OH	84.3 mA cm^{-2}	2.49	Graphene nanosheets were obtained through chemical reduction by using NaBH$_4$ of graphite oxide (GO), which was prepared by a modified Hummers and Offenmans method.	[179]
Pt-Cu/RGO	65.9 m^2 g^{-1} $_{Pt}$	Ethanol	0.5 M H$_2$SO$_4$ + 0.5 M ethanol	0.7 mA cm^{-2}	-	GO was prepared from purified natural graphite according to a modified Hummer's method	[162]
Ni@Au@Pd/rGO	230.0 m^2 g^{-1}	Ethanol	2.0 M KOH and 1.0 M high purity ethanol	8.9 mA cm^{-2}	-	Graphene oxide (GO) nanosheets were synthesized with the improved Hummers' method by liquid phase oxidation and followed with repeated washing to be neutral and freeze-drying.	[183]
PdRu/RGO	52.0 m^2 g^{-1}	Ethanol	1.0 M NaOH + 1.0 M ethanol	0.5 mA cm^{-2}	-	Pre-oxidized commercial graphite was further oxidized to graphite oxide according to Hummer's method.	[171]
Au@Pd/graphene Pd/Au ratio of 1:3	-	Ethanol	1 M NaOH and 1 M ethanol	11.6 A mg^{-1}	1.09	GO was prepared using commercial graphite powder by Hummers and Offeman method. Ascorbic acid solution and freeze-drying were using for synthesized catalysts.	[184]

45

Catalyst	Surface area	Fuel	Electrolyte	Activity		Synthesis method	Ref.
Cu@Pd/SnO$_2$-Gr	467.5 m^2 g^{-1}	Ethanol	1 M ethanol + 0.5 M NaOH	104.2 mA mg^{-1} 63.2 mA cm^{-2}	0.86	A modified Hummer's method was employed to synthesize graphene oxide from natural graphite powder. Graphene oxide was then reduced using sodium borohydride. The fabricated powder was then separated by filtration, drying, and calcination.	[192]
Pd/MnO$_2$/GNRs	114.7 m^2 g^{-1}	Ethanol	1.0 M NaOH + 1.0 M C$_2$H$_5$OH	2510.0 mA mg^{-1}	-	The GNRs were synthesized by step-oxidation unzipping of commercial MWCNTs.	[127]
Pd-Cd/rGO/GCE	33.2 m^2 g^{-1}	Ethanol	1.0 M KOH + 1.0 M C$_2$H$_5$OH	179.0 mA cm^{-2}	1.63	Graphene oxide (GO) is synthesized from natural graphite powder by Hummers method. GO solution is dropped onto the surface of the Glassy carbon electrode (GCE) and was dried at room temperature. The GO/GCE is reduced to rGO/GCE via galvanic reduction by Zn/HCl system.	[186]
Pt$_1$Pd$_5$/RGO	72.0 m^2 g^{-1}	Ethanol	0.5 M H$_2$SO$_4$ + 0.5 M ethanol	1080.0 mA mg^{-1}	-	Graphene oxides were prepared according to Hummer's method. PtPd nanocubes (NCs) were deposited on the reduced graphene oxides (rGOs) via a one-pot solvothermal reduction (into a polytetrafluoroethylene reactor).	[188]
Pd-Cu$_{(F)}$/ RGO	151.9 m^2 g^{-1}	Ethanol	0.5 M NaOH + 0.5 M C$_2$H$_5$OH	2416.3 mA mg^{-1} $_{Pd}$	-	GO was prepared from graphite powder according to a modified Hummers method. Catalysts synthesized via a facile one-pot hydrothermal approach.	[181]
Pd–PEDOT/Graphene	13.2 m^2 g^{-1} $_{Pd}$	Ethanol	1.0 M KOH + 1.0 M C$_2$H$_5$OH	458.5 A g^{-1} $_{Pd}$	0.92	The GO was synthesized using the classical modified Hummer's method. Firstly Pd nanoparticles dispersed on the PEDOT nanospheres and then Pd/PEDOT and GO were used for Pd-PEDOT/Graphene synthesis by adding NaBH$_4$.	[197]
AuPt@Pt NCs/rGO	-	EG	0.5M H$_2$SO$_4$ + 0.5 M EG	600.0 mA mg^{-1} $_{Pt}$	1.58	A modified Hummer's method was used to prepare graphene oxide (GO). Next, poly(ViEtImBr), hydrazine solution and metal precursors were used for a facile one-step aqueous method.	[199]
			0.5 M KOH + 0.5 M EG	1850.0 mA mg^{-1} $_{Pt}$	4.37		

	Support material surface area (m²/g)	solvent	Media	current density			Ref
Ag@Pt/RGO	287.7 m²g⁻¹	EG	1.0 M KOH + 0.75 M EG	120.3 mA cm²	-	GO was firstly prepared from graphite powder according to the Hummers' method with some modifications. Then, the one-pot solvothermal method was developed for the preparation of the catalysts.	[200]
Pt–CeO₂/GNS	68.6 m² mg⁻¹$_{Pt}$	EG	1M H₂SO₄ + 1 M EG	250.0 mA mg⁻¹$_{Pt}$	-	The GNS support was prepared by chemical vapor deposition.	[204]
PdRu/RGO	52.0 m²g⁻¹	1-Propanol	1M NaOH + 1.0 M 1-Propanol	0.5 mA m⁻²	-	Pre-oxidized commercial graphite was further oxidized to graphite oxide according to Hummer's method	[171]
Pt(0)/DPA@GO	73.1 m²g⁻¹	2-propanol	0.1M HClO4 0.5 M 2- propanol	1.00 A/mg$_{Pt}$	-	Graphene oxide (GO) was synthesized from graphite powder using a modified Hummer's method. Pt(0)/DPA@GO NPs was produced by sonochemical double solvent reduction method.	[180]

Table 7. Comparison of electrocatalytic activities of mesoporous carbon supports for alcohol oxidation.

Catalyst	Support material surface area (m²/g)	Pore size (nm)	ECSA	Alcohol	Media	Forward peak (current density or mass current density)	I$_f$/I$_b$	Ref
Pt/CMK	1206.0	4.8	208.2 m²/g$_{Pt}$	Methanol	0.5 M H₂SO₄ + 1.0 M CH₃OH	37.3 mA/cm²	1.07	[215]
Pt/CMK	1312.3	4.4	39.9 m²/g	Methanol	0.5 M H₂SO₄ + 1.0 M CH₃OH	about 87 mA/cm²	about 0.79	[208]
Pt/OMC	526.0	5	117.8 m²/g	Methanol	0.5 M H₂SO₄ + 0.5 M CH₃OH	449.1 mA mg⁻¹ $_{Pt}$	1.12	[207]
Pt/CMK-3	997.0	4.0	84.0 m²/g	Methanol	1.0 M H₂SO₄ + 1.0 M CH₃OH	185.0 mA(mg.$_{Pt}$)⁻¹	1.33	[214]
Pt/OMCs-FT	699.0	6.5	62.9 m²g⁻¹	Methanol	0.5 M HClO₄ + 0.5 M CH₃OH	86.2 mA mg⁻¹	1.21	[216]
PtRu@WMC-F7	659.0	8.5	-	Methanol	0.5 M H₂SO₄ + 1.0 M CH₃OH	37.5 mA/cm²	1.13	[217]

PtRu/CMK3-CPDM	1066.0	3.8	12.3 m^2/g	Methanol	0.1 M HClO$_4$ and 1.0 M Methanol	166.0 mA mg^{-1}	About 2.76	[218]
Pt/OMC	526.0	5.0	117.8 m^2/g	Ethanol	0.5 M H$_2$SO$_4$ + 0.5M C$_2$H$_5$OH	350.2 mA mg^{-1} Pt	1.58	[207]
Pt$_{20}$Ru$_{10}$Sn$_{15}$@CMK	-	3.5	26.2 m^2 g^{-1}	Ethanol	0.5 M H$_2$SO$_4$ 1.0 M Ethanol	69.0 mA/g	1.00	[223]
Pt/GMC	232.0	3.6	117.0 m^2 g^{-1} Pt	Ethanol	0.5 M H$_2$SO$_4$ + 1.0 M CH$_3$CH$_2$OH	About 730 mA mg^{-1} Pt	About 1.30	[219]
Pt/CMK-3	1303.0	3.9	52.1 m^2 g^{-1} Pt	Ethanol	0.5 M H$_2$SO$_4$ + 1.0 M CH$_3$CH$_2$OH	About 480 mA mg^{-1} Pt	About 1.63	
Pt/MPC	1943.1	5.6	4.73% loss after 4000 potential cycles	Ethanol	0.1 M HClO$_4$ + 0.1 M Ethanol	690.0 mA cm^{-2}	About 1.32	[220]
Pd/HCH	1095.6	9.4	-	Ethanol	1.0 M KOH + 1.0 M Ethanol	About 43.0 mA cm^{-2}	About 0.88	[221]
Pt/OMC	526.0	5.0	117.8 m^2/g	Ethylene Glycol	0.5 M H$_2$SO$_4$ + 0.5M C$_2$H$_6$O$_2$	468.7 mA mg^{-1} Pt	1.55	[207]
Pt$_{20}$Ru$_{10}$Sn$_{15}$@CMK	-	3.5	26.2 m^2 g^{-1}	Ethylene Glycol	0.5 M H$_2$SO$_4$ 1.0 M Ethylene Glycol	61.0 mA/g	1.10	[223]
Pt/CMK	1312.3	4.4	39.9 m^2/g	Isopropanol	0.5 M H$_2$SO$_4$ +1.0 M Isopropanol	about 37.0 mA/cm^2	about 1.86	[208]

Table 8. Comparison of electrocatalytic activity of heteroatom carbon supports for alcohol oxidation.

Catalyst	BET Surface area ($m^2 g^{-1}$)	ECSA (Electrochemical surface area)	Alcohol	Media	Forward peak (current density or mass current density)	I_f/I_b	Explanation	Ref
Pt/HPNC900	755.0	641.0 $m^2 g^{-1}$	Methanol	0.5 M HClO$_4$ and 1.0 M CH$_3$OH	721 mA mg^{-1}	1.28	Chitosan/Cu(II) composites were freeze-dried, and then the powder was carbonized at different temperatures for the synthesis of hierarchical porous nitrogen-doped carbon material.	[37]
Pt1Ru1/PCN	1128.0	-	Methanol	2.0 M CH$_3$OH and 1.0 M H$_2$SO$_4$	420 mA mg^{-1}	-	N-doped porous carbon nanospheres (PCNs) were prepared by chemical activation of nonporous carbon nanospheres (CNs), which were obtained via carbonization of polypyrrole nanospheres (PNs).	[226]
Pt/HNPHCS	805.2	81.97 $m^2 g^{-1}$	Methanol	1.0 M CH$_3$OH + 0.1 M HClO$_4$	0.41 mA cm^{-2} or 336 mA mg^{-1}	1.04	The preparation of Hierarchical nitrogen-doped porous hollow carbon spheres was composed of four steps; (i) SiO$_2$ spheres with the diameter about 140 nm were synthesized, (ii) The SiO$_2$@PANI was prepared by oxidation polymerization of aniline, (iii) SiO$_2$@PANI composites were carbonized, (iv) macropores and mesopores were obtained via the removal of the silica template (etching).	[36]
Pt/MNC-900	787.0	97.90 $m^2 g^{-1}$	Methanol	0.5 M H$_2$SO$_4$ and methanol	10.2 mA cm^{-2}	-	Nitrogen-doped carbon material with honeycomb-like mesoporous structure was synthesized via carbonization of polyaniline coated silica nanoparticle.	[227]
Pt@NMC-22	447.0	44.40 $m^2 g^{-1}$	Methanol	0.5 M H$_2$SO$_4$ + 1.0 M CH$_3$OH	5.67 mA cm^{-2}	8.50	Colloidal silica nanospheres have different diameters (7, 12 or 22 nm) were used for porous structure, and then polymerization, carbonization, and etching were applied respectively.	[209]

Catalyst							Description	Ref.
PtOMC100	408.0	100.00 $m^2 g^{-1}$	Methanol	0.5 M H_2SO_4 and 1.0 M CH_3OH	505 $mA\ mg^{-1}$	1.50	Highly ordered mesoporous carbons obtained by the penetration of mesoporous silica template with pure acrylonitrile telomer (ANT), followed by the stabilization, carbonization, and removal of the template.	[228]
Pt/N-GNR	341.0	64.60 $m^2 g^{-1}$	Methanol	1.0 M H_2SO_4 and 2.0 M methanol	11.0 $mA\ cm^{-2}$	1.42	Hydrothermal self-assembly, freeze-drying, and thermal annealing approach are used for the fabrication of a catalyst made from Pt particles and three-dimensional (3D) nitrogen-doped graphene nanoribbons (NGNRs).	[229]
Pt/PCN	1010.0	24.60 m^2/g	Methanol	0.5 M H_2SO_4 and 1.0 M CH_3OH	343 $mA\ mg^{-1}$	-	PCNs were obtained by chemical activation of nonporous carbon nanospheres (CNs) that were prepared by carbonizing polypyrrole nanospheres (PNs) at high temperature.	[230]
PtNPs/3DNGA	144.3	42.17 $m^2 g^{-1}$	Methanol	0.5 M H_2SO_4 and 1.0 M CH_3OH	9.32 $mA\ cm^{-2}$	2.10	GO lost oxygenated functional groups under high temperatures and pressure at the hydrothermal process, and glycine was used as N-source; thus 3D hydrogel N doped GA structure is obtained.	[232]
Pt/NGA	401.3	60.60 $m^2 g^{-1}$	Methanol	0.5 M H_2SO_4 and 0.5 M CH_3OH	507.5 $mA\ mg^{-1}$	-	3D nitrogen-doped graphene was synthesized through a supramolecular melamine-cyanurate performed as N source by hydrothermal method.	[233]
Pt/CN$_x$-NTs-800	324.0	94.00 $m^2 g^{-1}$	Methanol	0.5 M H_2SO_4 and 0.5 M CH_3OH	870 $mA\ mg^{-1}$	-	PPy-NTs were synthesized using a self-degraded template method, and CN$_x$-NTs were prepared by pyrolysis of PPy-NTs.	[235]
Pt/S-MWCNT	-	161.40 $m^2 g^{-1}$	Methanol	0.5 M H_2SO_4 and 0.5 M CH_3OH	803.9 $mA\ mg^{-1}$	-	The S-MWCNTs were obtained by annealing poly(3,4-ethylenedioxythiophene) (PEDOT) functionalized commercial multi-walled carbon nanotubes at 800 °C.	[238]

Catalyst		Surface area	Fuel	Electrolyte	Current density		Synthesis description	Ref.
Pt/BN-GA	369.2	106.00 $m^2 g^{-1}$	Methanol	0.5 M H_2SO_4 and 1.0 M CH_3OH	1184.5 $mA\ mg^{-1}$	-	GO nanosheets were produced from Commercial graphite powders by a modified Hummers' method. For the incorporation of B and N atoms into the graphene sheets, ammonium fluoroborate (NH_4BF_4) was added into the synthesized GO solution,and then, the resulting mixture was transferred into a Teflon lined stainless steel autoclave for hydrothermal synthesis.	[242]
Pt-Pd/PPy/NGO	-	35.10 $m^2 g^{-1}$	Ethanol	KOH (0.5 M) and C_2H_5OH (1.0 M)	131.58 $mA\ cm^{-2}$	-	GO was synthesized by the Hummers method. The NGO was synthesized by pyrolysis method. PPy/NGO nanocomposite was prepared by in-situ chemical polymerization of pyrrole on NGO.	[250]
Pt@NMC-22	447.0	44.40 $m^2 g^{-1}$	Ethanol	0.5 M H_2SO_4 + 1.0 M Ethanol	About 4.75 $mA\ cm^{-2}$	-	Colloidal silica nanospheres have different diameters (7, 12 or 22 nm) were used for porous structure and then polymerization, carbonization,and etching were applied respectively.	[209]
Pd/ 3DNCNTs	195.0	20.10 $m^2 g^{-1}$	Ethanol	1.0 M KOH and 1.0 M C_2H_5OH	775.6 $mA\ mg^{-1}$	-	The acid oxidized carbon nanotubes (AO–CNTs) were obtained. 3DNCNTs were synthesized via a simple hydrothermal reaction in an autoclave with AO–CNTs and urea.	[246]
PtSn/Fe-C (1:1)	115.6	57.80 $m^2 g^{-1}$ Pt	Ethanol	0.5 M H_2SO_4 + 0.5 M C_2H_5OH	21.1 $mA\ cm^{-2}$	-	Melamine and Fe salt were used as the support precursors,and the sample was heated in a tube furnace under a nitrogen atmosphere. In addition to, the	[247]
PtSn/CNTs	-	22.10 $m^2 g^{-1}$ Pt	Ethanol	0.5 M H_2SO_4 + 0.5 M C_2H_5OH	10.6 $mA\ cm^{-2}$	-	commercial CNTs were treated with HNO_3 and H_2O_2 to produce carbonyl functional groups to act as active sites for anchoring PtSn nanoparticles.	[247]

Catalyst		Surface area	Solvent	Electrolyte			Description	Ref.
PtSn/NG-600	-	71.00 $m^2 g^{-1}$	Ethanol	0.5 M H_2SO_4 and 1.0 M C_2H_5OH	About 575 mA mg^{-1}	-	Graphene oxide (GO) was produced by the modified Hummers method, and NG was generated by thermal annealing of the GO and urea at different temperatures.	[249]
PtSn/INC	-	-	Ethanol	0.5 M C_2H_5OH 0.5 M H_2SO_4	424.5 mA mg^{-1} Pt	1.04	Iron nanoparticles wrapped in N-doped carbon (INC) were synthesized by a two-step method which includes hydrothermal and thermal treatments.	[77]
Pd/PMC	-	583.40 $cm^2 mg^{-1}$ Pd	EG	0.1 M KOH + 0.5 M $(CH_2OH)_2$	0.348 A mg^{-1} Pd	-	The carbonized phenanthroline modified carbon (PMC) support was prepared by heating the mixture of carbon (Vulcan XC72) and phenanthroline-nickel (II) complex.	[252]
Pd_1Au_1-NF/NG	-	136.52 $m^2 g^{-1}$	EG	1.0 M KOH + 1.0 M EG	8.7 A mg^{-1}	-	The polyvinylpyrrolidone (PVP), the N-Dimethylformamide (DMF) and $N_2H_4 \cdot H_2O$ as reducing agents were used via the ultrasonic-assisted procedure.	[251]
PdSn/NG	-	22.50 $m^2 g^{-1}$	EG	1M KOH and 0.5 M $(CH_2OH)_2$	118.05 mA cm^{-2}	-	N-doped graphene was synthesized by adding the appropriate amount of graphite oxide (GO), double-distilled water and ammonium hydroxide then transferred into a stainless steel autoclave for hydrothermal treatment, and then dried in the freezer dryer under vacuum.	[201]
Pt@NMC-22	447	44.40 $m^2 g^{-1}$	n-propanol	0.5 M H_2SO_4 + 1.0 M n-propanol	2.32 mA cm^{-2}	-	Colloidal silica nanospheres have different diameters (7, 12 or 22 nm) were used for porous structure and then polymerization, carbonization, and etching were applied respectively.	[209]

References

[1] E.H.L. Falcao, F. Wudl, Carbon allotropes: beyond graphite and diamond, J. Chem. Technol. Biot. 82 (2007) 524-531. https://doi.org/10.1002/jctb.1693

[2] B. McEnaney, Structure and bonding in carbon materials, in: T.D. burchell (Ed.) carbon materials for advanced technologies, Elsevier Science Ltd, Oxford, 1999, pp. 1-33. https://doi.org/10.1016/B978-008042683-9/50003-0

[3] C. Lamy, C. Coutanceau, Electrocatalysis of alcohol oxidation reactions at platinum group metals, catalysts for alcohol-fuelled direct oxidation fuel cells, The Royal Society of Chemistry, 2012, pp. 1-70.

[4] B.H.R. Suryanto, C. Zhao, Surface-oxidized carbon black as a catalyst for the water oxidation and alcohol oxidation reactions, Chem. Commun. 52 (2016) 6439-6442. https://doi.org/10.1039/C6CC01319H

[5] A. Lavacchi, H. Miller, F. Vizza, Carbon-Based Nanomaterials, in: A. Lavacchi, H. Miller, F. Vizza (Eds.) Nanotechnology in Electrocatalysis for Energy, Springer New York, New York, 2013, pp. 115-144. https://doi.org/10.1007/978-1-4899-8059-5_5

[6] D. Gervasio, Fuel cells – direct alcohol fuel cells, new materials, in: J. Garche (Ed.) Encyclopedia of Electrochemical Power Sources, Elsevier, Amsterdam, 2009, pp. 420-427. https://doi.org/10.1016/B978-044452745-5.00247-1

[7] V.K. Puthiyapura, D.J.L. Brett, A.E. Russell, W.F. Lin, C. Hardacre, Biobutanol as fuel for direct alcohol fuel cells—investigation of Sn-modified Pt catalyst for butanol electro-oxidation, ACS Appl.Mater. Interfaces. 8 (2016) 12859-12870. https://doi.org/10.1021/acsami.6b02863

[8] C. Bianchini, P.K. Shen, Palladium-based electrocatalysts for alcohol oxidation in half cells and in direct alcohol fuel cells, Chem. Rev. 109 (2009) 4183-4206. https://doi.org/10.1021/cr9000995

[9] B. Braunchweig, D. Hibbitts, M. Neurock, A. Wieckowski, Electrocatalysis: A direct alcohol fuel cell and surface science perspective, Catal. Today. 202 (2013) 197-209. https://doi.org/10.1016/j.cattod.2012.08.013

[10] O.O. Fashedemi, H.A. Miller, A. Marchionni, F. Vizza, K.I. Ozoemena, Electro-oxidation of ethylene glycol and glycerol at palladium-decorated FeCo@Fe core-shell nanocatalysts for alkaline direct alcohol fuel cells: functionalized MWCNT supports and impact on product selectivity, J. Mater. Chem. A. 3 (2015) 7145-7156. https://doi.org/10.1039/C5TA00076A

[11] Y.X. Chen, M. Bellini, M. Bevilacqua, P. Fornasiero, A. Lavacchi, H.A. Miller, L.Q. Wang, F. Vizza, Direct alcohol fuel cells: toward the power densities of hydrogen-fed

proton exchange membrane fuel cells, ChemSusChem. 8 (2015) 524-533.
https://doi.org/10.1002/cssc.201402999

[12] J.E. Solis-Tobias, J.A. Diaz-Guillen, P.C. Melendez-Gonzalez, N.M. Sanchez-
Padilla, R. Perez-Hernandez, I.L. Alonso-Lemus, F.J. Rodriguez-Varela, Enhanced
catalytic activity of supported nanostructured Pd for the oxidation of organic
molecules using gamma-Fe_2O_3 and Fe_3O_4 as co-electrocatalysts, Int. J. Hydrogen
Energ. 42 (2017) 30301-30309. https://doi.org/10.1016/j.ijhydene.2017.08.112

[13] C. Lamy, A. Lima, V. LeRhun, F. Delime, C. Coutanceau, J.M. Leger, Recent
advances in the development of direct alcohol fuel cells (DAFC), J. Power Sources.
105 (2002) 283-296. https://doi.org/10.1016/S0378-7753(01)00954-5

[14] C.C. Jin, Z. Zhang, Z.D. Chen, Q. Chen, High catalytic activity of Pt-modified ag
electrodes for oxidation of glycerol and allyl alcohol, Int. J. Electrochem. Sc. 8 (2013)
4215-4224.

[15] C.C. Jin, C.C. Wan, R.L. Dong, High activity of Pd deposited on Ag/C for allyl
alcohol oxidation, Electrochim. Acta. 262 (2018) 319-325.
https://doi.org/10.1016/j.electacta.2018.01.021

[16] M.Z.F. Kamarudin, S.K. Kamarudin, M.S. Masdar, W.R.W. Daud, Review: Direct
ethanol fuel cells, Int. J. Hydrogen Energ. 38 (2013) 9438-9453.
https://doi.org/10.1016/j.ijhydene.2012.07.059

[17] J.S. Spendelow, A. Wieckowski, Electrocatalysis of oxygen reduction and small
alcohol oxidation in alkaline media, Phys. Chem. Chem. Phys. 9 (2007) 2654-2675.
https://doi.org/10.1039/b703315j

[18] A. Santasalo-Aarnio, S. Tuomi, K. Jalkanen, K. Kontturi, T. Kallio, The correlation
of electrochemical and fuel cell results for alcohol oxidation in acidic and alkaline
media, Electrochim. Acta. 87 (2013) 730-738.
https://doi.org/10.1016/j.electacta.2012.09.100

[19] E. Antolini, E.R. Gonzalez, Alkaline direct alcohol fuel cells, J. Power Sources. 195
(2010) 3431-3450. https://doi.org/10.1016/j.jpowsour.2009.11.145

[20] J.L. Cohen, D.J. Volpe, H.D. Abruna, Electrochemical determination of activation
energies for methanol oxidation on polycrystalline platinum in acidic and alkaline
electrolytes, Phys. Chem. Chem. Phys. 9 (2007) 49-77.
https://doi.org/10.1039/B612040G

[21] T.S. Zhao, Z.X. Liang, J.B. Xu, Fuel cells – direct alcohol fuel cells overview, in: J.
Garche (Ed.) Encyclopedia of Electrochemical Power Sources, Elsevier, Amsterdam,
2009, pp. 362-369. https://doi.org/10.1016/B978-044452745-5.00240-9

[22] A. Ashok, A. Kumar, J. Ponraj, S.A. Mansour, F. Tarlochan, Single step synthesis of porous $NiCoO_2$ for effective electrooxidation of glycerol in alkaline medium, J. Electrochem. Soc. 165 (2018) J3301-J3309. https://doi.org/10.1149/2.0401815jes

[23] P.V. Samant, C.M. Rangel, M.H. Romero, J.B. Fernandes, J.L. Figueiredo, Carbon supports for methanol oxidation catalyst, J. Power Sources. 151 (2005) 79-84. https://doi.org/10.1016/j.jpowsour.2005.02.083

[24] Y.L. Guo, D.Q. Zheng, H.Y. Liu, A. Friedrich, J. Garche, Investigations of bifunctional mechanism in methanol oxidation on carbon-supported Pt and Pt-Ru catalysts, J. New Mat. Electr. Sys. 9 (2006) 33-39.

[25] S.C.S. Lai, S.E.F. Kleijn, F.T.Z. Ozturk, V.C.V. Vellinga, J. Koning, P. Rodriguez, M.T.M. Koper, Effects of electrolyte pH and composition on the ethanol electro-oxidation reaction, Catal. Today. 154 (2010) 92-104. https://doi.org/10.1016/j.cattod.2010.01.060

[26] J. Melke, A. Schoekel, D. Dixon, C. Cremers, D.E. Ramaker, C. Roth, Ethanol oxidation on carbon-supported Pt, PtRu, and PtSn catalysts studied by operando X-ray absorption spectroscopy, J. Phys. Chem. C. 114 (2010) 5914-5925. https://doi.org/10.1021/jp909342w

[27] A. Serov, C. Kwak, Recent achievements in direct ethylene glycol fuel cells (DEGFC), Appl. Catal. B-Environ. 97 (2010) 1-12. https://doi.org/10.1016/j.apcatb.2010.04.011

[28] M.S. Cogenli, A.B. Yurtcan, Catalytic activity, stability and impedance behavior of PtRu/C, PtPd/C and PtSn/C bimetallic catalysts toward methanol and formic acid oxidation, Int. J. Hydrogen. Energ. 43 (2018) 10698-10709. https://doi.org/10.1016/j.ijhydene.2018.01.081

[29] M.M.O. Thotiyl, S. Sampath, Electrochemical oxidation of ethanol in acid media on titanium nitride supported fuel cell catalysts, Electrochim. Acta. 56 (2011) 3549-3554. https://doi.org/10.1016/j.ijhydene.2018.01.081

[30] Z.L. Liu, X.Y. Ling, X.D. Su, J.Y. Lee, Carbon-supported Pt and PtRu nanoparticles as catalysts for a direct methanol fuel cell, J. Phys. Chem. B. 108 (2004) 8234-8240. https://doi.org/10.1016/j.ijhydene.2018.01.081

[31] H. Li, D.L. Kang, H. Wang, R.F. Wang, Carbon-supported Pt-RuCo nanoparticles with low-noble-metal content and superior catalysis for ethanol oxidization, Int. J. Electrochem. Soc. 6 (2011) 1058-1065.

[32] Y.J. Gu, W.T. Wong, Electro-oxidation of methanol on Pt particles dispersed on RuO_2 nanorods, J. Electrochem. Soc. 153 (2006) A1714-A1718. https://doi.org/10.1149/1.2217327

[33] R. Rizo, D. Sebastian, M.J. Lazaro, E. Pastor, On the design of Pt-Sn efficient catalyst for carbon monoxide and ethanol oxidation in acid and alkaline media, Appl. Catal. B-Environ. 200 (2017) 246-254. https://doi.org/10.1149/1.2217327

[34] L. Jiang, A. Hsu, D. Chu, R. Chen, Ethanol electro-oxidation on Pt/C and PtSn/C catalysts in alkaline and acid solutions, Int. J. Hydrogen Energ. 35 (2010) 365-372. https://doi.org/10.1016/j.ijhydene.2009.10.058

[35] C.G. Lee, M. Umeda, I. Uchida, Cyclic voltammetric analysis of C-1-C-4 alcohol electrooxidations with Pt/C and Pt-Ru/C microporous electrodes, J. Power Sources. 160 (2006) 78-89. https://doi.org/10.1016/j.jpowsour.2006.01.068

[36] J. Zhang, L. Ma, M.Y. Gan, F.F. Yang, S.N. Fu, X. Li, Well-dispersed platinum nanoparticles supported on hierarchical nitrogen-doped porous hollow carbon spheres with enhanced activity and stability for methanol electrooxidation, J.. Power Sources. 288 (2015) 42-52. https://doi.org/10.1016/j.jpowsour.2015.04.109

[37] A.N. Jiang, B.H. Zhang, Y.G. Xue, Y. Cheng, Z.H. Li, J.C. Hao, Pt electrocatalyst supported on metal ion-templated hierarchical porous nitrogen-doped carbon from chitosan for methanol electrooxidation, Micropor. Mesopor. Mat. 248 (2017) 99-107. https://doi.org/10.1016/j.micromeso.2017.04.025

[38] M.S. Ahmed, S. Jeon, Electrochemical activity evaluation of chemically damaged carbon nanotube with palladium nanoparticles for ethanol oxidation, J. Power. Sources. 282 (2015) 479-488. https://doi.org/10.1016/j.jpowsour.2015.02.072

[39] H.P. Boehm, Some aspects of the surface-chemistry of carbon-blacks and other carbons, Carbon. 32 (1994) 759-769. https://doi.org/10.1016/0008-6223(94)90031-0

[40] E. Antolini, Carbon supports for low-temperature fuel cell catalysts, Appl. Catal. B-Environ. 88 (2009) 1-24. https://doi.org/10.1016/j.apcatb.2008.09.030

[41] E. Auer, A. Freund, J. Pietsch, T. Tacke, Carbons as supports for industrial precious metal catalysts, Appl. Catal.A-Gen. 173 (1998) 259-271. https://doi.org/10.1016/S0926-860X(98)00184-7

[42] C.H.A. Wong, A. Ambrosi, M. Pumera, Thermally reduced graphenes exhibiting a close relationship to amorphous carbon, Nanoscale. 4 (2012) 4972-4977. https://doi.org/10.1039/c2nr30989k

[43] M. Uchida, Y. Aoyama, M. Tanabe, N. Yanagihara, N. Eda, A. Ohta, Influences of both carbon supports and heat-treatment of supported catalyst on electrochemical oxidation of methanol, J. Electrochem. Soc. 142 (1995) 2572-2576. https://doi.org/10.1149/1.2050055

[44] L. Calvillo, V. Celorrio, R. Moliner, A.B. Garcia, I. Camean, M.J. Lazaro, Comparative study of Pt catalysts supported on different high conductive carbon

materials for methanol and ethanol oxidation, Electrochim. Acta. 102 (2013) 19-27. https://doi.org/10.1016/j.electacta.2013.03.192

[45] T.R. Maumau, R.M. Modibedi, M.K. Mathe, Electro-oxidation of alcohols using carbon supported gold, palladium catalysts in alkaline media, Mater. Today-Proc. 5 (2018) 10542-10550. https://doi.org/10.1016/j.matpr.2017.12.386

[46] L.M. Zhang, Z.B. Wang, J.J. Zhang, X.L. Sui, L. Zhao, J.C. Han, investigation on electrocatalytic activity and stability of pt/c catalyst prepared by facile solvothermal synthesis for direct methanol fuel cell, Fuel Cell. 15 (2015) 619-627. https://doi.org/10.1002/fuce.201400172

[47] S. Chandravathanam, B. Kavitha, B. Viswanathan, Y.Y. Thangam, Study of sulphonic acid functionalization of Vulcan XC-72 carbon black support of Pt/Vulcan XC-72 catalyst for methanol electrooxidation, Indian. J. Chem. A. 51 (2012) 704-707.

[48] Y.Y. Chu, Z.B. Wang, D.M. Gu, G.P. Yin, Performance of Pt/C catalysts prepared by microwave-assisted polyol process for methanol electrooxidation, J. Power Sources. 195 (2010) 1799-1804. https://doi.org/10.1016/j.jpowsour.2009.10.039

[49] C.Y. Hu, X. Wang, Highly dispersed palladium nanoparticles on commercial carbon black with significantly high electro-catalytic activity for methanol and ethanol oxidation, Int. J. Hydrogen Energ. 40 (2015) 12382-12391. https://doi.org/10.1016/j.ijhydene.2015.07.100

[50] Z.H. Yang, M.R. Berber, N. Nakashima, A polymer-coated carbon black-based fuel cell electrocatalyst with high CO-tolerance and durability in direct methanol oxidation, J. Mater. Chem. A. 2 (2014) 18875-18880. https://doi.org/10.1039/C4TA03185G

[51] R.J. Feng, M. Li, J.X. Liu, Synthesis of core-shell Au@Pt nanoparticles supported on Vulcan XC-72 carbon and their electrocatalytic activities for methanol oxidation, Colloid. Surface A. 406 (2012) 6-12. https://doi.org/10.1016/j.colsurfa.2012.04.030

[52] L. Khotseng, A. Bangisa, R.M. Modibedi, V. Linkov, Electrochemical evaluation of Pt-based binary catalysts on various supports for the direct methanol fuel cell, Electrocatalysis-Us. 7 (2016) 1-12. https://doi.org/10.1007/s12678-015-0282-x

[53] A.B. Kashyout, A.A.A. Nassr, L. Giorgi, T. Maiyalagan, B.A.B. Youssef, Electrooxidation of methanol on carbon supported Pt-Ru nanocatalysts prepared by ethanol reduction method, Int. J. Electrochem. Sci. 6 (2011) 379-393.

[54] Q. Wang, H.L. Tao, Z.Q. Li, S.S. Liu, L. Han, Enhanced activity for methanol electro-oxidation on PtRu/C catalyst by reduction treatment, Int. J. Electrochem. Sci. 12 (2017) 6211-6220. https://doi.org/10.20964/2017.07.62

[55] J. Qi, L.H. Jiang, M.Y. Jing, Q.W. Tang, G.Q. Sun, Preparation of Pt/C via a polyol process - Investigation on carbon support adding sequence, Int. J. Hydrogen Energ. 36 (2011) 10490-10501. https://doi.org/10.1016/j.ijhydene.2011.06.022

[56] M. Carmo, A.R. Dos Santos, J.G.R. Poco, M. Linardi, Physical and electrochemical evaluation of commercial carbon black as electrocatalysts supports for DMFC applications, J. Power Sources. 173 (2007) 860-866. https://doi.org/10.1016/j.jpowsour.2007.08.032

[57] Z.H. Yang, C. Kim, S. Hirata, T. Fujigaya, N. Nakashima, Facile enhancement in CO-tolerance of a polymer-coated pt electrocatalyst supported on carbon black: Comparison between vulcan and ketjenblack, Acs Appl. Mater. Interfaces. 7 (2015) 15885-15891. https://doi.org/10.1021/acsami.5b03371

[58] N. Nakagawa, Y. Suzuki, T. Watanabe, T. Takei, K. Kanamura, Preparation of Pt-Ru nanoparticles with a uniform size distribution on a mesoporous carbon and their activity towards methanol electro-oxidation, Electrochemistry. 75 (2007) 172-174. https://doi.org/10.5796/electrochemistry.75.172

[59] J.X. Cheng, X.L. Hu, J.B. Zhang, H.H. Huang, N. Su, H.K. Zhu, Fabrication of a composite of platinum, N-g-C_3N_4 and Ketjen black for photo-electrochemical methanol oxidation, J. Mater. Sci. 52 (2017) 8444-8454. https://doi.org/10.1007/s10853-017-1110-x

[60] Y. Zhang, Y.H. Wang, L.Y. Bian, R. Lu, J.B. Zang, Functional separation of oxidation-reduction reaction and electron transport: PtRu/undoped nanodiamond and acetylene black as a hybrid electrocatalyst in a direct methanol fuel cell, Int. J. Hydrogen Energ. 41 (2016) 4624-4631. https://doi.org/10.1016/j.ijhydene.2016.01.082

[61] M.R. Tarasevich, Z.R. Karichev, V.A. Bogdanovskaya, A.V. Kapustin, E.N. Lubnin, M.A. Osina, Oxidation of methanol and other low-molecular-weight alcohols on the RuNi catalysts in an alkaline environment, Russ. J. Electrochem. 41 (2005) 736-745. https://doi.org/10.1007/s11175-005-0133-9

[62] E.E. Said-Galiyev, A.Y. Nikolaev, E.E. Levin, E.K. Lavrentyeva, M.O. Gallyamov, S.N. Polyakov, G.A. Tsirlina, O.A. Petrii, A.R. Khokhlov, Structural and electrocatalytic features of Pt/C catalysts fabricated in supercritical carbon dioxide, J. Solid State Electr. 15 (2011) 623-633. https://doi.org/10.1007/s10008-010-1169-7

[63] Y.J. Luo, Y.H. Xiao, G.H. Cai, Y. Zheng, K.M. Wei, Complete methanol oxidation in carbon monoxide streams over Pd/CeO_2 catalysts: Correlation between activity and properties, Appl. Catal. B-Environ. 136 (2013) 317-324. https://doi.org/10.1016/j.apcatb.2013.02.020

[64] W. Wang, Y. Li, H. Wang, Improved methanol oxidation on a PtRu-RuO$_2$/C composite catalyst with close contact, React. Kinet. Mech. Cat. 108 (2013) 433-441. https://doi.org/10.1007/s11144-012-0532-3

[65] S. Jana, G. Mondal, B.C. Mitra, P. Bera, A. Mondal, Synthesis, characterization and electrocatalytic activity of SnO2, Pt-SnO2 thin films for methanol oxidation, Chem. Phys. 439 (2014) 44-48. https://doi.org/10.1016/j.chemphys.2014.05.003

[66] Y. Fan, J.H. Liu, H.T. Lu, P. Huang, D.L. Xu, Hierarchical structure SnO$_2$ supported Pt nanoparticles as enhanced electrocatalyst for methanol oxidation, Electrochim. Acta. 76 (2012) 475-479. https://doi.org/10.1016/j.electacta.2012.05.067

[67] H.Y. Sun, J.M. You, M.H. Yang, F.L. Qu, Synthesis of Pt/Fe$_3$O$_4$-CeO$_2$ catalyst with improved electrocatalytic activity for methanol oxidation, J. Power Sources. 205 (2012) 231-234. https://doi.org/10.1016/j.jpowsour.2012.01.014

[68] Y.H. Qin, Y.F. Li, R.L. Lv, T.L. Wang, W.G. Wang, C.W. Wang, 2 Enhanced methanol oxidation activity and stability of Pt particles anchored on carbon-doped TiO$_2$ nanocoating support, J. Power Sources. 278 (2015) 639-644. https://doi.org/10.1016/j.jpowsour.2014.12.096

[69] R.M.S. Rodrigues, R.R. Dias, C.A.L.G.O. Forbicini, M. Linardi, E.V. Spinace, A.O. Neto, Enhanced activity of PtRu/85%C+15% rare earth for methanol oxidation in acidic medium, Int. J. Electrochem. Sci. 6 (2011) 5759-5766.

[70] A. Bin Yousaf, M. Imran, N. Uwitonze, A. Zeb, S.J. Zaidi, T.M. Ansari, G. Yasmeen, S. Manzoor, Enhanced electrocatalytic performance of Pt$_3$Pd1 alloys supported on CeO$_2$/C for methanol oxidation and oxygen reduction reactions, J. Phys. Chem. C. 121 (2017) 2069-2079. https://doi.org/10.1021/acs.jpcc.6b11528

[71] H. Su, T.H. Chen, Preparation of PtSn$_2$-SnO$_2$/C nanocatalyst and its high performance for methanol electro-oxidation, Chinese. Chem. Lett. 27 (2016) 1083-1086. https://doi.org/10.1016/j.cclet.2016.03.010

[72] G.X. Wang, T. Takeguchi, Y. Zhang, E.N. Muhamad, M. Sadakane, S. Ye, W. Ueda, Effect of SnO$_2$deposition sequence in SnO$_2$-modified PtRu/C catalyst preparation on catalytic activity for methanol electro-oxidation, J. Electrochem. Soc. 156 (2009) B862-B869. https://doi.org/10.1149/1.3133249

[73] Y.Y. Chu, Z.H. Teng, B. Wu, Y.W. Tang, T.H. Lu, Y. Gao, Effect of mixed support of carbon black and nanographite on the activity of Pt catalyst for ethanol oxidation, J. Appl. Electrochem. 38 (2008) 1357-1362. https://doi.org/10.1007/s10800-008-9571-x

[74] X.Y. Ma, Y.F. Chen, H. Wang, Q.X. Li, W.F. Lin, W.B. Cai, Electrocatalytic oxidation of ethanol and ethylene glycol on cubic, octahedral and rhombic

dodecahedral palladium nanocrystals, Chem. Commun. 54 (2018) 2562-2565. https://doi.org/10.1039/C7CC08793D

[75] H. Chen, Y.Y. Huang, D. Tang, T. Zhang, Y.B. Wang, Ethanol oxidation on Pd/C promoted with $CaSiO_3$ in alkaline medium, Electrochim. Acta. 158 (2015) 18-23. https://doi.org/10.1016/j.electacta.2015.01.103

[76] F.M. Souza, J. Nandenha, B.L. Batista, V.H.A. Oliveira, V.S. Pinheiro, L.S. Parreira, A.O. Neto, M.C. Santos, PdxNby electrocatalysts for DEFC in alkaline medium: Stability, selectivity and mechanism for FOR, Int. J. Hydrogen Energ. 43 (2018) 4505-4516. https://doi.org/10.1016/j.ijhydene.2018.01.058

[77] H. Wang, Q.P. Zhao, Q. Ma, PtSn nanoparticles supported on iron nanoparticles wrapped inside nitrogen-doped carbon for ethanol oxidation, Ionics. 21 (2015) 1703-1709. https://doi.org/10.1007/s11581-014-1326-6

[78] J. Riberio, D.M. dos Anjos, K.B. Kokoh, C. Coutanceau, J.M. Leger, P. Olivi, A.R. de Andrade, G. Tremiliosi-Filho, Carbon-supported ternary PtSnIr catalysts for direct ethanol fuel cell, Electrochim. Acta. 52 (2007) 6997-7006. https://doi.org/10.1016/j.electacta.2007.05.017

[79] E. Higuchi, K. Miyata, H. Inoue, Preparation of nanoparticles of Pt and SnO_2highly dispersed on carbon black support and their activity for ethanol oxidation reaction, Electrochemistry. 78 (2010) 526-528. https://doi.org/10.5796/electrochemistry.78.526

[80] P.K. Shen, C.W. Xu, Alcohol oxidation on nanocrystalline oxide Pd/C promoted electrocatalysts, Electrochem. Commun. 8 (2006) 184-188. https://doi.org/10.1016/j.elecom.2005.11.013

[81] D.R.M. Godoi, H.M. Villullas, F.C. Zhu, Y.X. Jiang, S.G. Sun, J.S. Guo, L.L. Sun, R.R. Chen, A comparative investigation of metal-support interactions on the catalytic activity of Pt nanoparticles for ethanol oxidation in alkaline medium, J. Power Sources. 311 (2016) 81-90. https://doi.org/10.1016/j.jpowsour.2016.02.011

[82] R.F.B. De Souza, M.M. Tusi, M. Brandalise, R.R. Dias, M. Linardi, E.V. Spinace, M.C. dos Santos, A.O. Neto, Preparation of PtSn/C-Rh and PtSn/C-CeO_2 for Ethanol Electro-Oxidation, Int. J. Electrochem. Sci. 5 (2010) 895-902.

[83] R.F. Wang, Z.Y. Liu, Y.J. Ma, H. Wang, V. Linkov, S. Ji, Heterostructure core PdSn-SnO_2 decorated by Pt as efficient electrocatalysts for ethanol oxidation, Int. J. Hydrogen Energ. 38 (2013) 13604-13610. https://doi.org/10.1016/j.ijhydene.2013.08.044

[84] J.F. Pang, M.Y. Zheng, A.Q. Wang, T. Zhang, Catalytic hydrogenation of corn stalk to ethylene glycol and 1,2-propylene glycol, Ind. Eng. Chem. Res. 50 (2011) 6601-6608. https://doi.org/10.1021/ie102505y

[85] R.B. de Lima, V. Paganin, T. Iwasita, W. Vielstich, On the electrocatalysis of ethylene glycol oxidation, Electrochim. Acta. 49 (2003) 85-91. https://doi.org/10.1016/j.electacta.2003.05.004

[86] C. Coutanceau, L. Demarconnay, C. Lamy, J.M. Leger, Development of electrocatalysts for solid alkaline fuel cell (SAFC), J. Power Sources. 156 (2006) 14-19. https://doi.org/10.1016/j.jpowsour.2005.08.035

[87] M.T. Liu, L.X. Chen, D.N. Li, A.J. Wang, Q.L. Zhang, J.J. Feng, One-pot controlled synthesis of AuPd@Pd core-shell nanocrystals with enhanced electrocatalytic performances for formic acid oxidation and glycerol oxidation, J. Colloid Interf. Sci. 508 (2017) 551-558. https://doi.org/10.1016/j.jcis.2017.08.041

[88] Z.Z. Yang, L. Liu, A.J. Wang, J.H. Yuan, J.J. Feng, Q.Q. Xu, Simple wet-chemical strategy for large-scaled synthesis of snowflake-like PdAu alloy nanostructures as effective electrocatalysts of ethanol and ethylene glycol oxidation, Int. J. Hydrogen Energ. 42 (2017) 2034-2044. https://doi.org/10.1016/j.ijhydene.2016.08.088

[89] L. Demarconnay, S. Brimaud, C. Coutanceau, J.M. Leger, Ethylene glycol electrooxidation in alkaline medium at multi-metallic Pt based catalysts, J. Electroanal. Chem. 601 (2007) 169-180. https://doi.org/10.1016/j.jelechem.2006.11.006

[90] Y. Li, D. Lu, L.Q. Zhou, M.L. Ye, X. Xiong, K.Z. Yang, Y.X. Pan, M.H. Chen, P. Wu, T. Li, Y.T. Chen, Z. Wang, Q.H. Xia, Bi-modified Pd-based/carbon-doped TiO_2 hollow spheres catalytic for ethylene glycol electrooxidation in alkaline medium, J. Mater. Res. 31 (2016) 3712-3722. https://doi.org/10.1557/jmr.2016.429

[91] H.P. Liu, J.Q. Ye, C.W. Xu, S.P. Jiang, Y.X. Tong, Electro-oxidation of methanol, 1-propanol and 2-propanol on Pt and Pd in alkaline medium, J. Power Sources. 177 (2008) 67-70. https://doi.org/10.1016/j.jpowsour.2007.11.015

[92] E. Pastor, S. Wasmus, T. Iwasita, M.C. Arevalo, S. Gonzalez, A.J. Arvia, Spectroscopic investigations of C-3 primary alcohols on platinum-electrodes in acid-solutions .1. N-Propanol, J. Electroanal. Chem. 350 (1993) 97-116. https://doi.org/10.1016/0022-0728(93)80199-R

[93] A. Sadiki, P. Vo, S.Z. Hu, T.S. Copenhaver, L. Scudiero, S. Ha, J.L. Haan, Increased electrochemical oxidation rate of alcohols in alkaline media on palladium surfaces electrochemically modified by antimony, lead, and tin, Electrochim. Acta. 139 (2014) 302-307. https://doi.org/10.1016/j.electacta.2014.07.019

[94] C.W. Xu, Z.Q. Tian, Z.C. Chen, S.P. Jiang, Pd/C promoted by Au for 2-propanol electrooxidation in alkaline media, Electrochem. Commun. 10 (2008) 246-249. https://doi.org/10.1016/j.elecom.2007.11.036

[95] W.S. Su, T.C. Leung, C.T. Chan, Work function of single-walled and multiwalled carbon nanotubes: First-principles study, Phys. Rev. B. 76 (2007). https://doi.org/10.1103/PhysRevB.76.235413

[96] G. Wu, B.Q. Xu, Carbon nanotube supported Pt electrodes for methanol oxidation: A comparison between multi- and single-walled carbon nanotubes, J. Power Sources. 174 (2007) 148-158. https://doi.org/10.1016/j.jpowsour.2007.08.024

[97] R.R. Sanganna Gari, Z. Li, L. Dong, Effects of different carbon nanotube supported catalysts on methanol and ethanol electro-oxidation, MRS Proceedings. 1213 (2009) 1213-T1208-1217.

[98] M. Kumar, Y. Ando, Chemical vapor deposition of carbon nanotubes: A review on growth mechanism and mass production, J. Nanosci. Nanotechno. 10 (2010) 3739-3758. https://doi.org/10.1166/jnn.2010.2939

[99] L.C. Qin, Determination of the chiral indices (n,m) of carbon nanotubes by electron diffraction, Phys. Chem. Chem. Phys. 9 (2007) 31-48. https://doi.org/10.1039/B614121H

[100] C.H. Hsu, P.L. Kuo, The use of carbon nanotubes coated with a porous nitrogen-doped carbon layer with embedded Pt for the methanol oxidation reaction, J. Power Sources. 198 (2012) 83-89. https://doi.org/10.1016/j.jpowsour.2011.10.012

[101] J.H. Lehman, M. Terrones, E. Mansfield, K.E. Hurst, V. Meunier, Evaluating the characteristics of multiwall carbon nanotubes, Carbon. 49 (2011) 2581-2602. https://doi.org/10.1016/j.carbon.2011.03.028

[102] Q.F. Yi, L.Z. Sun, X.P. Liu, H.D. Nie, Palladium-nickel nanoparticles loaded on multi-walled carbon nanotubes modified with beta-cyclodextrin for electrooxidation of alcohols, Fuel. 111 (2013) 88-95. https://doi.org/10.1016/j.fuel.2013.04.051

[103] J.M. Sieben, M.M.E. Duarte, Methanol, ethanol and ethylene glycol electro-oxidation at Pt and Pt-Ru catalysts electrodeposited over oxidized carbon nanotubes, Int. J. Hydrogen Energ. 37 (2012) 9941-9947. https://doi.org/10.1016/j.ijhydene.2012.01.173

[104] V.S. Kumar, M. Satyanarayana, K.Y. Goud, K.V. Gobi, Pd nanoparticles-embedded carbon nanotube interface for electrocatalytic oxidation of methanol toward DMFC applications, Clean. Technol. Envir. 20 (2018) 759-768. https://doi.org/10.1007/s10098-017-1449-3

[105] Y.Y. Zhou, C.H. Liu, J. Liu, X.L. Cai, Y. Lu, H. Zhang, X.H. Sun, S.D. Wang, Self-decoration of PtNi alloy nanoparticles on multiwalled carbon nanotubes for highly efficient methanol electro-oxidation, Nano-Micro. Lett. 8 (2016) 371-380. https://doi.org/10.1007/s40820-016-0096-2

[106] M. Zhao, Y. Ji, N. Zhong, Fabrication of ultrafine amorphous Pd-Ni-P nanoparticles supported on carbon nanotubes as an effective catalyst for electro-oxidation of methanol, Int. J. Electrochem. Sci. 11 (2016) 10488-10497. https://doi.org/10.20964/2016.12.61

[107] Z.H. Yang, A. Nagashima, T. Fujigaya, N. Nakashima, Electrocatalyst composed of platinum nanoparticles deposited on doubly polymer-coated carbon nanotubes shows a high CO-tolerance in methanol oxidation reaction, Int. J. Hydrogen Energ. 41 (2016) 19182-19190. https://doi.org/10.1016/j.ijhydene.2016.08.198

[108] Z.H. Yang, F. Luo, Pt nanoparticles deposited on dihydroxy-polybenzimidazole wrapped carbon nanotubes shows a remarkable durability in methanol electro-oxidation, Int. J. Hydrogen Energ. 42 (2017) 507-514. https://doi.org/10.1016/j.ijhydene.2016.10.148

[109] K. Kardimi, T. Tsoufis, A. Tomou, B.J. Kooi, M.I. Prodromidis, D. Gournis, Synthesis and characterization of carbon nanotubes decorated with Pt and PtRu nanoparticles and assessment of their electrocatalytic performance, Int. J. Hydrogen Energ. 37 (2012) 1243-1253. https://doi.org/10.1016/j.ijhydene.2011.09.143

[110] M. Xu, Z. Sun, Q. Chen, B.K. Tay, Effect of chemical oxidation on the gas sensing properties of multi-walled carbon nanotubes, Int. J. Nanotechnol. 6 (2009) 735-744. https://doi.org/10.1504/IJNT.2009.025311

[111] J.J. Yang, Y.Y. Chu, L. Li, H.T. Wang, Z. Dai, X.Y. Tan, Effects of calcination temperature and CeO_2 contents on the performance of Pt/CeO_2-CNTs hybrid nanotube catalysts for methanol oxidation, J. Appl. Electrochem. 46 (2016) 369-377. https://doi.org/10.1007/s10800-016-0931-7

[112] C.M. Zhou, H.J. Wang, J.H. Liang, F. Peng, H. Yu, J. Yang, Effects of RuO_2content in Pt/RuO_2/CNTs nanocatalyst on the electrocatalytic oxidation performance of methanol, Chinese J. Catal. 29 (2008) 1093-1098. https://doi.org/10.1016/S1872-2067(09)60007-3

[113] M.H. Huang, J.S. Zhang, C.X. Wu, L.H. Guan, Pt Nanoparticles densely coated on SnO_2-covered multiwalled carbon nanotubes with excellent electrocatalytic activity and stability for methanol oxidation, Acs Appl. Mater. Interfaces. 9 (2017) 26921-26927. https://doi.org/10.1021/acsami.7b07866

[114] G. Liu, Z.C. Pan, W.Y. Li, K. Yu, G.W. Xia, Q.X. Zhao, S.K. Shi, G.H. Hu, C.M. Xiao, Z.G. Wei, The effect of titanium nickel nitride decorated carbon nanotubes-reduced graphene oxide hybrid support for methanol oxidation, Appl. Surf. Sci. 410 (2017) 70-78. https://doi.org/10.1016/j.apsusc.2017.03.075

[115] F. Peng, C.M. Zhou, H.J. Wang, H. Yu, J.H. Liang, J.A. Yang, The role of RuO_2 in the electrocatalytic oxidation of methanol for direct methanol fuel cell, Catal. Commun. 10 (2009) 533-537. https://doi.org/10.1016/j.catcom.2008.10.037

[116] L. Li, Y. Qian, J.J. Yang, X.Y. Tan, Z. Dai, Y.X. Jin, H.T. Wang, W.L. Qu, Y.Y. Chu, A novel structural design of hybrid nanotube with CNTs and CeO_2 supported Pt nanoparticles with improved performance for methanol electro-oxidation, Int. J. Hydrogen Energ. 41 (2016) 9284-9294. https://doi.org/10.1016/j.ijhydene.2016.04.069

[117] N. Kakati, J. Maiti, S.H. Jee, S.H. Lee, Y.S. Yoon, Hydrothermal synthesis of PtRu on CNT/SnO_2 composite as anode catalyst for methanol oxidation fuel cell, J. Alloy. Compd. 509 (2011) 5617-5622. https://doi.org/10.1016/j.jallcom.2011.02.087

[118] M.C. Tsai, T.K. Yeh, C.H. Tsai, Methanol oxidation efficiencies on carbon-nanotube-supported platinum and platinum-ruthenium nanoparticles prepared by pulsed electrodeposition, Int. J. Hydrogen Energ. 36 (2011) 8261-8266. https://doi.org/10.1016/j.ijhydene.2011.03.107

[119] Y.R. Wang, Q.L. He, K.Q. Ding, H.G. Wei, J. Guo, Q. Wang, R. O'Connor, X.H. Huang, Z.P. Luo, T.D. Shen, S.Y. Wei, Z.H. Guo, Multiwalled carbon nanotubes composited with palladium nanocatalysts for highly efficient ethanol oxidation, J. Electrochem. Soc. 162 (2015) F755-F763. https://doi.org/10.1149/2.0751507jes

[120] C. Wen, X. Zhang, Y. Wei, T. Zhang, C. Chen, A facile self-assembly approach to prepare palladium/carbon nanotubes catalyst for the electro-oxidation of ethanol, Mater. Res. Express. 5 (2018) 025013. https://doi.org/10.1088/2053-1591/aaa9f0

[121] J.M. Sieben, M.M.E. Duarte, Nanostructured Pt and Pt-Sn catalysts supported on oxidized carbon nanotubes for ethanol and ethylene glycol electro-oxidation, Int. J. Hydrogen Energ. 36 (2011) 3313-3321. https://doi.org/10.1016/j.ijhydene.2010.12.020

[122] S. Thilaga, S. Durga, V. Selvarani, S. Kiruthika, B. Muthukumaran, Multiwalled carbon nanotube supported Pt-Sn-M (M = Ru, Ni, and Ir) catalysts for ethanol electrooxidation, Ionics. 24 (2018) 1721-1731. https://doi.org/10.1007/s11581-017-2349-6

[123] Q.F. Yi, H. Chu, Q.H. Chen, Z. Yang, X.P. Liu, High performance Pd, PdNi, PdSn and PdSnNi nanocatalysts supported on carbon nanotubes for electrooxidation of C2-C4 alcohols, Electroanal. 27 (2015) 388-397. https://doi.org/10.1002/elan.201400423

[124] H.Q. Song, X.P. Qiu, F.S. Li, Promotion of carbon nanotube-supported Pt catalyst for methanol and ethanol electro-oxidation by ZrO2 in acidic media, Appl. Catal a-Gen. 364 (2009) 1-7. https://doi.org/10.1016/j.apcata.2009.04.046

[125] H.Q. Song, X.P. Qiu, F.S. Li, W.T. Zhu, L.Q. Chen, Ethanol electro-oxidation on catalysts with TiO_2 coated carbon nanotubes as support, Electrochem. Commun. 9 (2007) 1416-1421. https://doi.org/10.1016/j.elecom.2007.01.048

[126] H.L. Pang, J.P. Lu, J.H. Chen, C.T. Huang, B. Liu, X.H. Zhang, Preparation of SnO_2-CNTs supported Pt catalysts and their electrocatalytic properties for ethanol oxidation, Electrochim. Acta. 54 (2009) 2610-2615. https://doi.org/10.1016/j.electacta.2008.10.058

[127] Q. Liu, K. Jiang, J.C. Fan, Y. Lin, Y.L. Min, Q.J. Xu, W.B. Cai, Manganese dioxide coated graphene nanoribbons supported palladium nanoparticles as an efficient catalyst for ethanol electrooxidation in alkaline media, Electrochim. Acta. 203 (2016) 91-98. https://doi.org/10.1016/j.electacta.2016.04.021

[128] V. Selvaraj, M. Vinoba, M. Alagar, Electrocatalytic oxidation of ethylene glycol on Pt and Pt-Ru nanoparticles modified multi-walled carbon nanotubes, J. Colloid. Interf. Sci. 322 (2008) 537-544. https://doi.org/10.1016/j.jcis.2008.02.069

[129] N.W. Maxakato, C.J. Arendse, K.I. Ozoemena, Insights into the electro-oxidation of ethylene glycol at Pt/Ru nanocatalysts supported on MWCNTs: Adsorption-controlled electrode kinetics, Electrochem. Commun. 11 (2009) 534-537. https://doi.org/10.1016/j.elecom.2008.12.038

[130] Z.P. Sun, X.G. Zhang, Y.Y. Liang, H.L. Li, A facile approach towards sulfonate functionalization of multi-walled carbon nanotubes as Pd catalyst support for ethylene glycol electro-oxidation, J. Power Sources. 191 (2009) 366-370. https://doi.org/10.1016/j.jpowsour.2009.01.093

[131] T. Ramulifho, K.I. Ozoemena, R.M. Modibedi, C.J. Jafta, M.K. Mathe, Electrocatalytic oxidation of ethylene glycol at palladium-bimetallic nanocatalysts (PdSn and PdNi) supported on sulfonate-functionalised multi-walled carbon nanotubes, J. Electroanal. Chem. 692 (2013) 26-30. https://doi.org/10.1016/j.jelechem.2012.12.010

[132] R. Khaleghian-Moghadam, M. Noroozifar, M. Khorasani-Motlagh, M.S. Ekrami-Kakhki, Electrochemical activities of platinum-decorated multi-wall carbon nanotube/chitosan composites for the oxidations of alcohols, J. Solid. State. Electr. 17 (2013) 643-654. https://doi.org/10.1007/s10008-012-1908-z

[133] Y.y. Zhang, Q.f. Yi, H. Chu, H.d. Nie, Catalytic activity of Pd-Ag nanoparticles supported on carbon nanotubes for the electro-oxidation of ethanol and propanol, J. Fuel Chem. Tech. 45 (2017) 475-483. https://doi.org/10.1016/S1872-5813(17)30026-9

[134] S.I. Kim, T. Yamamoto, A. Endo, T. Ohmori, M. Nakaiwa, Preparation of platinum nanoparticles supported on resorcinol-formaldehyde carbon cryogel microspheres, J. Ind. Eng. Chem. 12 (2006) 769-776.

[135] M. Enterria, J.L. Figueiredo, Nanostructured mesoporous carbons: Tuning texture and surface chemistry, Carbon. 108 (2016) 79-102. https://doi.org/10.1016/j.carbon.2016.06.108

[136] R.W. Pekala, Organic aerogels from the polycondensation of resorcinol with formaldehyde, J. Mater. Sci. 24 (1989) 3221-3227. https://doi.org/10.1007/BF01139044

[137] H. Tamon, H. Ishizaka, T. Yamamoto, T. Suzuki, Influence of freeze-drying conditions on the mesoporosity of organic gels as carbon precursors, Carbon. 38 (2000) 1099-1105. https://doi.org/10.1016/S0008-6223(99)00235-3

[138] P.V. Samant, J.B. Fernandes, C.M. Rangel, J.L. Figueiredo, Carbon xerogel supported Pt and Pt-Ni catalysts for electro-oxidation of methanol in basic medium, Catal. Today. 102 (2005) 173-176. https://doi.org/10.1016/j.cattod.2005.02.039

[139] H. Zhu, Z.J. Guo, X.W. Zhang, K.F. Han, Y.B. Guo, F.H. Wang, Z.M. Wang, Y.S. Wei, Methanol-tolerant carbon aerogel-supported Pt-Au catalysts for direct methanol fuel cell, Int. J. Hydrogen Energ. 37 (2012) 873-876. https://doi.org/10.1016/j.ijhydene.2011.04.032

[140] S.L. Wei, D.C. Wu, X.L. Shang, R.W. Fu, Studies on the structure and electrochemical performance of pt/carbon aerogel catalyst for direct methanol fuel cells, Energ. Fuel. 23 (2009) 908-911. https://doi.org/10.1021/ef8006432

[141] Y. Liu, M. Wei, J.Y. Qu, L.Q. Mao, Carbon aerogels supported Pt nanoparticles as electrocatalysts for methanol oxidation in alkaline media, J. Chin. Chem. Soc-Taip. 61 (2014) 404-408. https://doi.org/10.1002/jccs.201300637

[142] X.L. Wang, C. Li, G.Q. Shi, A high-performance platinum electrocatalyst loaded on a graphene hydrogel for high-rate methanol oxidation, Phys. Chem. Chem. Phys. 16 (2014) 10142-10148. https://doi.org/10.1039/c3cp54058h

[143] J.L. Duan, X.L. Zhang, W.J. Yuan, H.L. Chen, S. Jiang, X.W. Liu, Y.F. Zhang, L.M. Chang, Z.Y. Sun, J. Du, Graphene oxide aerogel-supported Pt electrocatalysts for methanol oxidation, J. Power Sources. 285 (2015) 76-79. https://doi.org/10.1016/j.jpowsour.2015.03.064

[144] L. Zhao, Z.B. Wang, J.L. Li, J.J. Zhang, X.L. Sui, L.M. Zhang, One-pot synthesis of a three-dimensional graphene aerogel supported Pt catalyst for methanol electrooxidation, Rsc Adv. 5 (2015) 98160-98165. https://doi.org/10.1039/C5RA20503D

[145] P. Kolla, K. Kerce, Y. Normah, H. Fong, A. Smirnova, Metal oxides modified mesoporous carbon supports as anode catalysts in DMFC, Ecs. Transactions. 45 (2013) 35-45. https://doi.org/10.1149/04521.0035ecst

[146] L. Zhao, Z.B. Wang, J.L. Li, J.J. Zhang, X.L. Sui, L.M. Zhang, Hybrid of carbon-supported Pt nanoparticles and three dimensional graphene aerogel as high stable electrocatalyst for methanol electrooxidation, Electrochim. Acta. 189 (2016) 175-183. https://doi.org/10.1016/j.electacta.2015.12.072

[147] C.H.A. Tsang, K.N. Hui, K.S. Hui, L. Ren, Deposition of Pd/graphene aerogel on nickel foam as a binder-free electrode for direct electro-oxidation of methanol and ethanol, J. Mater. Chem. A. 2 (2014) 17986-17993. https://doi.org/10.1039/C4TA03138E

[148] Y.J. Huo, F.F. Yao, Y.S. Ma, Catalytic Performance of graphite oxide supported Au nanoparticles in aerobic oxidation of benzyl alcohol: Support effect, Chinese J. Chem. Phys. 30 (2017) 90-96. https://doi.org/10.1063/1674-0068/30/cjcp1604088

[149] L. Ren, K.S. Hui, K.N. Hui, Self-assembled free-standing three-dimensional nickel nanoparticle/graphene aerogel for direct ethanol fuel cells, J. Mater. Chem. A. 1 (2013) 5689-5694. https://doi.org/10.1039/c3ta10657h

[150] X.F. Zhang, Z.Q. Tian, P.K. Shen, Composite of nanosized carbides and carbon aerogel and its supported Pd electrocatalyst for synergistic oxidation of ethylene glycol, Electrochem. Commun. 28 (2013) 9-12. https://doi.org/10.1016/j.elecom.2012.11.031

[151] A. Krittayavathananon, M. Sawangphruk, Electrocatalytic oxidation of ethylene glycol on palladium coated on 3D reduced graphene oxide aerogel paper in alkali media: Effects of carbon supports and hydrodynamic diffusion, Electrochim. Acta. 212 (2016) 237-246. https://doi.org/10.1016/j.electacta.2016.06.162

[152] A.H. Castro Neto, F. Guinea, N.M.R. Peres, K.S. Novoselov, A.K. Geim, The electronic properties of graphene, Rev. Mod. Phys. 81 (2009) 109-162. https://doi.org/10.1103/RevModPhys.81.109

[153] S. Park, R.S. Ruoff, Chemical methods for the production of graphenes, Nat. Nanotechnol. 4 (2009) 217-224. https://doi.org/10.1038/nnano.2009.58

[154] Y.W. Zhu, S. Murali, W.W. Cai, X.S. Li, J.W. Suk, J.R. Potts, R.S. Ruoff, Graphene and graphene oxide: synthesis, properties, and applications, Adv. Mater. 22 (2010) 3906-3924. https://doi.org/10.1002/adma.201001068

[155] H.L. Gao, L.L. He, Y. Zhang, S.L. Zhang, L.Z. Wang, Facile synthesis of Pt nanoparticles supported on graphene/Vulcan XC-72 carbon and their application for

methanol oxidation, Ionics. 23 (2017) 435-442. https://doi.org/10.1007/s11581-016-1861-4

[156] S. Woo, J. Lee, S.K. Park, H. Kim, T.D. Chung, Y. Piao, Enhanced electrocatalysis of PtRu onto graphene separated by Vulcan carbon spacer, J. Power Sources. 222 (2013) 261-266. https://doi.org/10.1016/j.jpowsour.2012.07.115

[157] Y.L. Ma, Q. Wang, Y.L. Miao, Y. Lin, R.Y. Li, Plasma synthesis of Pt nanoparticles on 3D reduced graphene oxidecarbon nanotubes nanocomposites towards methanol oxidation reaction, Appl. Surf. Sci. 450 (2018) 413-421. https://doi.org/10.1016/j.apsusc.2018.04.094

[158] S.H. Yang, F.F. Zhang, C.L. Gao, J.F. Xia, L. Lu, Z.H. Wang, A sandwich-like PtCo-graphene/carbon dots/graphene catalyst for efficient methanol oxidation, J. Electroanal. Chem. 802 (2017) 27-32. https://doi.org/10.1016/j.jelechem.2017.08.027

[159] R. Zhang, W.F. Xia, W.J. Kang, R. Li, K.G. Qu, Y.T. Zhang, B.L. Chen, H.S. Wang, Y.F. Sun, H.B. Li, Methanol oxidation reaction performance on graphene-supported ptag alloy nanocatalyst: Contrastive study of electronic and geometric effects induced from Ag doping, Chemistryselect. 3 (2018) 3615-3620. https://doi.org/10.1002/slct.201800010

[160] J.L. Xie, X.J. Yang, X.Y. Xu, C. Yang, Microwave synthesis of reduced graphene oxide-supported platinum nanocomposite with high electrocatalytic activity for methanol oxidation, Int. J. Electrochem. Sci. 12 (2017) 466-474. https://doi.org/10.20964/2017.01.42

[161] A. Eshghi, M. Kheirmand, M.M. Sabzehmeidani, platinum-iron nanoparticles supported on reduced graphene oxide as an improved catalyst for methanol electro oxidation, Int. J. Hydrogen Energ. 43 (2018) 6107-6116. https://doi.org/10.1016/j.ijhydene.2018.01.206

[162] F.H. Li, Y.Q. Guo, M.X. Chen, H.X. Qiu, X.Y. Sun, W. Wang, Y. Liu, J.P. Gao, Comparison study of electrocatalytic activity of reduced graphene oxide supported Pt-Cu bimetallic or Pt nanoparticles for the electrooxidation of methanol and ethanol, Int. J. Hydrogen Energ. 38 (2013) 14242-14249. https://doi.org/10.1016/j.ijhydene.2013.08.093

[163] J. Florez-Montano, A. Calderon-Cardenas, W. Lizcano-Valbuena, J.L. Rodriguez, E. Pastor, Ni@Pt nanodisks with low Pt content supported on reduced graphene oxide for methanol electrooxidation in alkaline media, Int. J. Hydrogen Energ. 41 (2016) 19799-19809. https://doi.org/10.1016/j.ijhydene.2016.06.166

[164] M.S. Ahmed, S. Jeon, Synthesis and electrocatalytic activity evaluation of nanoflower shaped Ni-Pd on alcohol oxidation reaction, J. Electrochem. Soc. 161 (2014) F1300-F1306. https://doi.org/10.1149/2.1041412jes

[165] Y. Yang, L.M. Luo, Y.F. Guo, Z.X. Dai, R.H. Zhang, C.H. Sun, X.W. Zhou, In situ synthesis of PtPd bimetallic nanocatalysts supported on graphene nanosheets for methanol oxidation using triblock copolymer as reducer and stabilizer, J. Electroanal. Chem. 783 (2016) 132-139. https://doi.org/10.1016/j.jelechem.2016.11.034

[166] H.X. Wang, L.M. Sheng, X.L. Zhao, K. An, Z.M. Ou, Y.H. Fang, One-step synthesis of Pt-Pd catalyst nanoparticles supported on few-layer graphene for methanol oxidation, Curr. Appl. Phys. 18 (2018) 898-904. https://doi.org/10.1016/j.cap.2018.04.006

[167] X.Y. Yan, T. Liu, J. Jin, S. Devaramani, D.D. Qin, X.Q. Lu, Well dispersed Pt-Pd bimetallic nanoparticles on functionalized graphene as excellent electro-catalyst towards electro-oxidation of methanol, J. Electroanal. Chem. 770 (2016) 33-38. https://doi.org/10.1016/j.jelechem.2016.03.033

[168] S. Themsirimongkon, K. Ounnunkad, S. Saipanya, Electrocatalytic enhancement of platinum and palladium metal on polydopamine reduced graphene oxide support for alcohol oxidation, J. Colloid Interf. Sci. 530 (2018) 98-112. https://doi.org/10.1016/j.jcis.2018.06.072

[169] R.N. Singh, R. Awasthi, Graphene support for enhanced electrocatalytic activity of Pd for alcohol oxidation, Catal. Sci. Technol. 1 (2011) 778-783. https://doi.org/10.1039/c1cy00021g

[170] S. Themsirimongkon, A. Khammamung, A. Pinithchaisakula, K. Ounangkad, S. Saipanya, Determination of PtAuPd metal sequences for electrodeposition on graphene oxide for anode catalyst improvement in methanol oxidation, Mol. Cryst. Liq. Cryst. 653 (2017) 164-176. https://doi.org/10.1080/15421406.2017.1351281

[171] E.J. Lim, Y. Kim, S.M. Choi, S. Lee, Y. Noh, W.B. Kim, Binary PdM catalysts (M = Ru, Sn, or Ir) over a reduced graphene oxide support for electro-oxidation of primary alcohols (methanol, ethanol, 1-propanol) under alkaline conditions, J. Mater. Chem. A. 3 (2015) 5491-5500. https://doi.org/10.1039/C4TA06893A

[172] H.Q. Ye, Y.M. Li, J.H. Chen, J.L. Sheng, X.Z. Fu, R. Sun, C.P. Wong, PdCu alloy nanoparticles supported on reduced graphene oxide for electrocatalytic oxidation of methanol, J. Mater. Sci. 53 (2018) 15871-15881. https://doi.org/10.1007/s10853-018-2759-5

[173] F. Yang, B. Zhang, S. Dong, Y.S. Tang, L.Q. Hou, Z. Chen, Z.H. Li, W. Yang, C. Xu, M.J. Wang, Y. Li, Y.F. Li, Silica nanosphere supported palladium nanoparticles

encapsulated with graphene: High-performance electrocatalysts for methanol oxidation reaction, Appl. Surf. Sci. 452 (2018) 11-18.
https://doi.org/10.1016/j.apsusc.2018.05.022

[174] T.H.T. Vu, T.T.T. Tran, H.N.T. Le, L.T. Tran, P.H.T. Nguyen, H.T. Nguyen, N.Q. Bui, Solvothermal synthesis of Pt -SiO$_2$/graphene nanocomposites as efficient electrocatalyst for methanol oxidation, Electrochim. Acta. 161 (2015) 335-342.
https://doi.org/10.1016/j.electacta.2015.02.100

[175] H.Q. Zhang, X. Han, Y. Zhao, Pd-TiO$_2$ nanoparticles supported on reduced graphene oxide: Green synthesis and improved electrocatalytic performance for methanol oxidation, J. Electroanal. Chem. 799 (2017) 84-91.
https://doi.org/10.1016/j.jelechem.2017.05.026

[176] L.T. Ye, Z.S. Li, L. Zhang, F.L. Lei, S. Lin, A green one-pot synthesis of Pt/TiO$_2$/graphene composites and its electro-photo-synergistic catalytic properties for methanol oxidation, J. Colloid Interf. Sci. 433 (2014) 156-162.
https://doi.org/10.1016/j.jcis.2014.06.012

[177] W. Zhuang, L.J. He, J.H. Zhu, R. An, X.B. Wu, L.W. Mu, X.H. Lu, L.H. Lu, X.J. Liu, H.J. Ying, TiO$_2$ nanofibers heterogeneously wrapped with reduced graphene oxide as efficient Pt electrocatalyst supports for methanol oxidation, Int. J. Hydrogen. Energ. 40 (2015) 3679-3688. https://doi.org/10.1016/j.ijhydene.2015.01.042

[178] R. Liu, H.H. Zhou, J. Liu, Y. Yao, Z.Y. Huang, C.P. Fu, Y.F. Kuang, Preparation of Pd/MnO$_2$-reduced graphene oxide nanocomposite for methanol electro-oxidation in alkaline media, Electrochem. Commun. 26 (2013) 63-66.
https://doi.org/10.1016/j.elecom.2012.10.019

[179] C.S. Sharma, A.S.K. Sinha, R.N. Singh, Use of graphene-supported manganite nano-composites for methanol electrooxidation, Int. J. Hydrogen Energ. 39 (2014) 20151-20158. https://doi.org/10.1016/j.ijhydene.2014.10.019

[180] B. Celik, G. Baskaya, H. Sert, O. Karatepe, E. Erken, F. Sen, Monodisperse Pt(0)/DPA@GO nanoparticles as highly active catalysts for alcohol oxidation and dehydrogenation of DMAB, Int. J. Hydrogen Energ. 41 (2016) 5661-5669.
https://doi.org/10.1016/j.ijhydene.2016.02.061

[181] J. Zhang, A. Feng, J. Bai, Z. Tan, W. Shao, Y. Yang, W. Hong, Z. Xiao, One-pot synthesis of hierarchical flower-like Pd-Cu alloy support on graphene towards ethanol oxidation, Nanoscale Res. Lett. 12 (2017) 521. https://doi.org/10.1186/s11671-017-2290-7

[182] Q. Dong, Y. Zhao, X. Han, Y. Wang, M.C. Liu, Y. Li, Pd/Cu bimetallic nanoparticles supported on graphene nanosheets: Facile synthesis and application as

novel electrocatalyst for ethanol oxidation in alkaline media, Int. J. Hydrogen Energ. 39 (2014) 14669-14679. https://doi.org/10.1016/j.ijhydene.2014.06.139

[183] F. Wang, J.S. Qiao, J. Wang, H.T. Wu, X.Y. Yue, Z.H. Wang, W. Sun, K.N. Sun, Reduced graphene oxide supported Ni@Au@Pd core@bishell nanoparticles as highly active electrocatalysts for ethanol oxidation reactions and alkaline direct bioethanol fuel cells applications, Electrochim. Acta. 271 (2018) 1-9. https://doi.org/10.1016/j.electacta.2018.03.013

[184] G.H. Jeong, D. Choi, M. Kang, J. Shin, J.G. Kang, S.W. Kim, One-pot synthesis of Au@ Pd/graphene nanostructures: electrocatalytic ethanol oxidation for direct alcohol fuel cells (DAFCs), Rsc Adv. 3 (2013) 8864-8870. https://doi.org/10.1039/c3ra40505b

[185] Q.F. Zhang, X.F. Wu, M.Y. Gao, H.F. Qiu, J. Hu, K.K. Huang, S.H. Feng, Y. Yang, T.T. Wang, B. Zhao, Z.L. Liu, Highly active electrocatalyst of 3D Pd/reduced graphene oxide nanostructure for electro-oxidation of methanol and ethanol, Inorg. Chem. Commun. 94 (2018) 43-47. https://doi.org/10.1016/j.inoche.2018.05.028

[186] N. Alfi, M.Z. Yazdan-Abad, A. Rezvani, M. Noroozifar, M. Khorasani-Motlagh, Three-dimensional Pd-Cd nanonetwork decorated on reduced graphene oxide by a galvanic method as a novel electrocatalyst for ethanol oxidation in alkaline media, J. Power Sources. 396 (2018) 742-748. https://doi.org/10.1016/j.jpowsour.2018.06.080

[187] H.V. Hien, T.D. Thanh, N.D. Chuong, D. Hui, N.H. Kim, J.H. Lee, Hierarchical porous framework of ultrasmall PtPd alloy-integrated graphene as active and stable catalyst for ethanol oxidation, Compos. Part B-Eng. 143 (2018) 96-104. https://doi.org/10.1016/j.compositesb.2018.02.013

[188] Y.Y. Zheng, J.H. Qiao, J.H. Yuan, J.F. Shen, A.J. Wang, S.T. Huang, Controllable synthesis of PtPd nanocubes on graphene as advanced catalysts for ethanol oxidation, Int. J. Hydrogen Energ. 43 (2018) 4902-4911. https://doi.org/10.1016/j.ijhydene.2018.01.131

[189] T. Oznuluer, U. Demir, H.O. Dogan, Fabrication of underpotentially deposited Cu monolayer/electrochemically reduced graphene oxide layered nanocomposites for enhanced ethanol electro-oxidation, Appl. Catal. B-Environ. 235 (2018) 56-65. https://doi.org/10.1016/j.apcatb.2018.04.065

[190] D. Puthusseri, S. Ramaprabhu, Platinum and SnO_2 Decorated graphene sheets as ethanol oxidation electrocatalyst in acidic medium, Graphene. 3 (2015) 29-33. https://doi.org/10.1166/graph.2015.1051

[191] Y.T. Qu, Y.Z. Gao, L. Wang, J.C. Rao, G.P. Yin, Mild synthesis of Pt/SnO2/graphene nanocomposites with remarkably enhanced ethanol electro-

oxidation activity and durability, Chem. Eur. J. 22 (2016) 193-198.
https://doi.org/10.1002/chem.201503867

[192] A.E. Fahim, R.M.A. Hameed, N.K. Allam, Synthesis and characterization of core-shell structured M@Pd/SnO_2-graphene [M = Co, Ni or Cu] electrocatalysts for ethanol oxidation in alkaline solution, New J. Chem. 42 (2018) 6144-6160.
https://doi.org/10.1039/C8NJ01078A

[193] Q. He, S. Mukerjee, B. Shyam, D. Ramaker, S. Parres-Esclapez, M.J. Illan-Gomez, A. Bueno-Lopez, Promoting effect of CeO_2 in the electrocatalytic activity of rhodium for ethanol electro-oxidation, J. Power Sources. 193 (2009) 408-415.
https://doi.org/10.1016/j.jpowsour.2009.03.056

[194] K. Kakaei, A. Rahimi, S. Husseindoost, M. Hamidi, H. Javan, A. Balavandi, Fabrication of Pt-CeO_2 nanoparticles supported sulfonated reduced graphene oxide as an efficient electrocatalyst for ethanol oxidation, Int. J. Hydrogen Energ. 41 (2016) 3861-3869. https://doi.org/10.1016/j.ijhydene.2016.01.013

[195] D.V. Kosynkin, A.L. Higginbotham, A. Sinitskii, J.R. Lomeda, A. Dimiev, B.K. Price, J.M. Tour, Longitudinal unzipping of carbon nanotubes to form graphene nanoribbons, Nature. 458 (2009) 872. https://doi.org/10.1038/nature07872

[196] D. Rajesh, P.I. Neel, A. Pandurangan, C. Mahendiran, Pd-NiO decorated multiwalled carbon nanotubes supported on reduced graphene oxide as an efficient electrocatalyst for ethanol oxidation in alkaline medium, Appl. Surf. Sci. 442 (2018) 787-796. https://doi.org/10.1016/j.apsusc.2018.02.174

[197] R.R. Yue, H.W. Wang, D. Bin, J.K. Xu, Y.K. Du, W.S. Lu, J. Guo, Facile one-pot synthesis of Pd-PEDOT/graphene nanocomposites with hierarchical structure and high electrocatalytic performance for ethanol oxidation, J. Mater. Chem. A. 3 (2015) 1077-1088. https://doi.org/10.1039/C4TA05131A

[198] S. Shahrokhian, S. Rezaee, Fabrication of trimetallic Pt-Pd-Co porous nanostructures on reduced graphene oxide by galvanic replacement: Application to electrocatalytic oxidation of ethylene glycol, Electroanal. 29 (2017) 2591-2601.
https://doi.org/10.1002/elan.201700355

[199] F.Q. Shao, X.X. Lin, J.J. Feng, J. Yuan, J.R. Chen, A.J. Wang, Simple fabrication of core-shell AuPt@Pt nanocrystals supported on reduced graphene oxide for ethylene glycol oxidation and hydrogen evolution reactions, Electrochim. Acta. 219 (2016) 321-329. https://doi.org/10.1016/j.electacta.2016.09.158

[200] J.N. Zheng, J.J. Lv, S.S. Li, M.W. Xue, A.J. Wang, J.J. Feng, One-pot synthesis of reduced graphene oxide supported hollow Ag@Pt core-shell nanospheres with

enhanced electrocatalytic activity for ethylene glycol oxidation, J. Mater. Chem. A. 2 (2014) 3445-3451. https://doi.org/10.1039/c3ta13935b

[201] P. Wu, Y.Y. Huang, L.T. Kang, M.X. Wu, Y.B. Wang, Multisource synergistic electrocatalytic oxidation effect of strongly coupled PdM (M = Sn, Pb)/N-doped graphene nanocomposite on small organic molecules, Sci. Rep-Uk. 5 (2015) 14173. https://doi.org/10.1038/srep14173

[202] C.W. Xu, P.K. Shen, Novel Pt/CeO$_2$/C catalysts for electrooxidation of alcohols in alkaline media, Chem. Commun. (2004) 2238-2239. https://doi.org/10.1039/b408589b

[203] Q. He, Y. Shen, K.J. Xiao, J.Y. Xi, X.P. Qiu, Alcohol electro-oxidation on platinum-ceria/graphene nanosheet in alkaline solutions, Int. J. Hydrogen Energ. 41 (2016) 20709-20719. https://doi.org/10.1016/j.ijhydene.2016.07.205

[204] L. Zhang, Y. Shen, One-pot synthesis of platinum-ceria/graphene nanosheet as advanced electrocatalysts for alcohol oxidation, Chemelectrochem. 2 (2015) 887-895. https://doi.org/10.1002/celc.201402432

[205] J.R.C. Salgado, V.A. Paganin, E.R. Gonzalez, M.F. Montemor, I. Tacchini, A. Anson, M.A. Salvador, P. Ferreira, F.M.L. Figueiredo, M.G.S. Ferreira, Characterization and performance evaluation of Pt-Ru electrocatalysts supported on different carbon materials for direct methanol fuel cells, Int. J. Hydrogen Energ. 38 (2013) 910-920. https://doi.org/10.1016/j.ijhydene.2012.10.079

[206] J.B. Joo, P. Kim, W. Kim, J. Yi, Preparation and application of mesocellular carbon foams to catalyst support in methanol electro-oxidation, Catal. Today. 131 (2008) 219-225. https://doi.org/10.1016/j.cattod.2007.10.086

[207] D. Morales-Acosta, F.J. Rodriguez-Varela, R. Benavides, Template-free synthesis of ordered mesoporous carbon: Application as support of highly active Pt nanoparticles for the oxidation of organic fuels, Int. J. Hydrogen Energ. 41 (2016) 3387-3398. https://doi.org/10.1016/j.ijhydene.2015.10.114

[208] S.Q. Song, K. Wang, Y.H. Liu, C.X. He, Y.R. Liang, R.W. Fu, D.C. Wu, Y. Wang, Highly ordered mesoporous carbons as the support for Pt catalysts towards alcohol electrooxidation: The combined effect of pore size and electrical conductivity, Int. J. Hydrogen Energ. 38 (2013) 1405-1412. https://doi.org/10.1016/j.ijhydene.2012.11.029

[209] L.R. Nan, W.B. Yue, Exceptional electrocatalytic activity and selectivity of platinum@nitrogen-doped mesoporous carbon nanospheres for alcohol oxidation, Acs Appl. Mater. Inter.10 (2018) 26213-26221. https://doi.org/10.1021/acsami.8b06347

[210] J.R.C. Salgado, F. Alcaide, G. Alvarez, L. Calvillo, M.J. Lazaro, E. Pastor, Pt-Ru electrocatalysts supported on ordered mesoporous carbon for direct methanol fuel cell,

J. Power Sources. 195 (2010) 4022-4029.
https://doi.org/10.1016/j.jpowsour.2010.01.001

[211] S. Sharma, B.G. Pollet, Support materials for PEMFC and DMFC electrocatalysts-A review, J. Power Sources. 208 (2012) 96-119.
https://doi.org/10.1016/j.jpowsour.2012.02.011

[212] R. Ryoo, S.H. Joo, S. Jun, Synthesis of highly ordered carbon molecular sieves via template-mediated structural transformation, J. Phys. Chem. B. 103 (1999) 7743-7746.
https://doi.org/10.1021/jp991673a

[213] R. Ryoo, S.H. Joo, M. Kruk, M. Jaroniec, Ordered mesoporous carbons, Adv. Mater. 13 (2001) 677-681. https://doi.org/10.1002/1521-4095(200105)13:9<677::AID-ADMA677>3.0.CO;2-C

[214] B. Kuppan, P. Selvam, Platinum-supported mesoporous carbon (Pt/CMK-3) as anodic catalyst for direct methanol fuel cell applications: The effect of preparation and deposition methods, Prog. Nat. Sci-Mater. 22 (2012) 616-624.
https://doi.org/10.1016/j.pnsc.2012.11.005

[215] L.B. Kong, H. Li, J. Zhang, Y.C. Luo, L. Kang, Platinum catalyst on ordered mesoporous carbon with controlled morphology for methanol electrochemical oxidation, Appl. Surf. Sci. 256 (2010) 6688-6693.
https://doi.org/10.1016/j.apsusc.2010.04.071

[216] Z.P. Sun, X.G. Zhang, Y.Y. Liang, H. Tong, R.L. Xue, S.D. Yang, H.L. Li, Ordered mesoporous carbons (OMCs) as supports of electrocatalysts for direct methanol fuel cells (DMFCs): Effect of the pore characteristics of OMCs on DMFCs, J. Electroanal. Chem. 633 (2009) 1-6. https://doi.org/10.1016/j.jelechem.2009.04.013

[217] W.F. Liu, X.P. Qin, X.F. Zhang, Z.G. Shao, B.L. Yi, Wormholelike mesoporous carbon supported PtRu catalysts toward methanol electrooxidation, J. Energy Chem. 26 (2017) 200-206. https://doi.org/10.1016/j.jechem.2016.08.003

[218] F.J. Li, K.Y. Chan, H. Yung, C.Z. Yang, S.W. Ting, Uniform dispersion of 1: 1 PtRu nanoparticles in ordered mesoporous carbon for improved methanol oxidation, Phys. Chem. Chem. Phys. 15 (2013) 13570-13577. https://doi.org/10.1039/c3cp00153a

[219] D.S. Yuan, X.L. Yuan, W.J. Zou, F.L. Zeng, X.J. Huang, S.L. Zhou, Synthesis of graphitic mesoporous carbon from sucrose as a catalyst support for ethanol electro-oxidation, J. Mater. Chem. 22 (2012) 17820-17826.
https://doi.org/10.1039/c2jm33658h

[220] M.H. Chen, Y.X. Jiang, S.R. Chen, R. Huang, J.L. Lin, S.P. Chen, S.G. Sun, Synthesis and durability of highly dispersed platinum nanoparticles supported on

ordered mesoporous carbon and their electrocatalytic properties for ethanol oxidation, J. Phys. Chem. C. 114 (2010) 19055-19061. https://doi.org/10.1021/jp1091398

[221] Z.X. Yan, H. Meng, L. Shi, Z.H. Li, P.K. Shen, Synthesis of mesoporous hollow carbon hemispheres as highly efficient Pd electrocatalyst support for ethanol oxidation, Electrochem. Commun. 12 (2010) 689-692. https://doi.org/10.1016/j.elecom.2010.03.007

[222] M.A. Hogue, D.C. Higgins, F.M. Hassan, J.Y. Choi, M.D. Pritzker, Z.W. Chen, Tin oxide - mesoporous carbon composites as platinum catalyst supports for ethanol oxidation and oxygen reduction, Electrochim. Acta. 121 (2014) 421-427. https://doi.org/10.1016/j.electacta.2013.12.075

[223] A.Y. Lo, Y.C. Chung, W.H. Hung, Y.C. Hsu, C.M. Tseng, W.L. Zhang, F.K. Wang, C.Y. Lin, $Pt_2 0RuxSny$ nanoparticles dispersed on mesoporous carbon CMK-3 and their application in the oxidation of 2-carbon alcohols and fermentation effluent, Electrochim. Acta. 225 (2017) 207-214. https://doi.org/10.1016/j.electacta.2016.12.098

[224] S.K. Park, H. Lee, M.S. Choi, D.H. Suh, P. Nakhanivej, H.S. Park, Straightforward and controllable synthesis of heteroatom-doped carbon dots and nanoporous carbons for surface-confined energy and chemical storage, Energy Storage Mater. 12 (2018) 331-340. https://doi.org/10.1016/j.ensm.2017.10.008

[225] D.D. Zhu, L.J. Li, J.J. Cai, M. Jiang, J.B. Qi, X.B. Zhao, Nitrogen-doped porous carbons from bipyridine-based metal-organic frameworks: Electrocatalysis for oxygen reduction reaction and Pt-catalyst support for methanol electrooxidation, Carbon. 79 (2014) 544-553. https://doi.org/10.1016/j.carbon.2014.08.013

[226] Z.L. Liu, F.B. Su, X.H. Zhang, S.W. Tay, Preparation and characterization of PtRu nanoparticles supported on nitrogen-doped porous carbon for electrooxidation of methanol, Acs Appl. Mater. Inter. 3 (2011) 3824-3830. https://doi.org/10.1021/am2010515

[227] L.M. Zhang, Z.B. Wang, J.J. Zhang, X.L. Sui, L. Zhao, D.M. Gu, Honeycomb-like mesoporous nitrogen-doped carbon supported Pt catalyst for methanol electrooxidation, Carbon. 93 (2015) 1050-1058. https://doi.org/10.1016/j.carbon.2015.06.022

[228] Y.Q. Chang, F. Hong, J.X. Liu, M.S. Xie, Q.L. Zhang, C.X. He, H.B. Niu, J.H. Liu, Nitrogen/sulfur dual-doped mesoporous carbon with controllable morphology as a catalyst support for the methanol oxidation reaction, Carbon. 87 (2015) 424-433. https://doi.org/10.1016/j.carbon.2015.02.063

[229] H.J. Huang, G.L. Ye, S.B. Yang, H.L. Fei, C.S. Tiwary, Y.J. Gong, R. Vajtai, J.M. Tour, X. Wang, P.M. Ajayan, Nanosized Pt anchored onto 3D nitrogen-doped graphene nanoribbons towards efficient methanol electrooxidation, J. Mater. Chem. A. 3 (2015) 19696-19701. https://doi.org/10.1039/C5TA05372B

[230] F.B. Su, Z.Q. Tian, C.K. Poh, Z. Wang, S.H. Lim, Z.L. Liu, J.Y. Lin, Pt Nanoparticles supported on nitrogen-doped porous carbon nanospheres as an electrocatalyst for fuel cells, Chem. Mater. 22 (2010) 832-839. https://doi.org/10.1021/cm901542w

[231] B. Xiong, Y.K. Zhou, Y.Y. Zhao, J. Wang, X. Chen, R. O'Hayre, Z.P. Shao, The use of nitrogen-doped graphene supporting Pt nanoparticles as a catalyst for methanol electrocatalytic oxidation, Carbon. 52 (2013) 181-192. https://doi.org/10.1016/j.carbon.2012.09.019

[232] X. Zhang, N. Hao, X.Y. Dong, S.B. Chen, Z. Zhou, Y. Zhang, K. Wang, One-pot hydrothermal synthesis of platinum nanoparticle-decorated three-dimensional nitrogen-doped graphene aerogel as a highly efficient electrocatalyst for methanol oxidation, Rsc Adv. 6 (2016) 69973-69976. https://doi.org/10.1039/C6RA12562J

[233] L. Zhao, X.L. Sui, J.Z. Li, J.J. Zhang, L.M. Zhang, G.S. Huang, Z.B. Wang, Supramolecular assembly promoted synthesis of three-dimensional nitrogen doped graphene frameworks as efficient electrocatalyst for oxygen reduction reaction and methanol electrooxidation, Appl. Catal B-Environ. 231 (2018) 224-233. https://doi.org/10.1016/j.apcatb.2018.03.020

[234] L. Zhao, X.L. Sui, J.L. Li, J.J. Zhang, L.M. Zhang, Z.B. Wang, Ultra-fine Pt nanoparticles supported on 3D porous N-doped graphene aerogel as a promising electro-catalyst for methanol electrooxidation, Catal. Commun. 86 (2016) 46-50. https://doi.org/10.1016/j.catcom.2016.08.011

[235] L.M. Zhang, X.L. Sui, L. Zhao, J.J. Zhang, D.M. Gu, Z.B. Wang, Nitrogen-doped carbon nanotubes for high-performance platinum-based catalysts in methanol oxidation reaction, Carbon. 108 (2016) 561-567. https://doi.org/10.1016/j.carbon.2016.07.059

[236] S.Y. Wang, T. Cochell, A. Manthiram, Boron-doped carbon nanotube-supported Pt nanoparticles with improved CO tolerance for methanol electro-oxidation, Phys. Chem. Chem. Phys. 14 (2012) 13910-13913. https://doi.org/10.1039/c2cp42414b

[237] Z.W. Liu, Q.Q. Shi, F. Peng, H.J. Wang, R.F. Zhang, H. Yu, Pt supported on phosphorus-doped carbon nanotube as an anode catalyst for direct methanol fuel cells, Electrochem. Commun. 16 (2012) 73-76. https://doi.org/10.1016/j.elecom.2011.11.033

[238] J.J. Fan, Y.J. Fan, R.X. Wang, S. Xiang, H.G. Tang, S.G. Sun, A novel strategy for the synthesis of sulfur-doped carbon nanotubes as a highly efficient Pt catalyst support toward the methanol oxidation reaction, J. Mater. Chem. A. 5 (2017) 19467-19475. https://doi.org/10.1039/C7TA05102F

[239] X.M. Ning, Y.H. Li, B.Q. Dong, H.J. Wang, H. Yu, F. Peng, Y.H. Yang, Electron transfer dependent catalysis of Pt on N-doped carbon nanotubes: Effects of synthesis method on metal-support interaction, J. Catal. 348 (2017) 100-109. https://doi.org/10.1016/j.jcat.2017.02.011

[240] H.Y. Du, C.H. Wang, H.C. Hsu, S.T. Chang, U.S. Chen, S.C. Yen, L.C. Chen, H.C. Shih, K.H. Chen, Controlled platinum nanoparticles uniformly dispersed on nitrogen-doped carbon nanotubes for methanol oxidation, Diam. Relat. Mater. 17 (2008) 535-541. https://doi.org/10.1016/j.diamond.2008.01.116

[241] Y.R. Sun, C.Y. Du, M.C. An, L. Du, Q. Tan, C.T. Liu, Y.Z. Gao, G.P. Yin, Boron-doped graphene as promising support for platinum catalyst with superior activity towards the methanol electrooxidation reaction, J. Power Sources. 300 (2015) 245-253. https://doi.org/10.1016/j.jpowsour.2015.09.046

[242] M.M. Li, Q.G. Jiang, M.M. Yan, Y.J. Wei, J.B. Zong, J.F. Zhang, Y.P. Wu, H.J. Huang, Three-dimensional boron- and nitrogen-Codoped graphene aerogel-supported Pt nanoparticles as highly active electrocatalysts for methanol oxidation reaction, Acs Sustain. Chem. Eng. 6 (2018) 6644-6653. https://doi.org/10.1021/acssuschemeng.8b00425

[243] V. Deerattrakul, P. Puengampholsrisook, W. Limphirat, P. Kongkachuichay, Characterization of supported Cu-Zn/graphene aerogel catalyst for direct CO_2 hydrogenation to methanol: Effect of hydrothermal temperature on graphene aerogel synthesis, Catal. Today. 314 (2018) 154-163. https://doi.org/10.1016/j.cattod.2017.12.010

[244] D.X. Yu, A.J. Wang, L.L. He, J.H. Yuan, L. Wu, J.R. Chen, J.J. Feng, Facile synthesis of uniform AuPd@Pd nanocrystals supported on three-dimensional porous N-doped reduced graphene oxide hydrogels as highly active catalyst for methanol oxidation reaction, Electrochim. Acta. 213 (2016) 565-573. https://doi.org/10.1016/j.electacta.2016.07.141

[245] J. Wang, R. Huang, Y.J. Zhang, J.Y. Diao, J.Y. Zhang, H.Y. Liu, D.S. Su, Nitrogen-doped carbon nanotubes as bifunctional catalysts with enhanced catalytic performance for selective oxidation of ethanol, Carbon. 111 (2017) 519-528. https://doi.org/10.1016/j.carbon.2016.10.038

[246] H.L. Yang, X.Y. Zhang, H. Zou, Z.N. Yu, S.W. Li, J.H. Sun, S.D. Chen, J. Jin, J.T. Ma, Palladium nanoparticles anchored on three-dimensional nitrogen-doped carbon nanotubes as a robust electrocatalyst for ethanol oxidation, Acs Sustain. Chem. Eng. 6 (2018) 7918-7923. https://doi.org/10.1021/acssuschemeng.8b01157

[247] H. Wang, Y.J. Ma, W.Z. Lv, S. Ji, J.L. Key, R.F. Wang, Platinum-tin nanowires anchored on a nitrogen-doped nanotube composite embedded with iron/iron carbide particles as an ethanol oxidation electrocatalyst, J. Electrochem. Soc. 162 (2015) H79-H85. https://doi.org/10.1149/2.1031501jes

[248] J.C.M. Silva, I.C. de Freitas, A.O. Neto, E.V. Spinace, V.A. Ribeiro, Palladium nanoparticles supported on phosphorus-doped carbon for ethanol electro-oxidation in alkaline media, Ionics. 24 (2018) 1111-1119. https://doi.org/10.1007/s11581-017-2257-9

[249] J.G. Yu, M.M. Jia, T.M. Dai, F.M. Qin, Y.N. Zhao, Nitrogen-doped graphene supporting PtSn nanoparticles with a tunable microstructure to enhance the activity and stability for ethanol oxidation, J. Solid State. Electr. 21 (2017) 967-974. https://doi.org/10.1007/s10008-016-3449-3

[250] M. Boulaghi, H.G. Taleghani, M.S. Lashkenari, M. Ghorbani, Platinum-palladium nanoparticles-loaded on N-doped graphene oxide/polypyrrole framework as a high performance electrode in ethanol oxidation reaction, Int. J. Hydrogen Energ. 43 (2018) 15164-15175. https://doi.org/10.1016/j.ijhydene.2018.06.092

[251] H. Xu, B. Yan, K. Zhang, J. Wang, S.M. Li, C.Q. Wang, Y. Shiraishi, Y.K. Du, P. Yang, Ultrasonic-assisted synthesis of N-doped graphene-supported binary PdAu nanoflowers for enhanced electro-oxidation of ethylene glycol and glycerol, Electrochim. Acta. 245 (2017) 219-228. https://doi.org/10.1016/j.electacta.2017.05.146

[252] Y. Yang, W. Wang, Y.Q. Liu, F.X. Wang, D. Chai, Z.Q. Lei, Pd nanoparticles supported on phenanthroline modified carbon as high active electrocatalyst for ethylene glycol oxidation, Electrochim. Acta. 154 (2015) 1-8. https://doi.org/10.1016/j.electacta.2014.12.072

Materials Research Forum LLC
doi: https://doi.org/10.21741/9781644900192-2

Chapter 2

Carbon-Based Nanomaterials for Alcohol Oxidation

D. Yang

National Research Council of Canada

Dongfang.yang@nrc-cnrc.gc.ca

Abstract

Carbon-based porous materials with a high specific surface area are extensively used as catalyst support in direct alcohol fuel cells to improve the efficiency of catalyst nanoparticles. New types of carbon-based nanomaterials such as carbon nanotubes, mesoporous carbon, nanodiamond, and graphene are of great interest due to their tunable mesoporosity, high surface areas, and good electrical conductivity. Furthermore, they possess suitable anchoring sites for catalyst nanoparticles and have the potential to replace carbon blacks with improved stability. This chapter is devoted to reviewing the recent advances in the preparation, characterization and performance evaluation of these new types of the catalyst support.

Keywords

Alcohol Fuel Cells, Carbon Support, Electrocatalysts, Carbon Blacks, Graphene, Mesoporous Carbon, Platinum Nanoparticles, Carbon Nanotubes, PtRu Nanoparticles, Direct Methanol Fuel Cells

Contents

1.   Introduction

Direct alcohol fuel cells (DAFCs) such as direct methanol fuel cells (DMFCs) are considered as potential power sources for handheld electrical devices and electric vehicles owing to their high energy density, fuel availability, and portability. Pt and Pt-Ru are by far the most promising catalysts for electrochemical oxidation of alcohols such as methanol and ethanol in DAFCs[1]. In DAFCs, catalyst nanoparticles are deposited on the catalyst support using either physical vapor deposition methods such as sputtering, thermal evaporation and electron beam deposition, or by wet chemical methods such as electrochemical reduction, impregnation, and ion-exchange. The catalyst layer that consists of catalyst nanoparticles with their support were placed on carbon cloth to form the fuel cell electrode. The specific electrochemical activity of catalyst nanoparticles is improved owing to the existed interaction between the catalyst and its support. The catalyst layer structure is schematically shown in the Figure. 1, in which, the catalyst nanoparticles are uniformly dispersed on the surface of carbon support particles, and the polymer electrolyte/ionomers cover the catalyst nanoparticles. The electrochemical oxidation of alcohols takes place at the alcohol-electrolyte-support triple phase boundary. In DMFCs, for example, the electrochemical reaction occurred at the triple phase boundary at the anode follows,

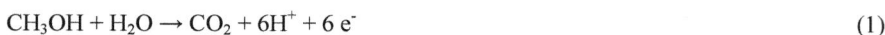

$$CH_3OH + H_2O \rightarrow CO_2 + 6H^+ + 6\ e^- \tag{1}$$

While at the cathode, $3/2\ O_2$ molecules are reduced to $3\ H_2O$,

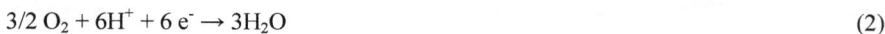

$$3/2\ O_2 + 6H^+ + 6\ e^- \rightarrow 3H_2O \tag{2}$$

The total reaction is described as,

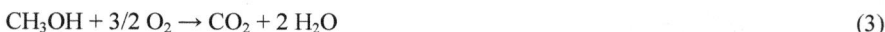

$$CH_3OH + 3/2\ O_2 \rightarrow CO_2 + 2\ H_2O \tag{3}$$

Materials Research Forum LLC

doi: https://doi.org/10.21741/9781644900192-2

A multi-step reaction mechanism is used to describe the methanol electrochemical oxidation reaction (3), in which reaction intermediates, such as HCHO, CO and HCOO, are produced, adsorbed on the catalyst surface and consumed in a subsequent step to ultimately generate the final reaction products, CO_2 and H_2O.

For the maximum utilization of catalyst nanoparticles, polymer electrolyte/ionomer should cover most of the catalyst particles so they will be active for electrochemical reactions. The amount of polymer electrolyte/ionomer is critical: not enough will handle the proton conductance, while too much will slow down the transport of alcohol and CO_2 molecules. The keys for achieving maximum utilization of catalyst nanoparticles are to increase the contact between the catalyst particles and their support, as well as between the polymer electrolyte/ionomers and catalyst particles.

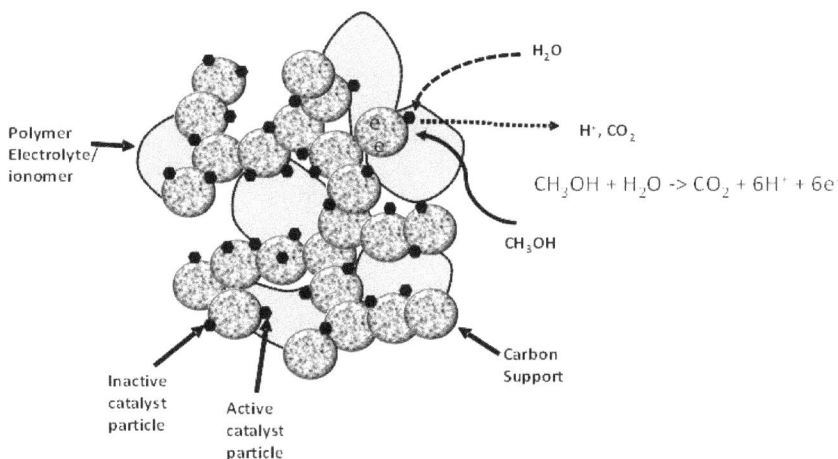

Figure 1. Schematic illustration of the catalyst layer that consists of the triple-phase-boundary (i.e., alcohol-electrolyte-catalyst/support) in a direct methanol fuel cell.

Porous conductive materials with a high specific surface area are usually used as the catalyst support. Catalyst support is used to improve the dispersion and particle size distribution of catalyst nanoparticles, as well as to increase their surface area and reduce their loading. Catalyst support affects the efficiency of catalyst and ultimately the performance and durability of the whole fuel cell device. Suitable catalyst support generally needs some features such as low cost, high surface area, high porosity, high electric conductivity, and good corrosion resistance. Specific surface properties such as suitable hydrophilicity and hydrophobicity are also required in order to handle water to prevent flooding. Ideal structure for the catalyst support is a mesoporous structure with a pore size ranging from 20 nm to 40 nm. For pores narrower than 20 nm, polymer electrolyte such as Nafion, needed in electrochemical reactions, may not be easy to enter or may be occluded by the carbon support. These catalyst nanoparticles chemically deposited inside such small pores are not in contact with the polymer proton conductor, therefore do not contribute to the oxidation of alcohols. Currently, the most commonly used catalyst supports in DAFCs are porous carbon materials such as carbon blacks (CBs). CBs are commercially available with low cost and possess relatively good electrical conductivity and high surface areas [2–3].

However, corrosion of carbon black at high electrode potentials reduces its durability in DAFCs. Corrosion of carbon support materials is the result of electrochemical oxidation of carbon to CO_2. At the electrode potentials higher than +0.207 V vs. RHE, the oxidation reaction of carbon becomes thermodynamically favorable. Fortunately, only at electrode potentials higher that +1.2 vs. RHE, the reaction occurs significantly[4]. Corrosion of carbon support materials results in agglomeration, Ostwald Ripening, and catalyst particle detachment. Agglomeration of carbon support densifies the catalyst layer and can severely reduce the mass transport of reactants in DAFCs and thus, deteriorate its performance[5]. Durability improvement of the catalyst support is one of the main challenges that need to be addressed in order to achieve good performance of DAFCs for long-term. As a general requirement, a fuel cell should maintain at least 90% of its performance after 5000-hour operation[4]. New types of nanocarbon such as carbon nanotubes, mesoporous carbon, nanodiamond, and graphene are materials of great interest in replacing carbon blacks as the catalyst support with improved stability. These materials are known for their tunable mesoporosity, high surface areas, good electrical conductivity, and possess suitable anchoring sites for catalyst nanoparticles. This chapter is devoted to reviewing the recent advances in the preparation, characterization and performance evaluation of these new types of nanocarbon catalyst support in DAFCs.

2.   Electrocatalyst support materials

CBs are the most commonly used catalyst support for DAFCs. They consist of primary particles, secondary particles, and secondary particle groups as schematically shown in Figure 2. Tens of primary carbon particles are linked together through covalent bonds to form clusters (aggregates). Such aggregates (secondary particles) are usually agglomerated through Van der Waals force to form secondary particle groups. In order to effectively increase the catalyst dispersion, the number of anchoring centers of CBs needs to be created through either chemical or physical activations, however, over-activation that can lead to structural collapse and should be avoided. As can be seen in figure 2, catalyst particles residing deep inside the primary CB particles cannot be covered by polymer electrolyte/ionomers, so they will not participate in electrochemical reactions. Reactants also have difficulty reaching these catalyst particles due to the small pore size of CBs. The presence of impurities in CBs affects its stability, which directly affects the durability of fuel cells [6]. Performance of various catalysts, such as Pt, Pt-SnO$_2$ and PtRu, that are supported by commercial CBs, i.e., Vulcan XC-72, are compared in DMFCs,and their evaluation results can be reviewed in the reference [7].

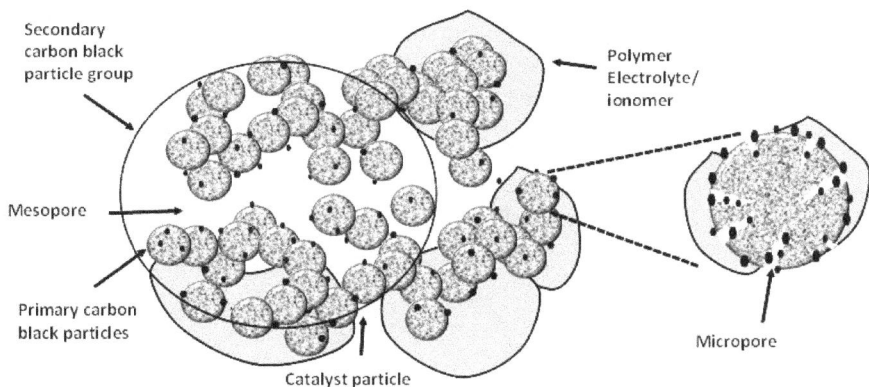

Figure 2. Schematic illustration of carbon black-supported catalyst for direct alcohol fuel cells

Recently, significant progress on the utilization of new carbon-based nanomaterials for catalyst supports in DAFCs has been made, which allows a better understanding of the effects of structures and properties of these nano-carbons on the activity and durability of catalysts in DAFCs. In the following subsection, the characteristics and performance of each type of carbon-based nanomaterials for alcohol oxidations in DAFCs will be summarized. The problems associated with each type of carbon-based support materials and future challenges will also be discussed.

2.1   Carbon nanomaterials of different shapes (tubes, fibers, horns, coils, walls, and spheres)

Carbon nanotubes (CNTs), nanofibers, nanohorns, nanocoils, nanowalls, and nanospheres represent a new class of carbon nanomaterials that have properties significantly different from other forms of carbon like CBs. These materials have been evaluated as the catalyst support for DAFCs, and among them, CNTs are the most studied materials due to their unique and tunable structures, high surface areas and sufficient good electrical conductivity, all contributing to enhanced performance as the catalyst support in DAFCs.

CNTs typically have an outer diameter of 10-50 nm, an inside diameter of 3-15 nm, and a length ranging from 10-50 mm, as shown in Figure 3. Beside good electronic conductivity, CNTs also have high thermal stability, due to the combination of their carbon atoms with the molecular perfection of buckytubes. Appropriate active sites for anchoring catalyst particles can be created on the surface of the tubular shaped CNTs. Uniformity and high dispersion of catalyst nanoparticles, which result from covalent and non-covalent bonds to active sites, leads to an increase of catalytic activity and efficiency. Functionalization of multiwalled CNTs using citric acid, for example, allows opening of the tubes and forming core-shell structures. If catalyst particles (i.e.,Pt with 2-5 nm size) deposit on the pore mouths of CNTs, they will participate in the electrochemical reactions, while if they exist inside the nanotube, they will not contribute to the electrochemical reaction. As the length of CNTs increase, the chance for Pt particles located inside the tube will increase, which will result in a decrease of the efficiency of the catalysts.

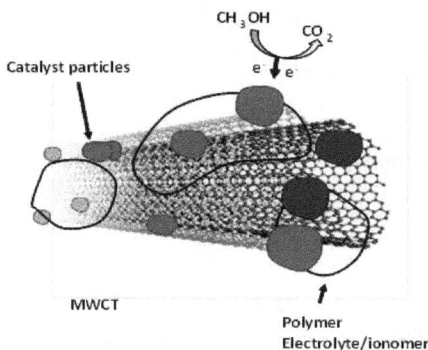

Figure 3. Schematic illustration of Pt nanoparticle bonded on CNTs used for methanol fuel cells.

Materials Research Forum LLC
doi: https://doi.org/10.21741/9781644900192-2

Conventional techniques that have been employed to fabricate carbon nanotubes include arc discharge, chemical vapor deposition (CVD), thermal evaporation and laser ablation. Wang et al. [8] grew CNTs directly on a carbon cloth using microwave plasma enhanced CVD. They then used a potentiostatic electrodeposition method to deposit Pt-Ru catalyst nanoparticles on the CNT surfaces in aqueous electrolytes. It was found that during the electrodeposition processes the concentrations of sulfuric acid, ethylene glycol, and Ru precursor salt affected the size and disperses of Pt-Ru nanoparticles on the surfaces of the CNTs that were directly grown on carbon cloths. The efficiency of methanol oxidation and the ability to suppress carbon monoxide adsorption (poisoning the catalyst) very much depend on how the Pt-Ru catalysts were dispersed on the CNT support. An electrochemical deposition method was also used by Zhao et al. [9] to grow 3D-flowerlike Pt catalyst onto multi-walled CNTs by employing a three electrochemical step process including a potential pulse sequence. For comparison, Pt nanoparticle dispersed uniformly in two dimensions on multi-walled CNTs was also prepared by similar three electrochemical steps including a cyclic voltammetry step rather than a potential-step. It was found that the 3D-flowerlike morphology exhibits much higher catalytic activity and better stability than the 2D morphology for the oxidation of methanol. The authors attributed the performance enhancement of the Pt catalyst to the high surface areas of the 3D-flowerlike structures deposited on the surface of CNTs.

In order to improve the adhesion of catalyst nanoparticles to CNTs support and therefore, the durability of the electrode, surface activation of CNTs is an important step. Kumara et al.[10] used a microwave to create surface active sites on multiwall carbon nanotubes to immobilize platinum and palladium nanoparticles. The hybrid materials were mixed with graphite paste and used as catalysts for electrochemical oxidation of methanol, ethanol and formic acid in acidic and alkaline media. Relatively low activation energy of 7.64 kJ mol^{-1} was achieved by the Pt-Pd catalyst for ethanol electrochemical oxidation.

Attempting to overcome the defect structures and disruption of electron structures caused by acid treatment, substantial efforts have been devoted to developing noncovalent functionalization of CNTs. Noncovalent functionalization of CNTs preserves the intrinsic structure of CNTs. Xi Genget al.[11] used a procedure adapted from Hu et al.[12] to perform noncovalent functionalization of multi-walled CNTs by using polyethyleneimine (PEI). Pt-Ru nanoparticles were uniformly deposited on the polyethyleneimine-functionalized multi-walled CNTs. It was found that the Pt-Ru supported on the PEI-multi-walled CNTs had a large surface area and exhibited enhanced electrocatalytic activity towards methanol oxidation in comparison with that of Pt-Ru catalyst supported on only oxidized multi-walled CNTs.

Besides CNTs, other shapes of carbon nanomaterials also have their unique features, for example, carbon nanofiber/nanofilaments possess high surface area and good electrical conductivity, and their entire surface is accessible, and all can be utilized to interact with reactants as the surface is external[13]. Catalysts, such as Pt and Pt-Ru, supported on carbon nanofilaments, exhibit higher electrochemical performance than carbon blacks such as Vulcan. Masataka Okada et al.[14] applied carbon nanofiber (CNF) on the top of the carbon paper following by precipitation of the Pt-Ru nanoparticles catalyst layer. The CNF interlayer was able to form a dense and smooth surface on the porous carbon paper. DMFC anode utilized the CNF coated carbon paper as the catalyst support exhibited higher power density than that of electrodes using carbon black covered carbon cloth. It is believed that the high density and cross-linkage of the CNF layer protects the catalyst particles from delamination, and also increases the active reaction sites owing to disperse of Pt-Ru on the CNF surface highly. CNF can also be decorated with a conductive layer such as graphite to form core/shell nanostructured carbon materials. Zhou et al.[15] heat-treated a CNF/polyaniline (CNF/PANI) composite at 900°C to form a core-shell structure in which nitrogen (N)-doped graphitic is the shell and the carbon nanofiber is the core. Pt nanoparticles were deposited on such supports by formic acid reduction. Pt nanoparticles supported on the carbonized composites showed significant performance improvement for methanol electrochemical oxidation as compared to that supported on the untreated CNF/PANI composites (current density for methanol oxidation is approximately seven times higher). Owing to its hollow-free structure and a larger diameter, CNF offers better durability and stability as the catalyst support than CNTs. Furthermore, CNFs can be used as the catalytic support without any activation treatment due to their peculiar structure.

Other less common shapes of carbon like carbon nanohorns and carbon nanocoils have also been demonstrated to be suitable supports for catalysts in DAFCs. Just like CNFs, these materials have high crystallinity, good electrical conductivity due to their graphitic structure, and a large surface area that is accessible by reactants/products. Single-wall carbon nanohorn (SWCNH) was used by Yoshitake et al. [16] to support platinum catalyst. Uniformly disperse of small Pt particles (~ 2 nm) on the SWCNH was accounted for better performance in a fuel cell than Pt supported on carbon black. Carbon nanocoils were used by Park et al.[17] and Sevilla et al.[18] as supports for Pt-Ru catalysts. Both research works found that catalysts supported on the carbon nanocoils exhibited better electrocatalytic performance than the same catalyst supported on Vulcan XC-72 carbon. Different loadings of Pt (0.19, 0.2, 0.4, 0.64, and 0.69 mg Pt cm^2) were deposited onto carbon nanohorn surface and investigated for its catalytic behavior in methanol oxidation reaction by Hamoudiet al.[19]. It was found that the loading of Pt films affects their morphologies. From low to high loadings, Pt film changes from spread nanoparticles in islands to ultrathin-films of smooth surface and finally thick porous films. Among the electrodes tested, the performance of high Pt loading electrode (0.69 mgPtcm^2) for methanol oxidation is the best due to its increased porosity, which allows ease of access by the reactants and electrolyte. Hollow carbon spheres (HCSs), prepared by combining the hydrothermal method and an intermittent microwave heating technique, were also investigated as a support for Pt nanoparticles and was found to possess high catalytic activity for methanol oxidation owing to the existence of nanochannels with open microspores [20].

Catalyst support basing on carbon nanomaterials of different shapes such as CNTs and CNFs show significant performance improvement in DAFCs, however, for commercial applications of those carbon nanomaterials, simple and cost-effective methods for their large-scale production are required. One of the promising techniques for the manufacturing of carbon nanofibers with high surface areas and high porosity is electrospun. However, the suitability of using electrospun nanofibers in DAFCs still requires further investigation [6].

2.2   Mesoporous carbon

Mesoporous carbons are nanostructured carbons which have pore sizes ranging from 2 nm to 50 nm, as illustrated in Figure 4. A large number of mesopores existed in ordered mesoporous carbon (OMC) makes the materials of great interest for DAFCs applications. Catalyst nanoparticles are relatively easy to disperse uniformly on the OMC. The unique mesoporous structures of OMC also facilitate the transport of reactants to the catalysts showing low charge-transfer resistance [21–22]. OMC supported catalysts exhibited greater catalytic activity compared to the same catalyst supported on carbon black which consists of mixed microporous and macroporous structures. Structural parameters of OMC (e.g., particle size, pore size and structure, and wall thickness) affect its performance as a catalyst support. The optimized structural parameters of OMC were found to have a three-dimensional structure with particles smaller than 100 nm and pore size in the range of 20–40 nm. The stability of OMC-supported catalyst nanoparticles are similar to that of catalysts supported on carbon black but can be improved by the graphitization of the OMC. Graphitization could also reduce the carbon corrosion rate of OMC in DAFCs. Another advantage of graphitization is the ability to increase the electrical conductivity of OMC which could be done through doping with nitrogen [23–24]. However, graphitization decreases the surface area and porosity of OMC. This shortcoming can be addressed through composing it with other tunable carbon nanostructures or polymers during fabrication. Activation of OMC with oxidizing agents such as HNO_3 results in modifying its surface chemistry and affects its bonding with catalysts such as Pt. Also interesting to note, is that OMC can be composited with metal oxides such as WO_3. In the work of Zeng et al. [25], a hard-template method, in which mesoporous SBA-15 silica was adapted as the template, was used to produce the OMC through carbonization of sucrose. The OMC–WO_3 composites were synthesized by immersing OMC in phosphotungstic acid, and followed by thermal decomposition in N_2 atmosphere at 600 °C. Such composites were investigated as the catalysts support in DMFCs. Pt catalysts were deposited on the OMC–WO_3 support by impregnation in chloroplatinicacid hexahydrate acetone solution. It was found that the specific activity of Pt catalyst increases and the ohmic resistance decreases with the presence of WO_3.

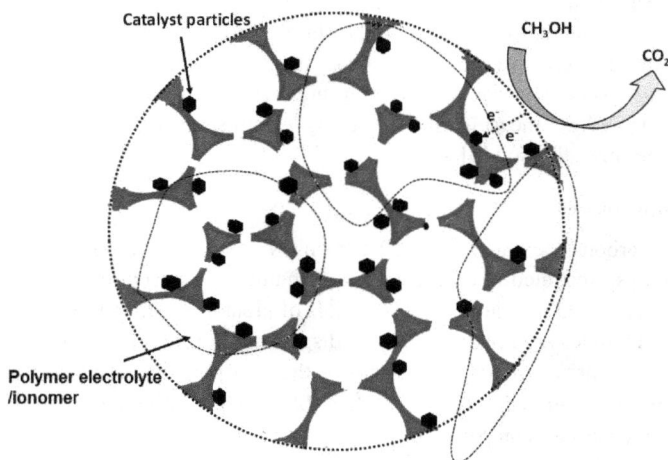

Figure 4. Schematic depiction of OMC supported catalyst for alcohol fuel cells

Mesopore structural carbon xerogels are also attracting for their use as the catalyst support due to their high purity, large surface area, and narrow and controllable pore size distribution. Preparation of carbon xerogels consists of polycondensation of resorcinol ($C_6H_6O_2$) with formaldehyde (CH_2O) followed by drying and pyrolysis [26]. Pt and Pt-Ru nanoparticles supported on functionalized xerogels were found to perform better than the catalysts supported on Vulcan XC-72R, as indicated by higher current density values for methanol oxidation. Researchers also found that catalyst support using S-doped mesoporous carbon xerogels enhanced the catalytic activity of Pt towards methanol

Nanomaterials for Alcohol Fuel Cells
Materials Research Foundations **49** (2019) 79-102

Materials Research Forum LLC
doi: https://doi.org/10.21741/9781644900192-2

electrochemical oxidation [27], exceeding the performance of Pt catalyst supported on both the bare xerogel and commercial carbon blacks.

2.3   Graphene oxide and reduced graphene oxide

Graphene is composed of a single layer of carbon atoms arranged in a hexagonal lattice [28], as shown in Figure 5. The combination of its large surface area, good electric conductivity, high chemical stability, and unique graphitized basal plane structure makes it an interesting catalyst support for DAFCs [29]. By using various methods of functionalization of graphene, more anchoring sites can be created for a better and more homogeneous catalyst dispersion, and thereby improve the durability and activity of the catalyst. Furthermore, doped graphene sheets with elements such as N, have demonstrated better electronic conductivity compared to that of pure graphene sheets. However, for graphene to be more cost-effective, novel methods for the synthesis of high-quality graphene in mass production scales are needed. When graphene nanosheets are used as a support, instead of other carbon supports, smaller sized catalyst particles and higher metal loadings can be obtained. Graphene is typically prepared from graphene oxide (GO) by chemical, thermal and light reduction. Due to incomplete reduction, oxygen-containing functional groups (e.g., epoxy, carboxyl,and hydroxyl) remain onto both sides of the graphene sheet [30–31]. These functional groups act as anchoring sites for catalyst nanoparticles that allow more homogeneous dispersion of catalysts on the graphene support, therefore increases its electrocatalytic activity and durability. Furthermore, the presence of functional groups on the surface of graphene nanosheets helps catalysts to oxidize adsorbed poisoning species originating from partial oxidation of alcohols. Figure 5 provides the schematic illustration of graphene supported Pt catalyst for DMFCs.

W. Zhao et al. [32] used an electrochemical method to deposit Pt nanoparticles on graphene nanosheets. It was found that Pt nanoparticles were well-dispersion on the surface of graphene nanosheets and no aggregation of Pt nanoparticles was found. The electrochemical measurement shows that Pt catalyst supported on graphene nanosheets has a better electrocatalytic activity for the methanol oxidation than that of Pt catalyst supported on Vulcan XC-72 carbon blacks. Magnetron sputtering method was used by N. Soin et al. [33] to deposit Pt nanoparticles on vertically aligned a few layered graphene nanoflakes, which were synthesized by microwave plasma enhanced chemical vapor deposition. The Pt nanoparticles were found to be around 6 nm in diameter, and when they were used as catalysts for DMFCs, relatively high specific peak current density of ~ 62 mA mg^{-1} cm^{-2} for methanol oxidation was obtained. Pt catalyst supported by graphene nanoflakes was also found to possess high resistance to carbon monoxide poisoning. The good catalytic activity of graphene support Pt catalyst was also demonstrated by the work of S. Liu et al. [34], in which Pt/graphene nanosheet composites were fabricated by microwave-assisted reduction of H_2PtCl_6 in a graphene oxide suspension using ethylene glycol as a reducing agent.

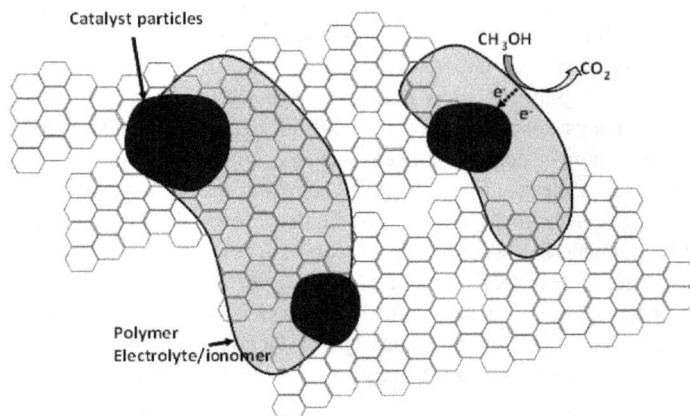

Figure 5. Schematic illustration of graphene supported Pt catalyst for methanol fuel cells

To enhance the uniform deposition of catalyst on the graphene support, functionalization of its surface is a good approach. Poly(diallyldimethylammonium chloride) (PDDA) was used by B. Luo et al. [35] to functionalize graphene nanosheets. They found that PDDA-functionalization introduces positive charges on the surfaces of graphene nanosheets, allowing self-assembly of negatively charged $PtCl_6^{2-}$ ions on their surface. The $PtCl_6^{2-}$ ions were reduced by ethylene glycol to form Pt nanoparticles uniformly depositing on the PDDA-functionalized graphene nanosheets. It was found that Pt catalyst supported on PDDA-graphene exhibited higher electrocatalytic activity for the methanol oxidation as compared with both Pt supported on graphene and PDDA-functionalized carbon black. 2-thiol benzimidazole was also used by OnurAkyıldırım et al. [36] to functionalize graphene nanosheets. It was found that the bimetallic Pt-Pd catalyst supported on such functionalized graphene nanosheets exhibited higher peak currents for methanol oxidation than those of single Pd or Pt catalyst supported on the same functionalized graphene nanosheets.

A single layer of graphene can be rolled up to form graphene nanoscrolls, which possess a spiral structure with open edges and ends. Such a spiral structure prevents aggregation of graphene nanosheets, and their interlayer space can be adjusted for better transportation of reactants and the electrolyte ions. Yu Liu et al.[37] developed a method to in situ roll-up of Pt/ graphene sheets with the help of hydrogen peroxide and ultrasonication. The Pt/graphene nanosheets were first fabricated by the co-reduction of graphite oxide and Pt salt using ethylene glycol. Pt nanoparticles are found to disperse on such graphene nanoscroll uniformly and exhibited a higher specific surface area, and broader range pore size (188 m^2 g^{-1}, 25–45 nm) than that of Pt supported on graphene nanosheets (122 m^2 g^{-1}, 30–38 nm). Pt catalyst supported on the nanoscrolls also exhibited significantly higher electrocatalytic activity and stability. If the size of graphene sheets is reduced to be ~10 nm, graphene quantum dots (GQDs) are produced. Due to the small size, quantum dots possess strong quantum confinement and edge effects showing unique physical properties. The electronic characteristics of graphene quantum dots can be further improved by nitrogen-doping. Z. Li et al. [38] used a hydrothermal method to prepare nitrogen-doped graphene quantum dots and used them as the supported for Pd nanoparticles. The Pd catalyst demonstrates an excellent electrocatalytic activity for the methanol oxidation reaction in an alkaline medium. Its performance is better than the Pd catalyst supported on graphene sheets as well as un-doped graphene quantum dots. The excellent performance of Pd catalyst supported on N-doped graphene quantum dots was attributed to good dispersion, and small and narrow size distribution of Pd nanoparticles. With the help of graphene oxide nanosheets, Pt nanoparticles can be assembled into well-controlled three-dimensional nanostructures. Such structure possesses porous structure, high surface area, and high catalytic activity. X. Chen et al. [39] created three-dimensional Pt nanoflowers supported on graphene oxide nanosheets by chemical reduction using ethanol. Pt nanoflowers of around 30 nm, which are composed of numerous small (4 nm) Pt nanoparticles, exhibited high activity for the methanol oxidation when compared to that supported on the commercial Pt black.

Besides creating various nanostructures of Pt catalyst on graphene support, forming alloys with other metal also allows alternating its catalyst properties. Pt-Au was co-electrodeposited on the surface of graphene nanosheets [40], and its electrocatalytic property for methanol oxidation was systematically investigated. It was found that Pt-Au catalysts supported on graphene nanosheets exhibit higher catalytic activity and stability than pure Pt or Au catalyst supported on the same graphene nanosheets. When Pt to Au molar ratio is 2:1, the highest catalyst activity was obtained. Owing to the synergistic catalytic effect of Au and the existence of the graphene support, carbonaceous species like CO formed during the oxidation of methanol can be easily removed from the surface of catalysts avoiding loss of their activity. Pt-Co alloy catalyst support by graphene was also used by S. Yang et al. [41] for DMFCs. To avoid irreversible agglomeration of graphene nanosheets because of the Van der Waals force, graphite dots were introduced into the space between graphene layers. Pt-Co nanoparticles were then deposited on the graphene/graphite dots/graphene support. It was found that Pt-Co nanoparticles were uniformly dispersed between the graphene layers and exhibited much better electrocatalytic activity and stability toward methanol oxidation when compared to Pt or Pt-Co supported on graphene nanosheets. The authors attributed the performance improvement to the synergistic effects between Pt and Co nanoparticles as well as the embedding of graphite dots between graphene sheets that increase the electrical conductivity and enhance the dispersion of Pt-Co nanoparticles.

2.4   Nanodiamond

Pure diamond is an insulator and therefore, is unsuitable as the catalyst support for DAFCs. The conductivity of the diamond, however, can be improved by elemental doping, which increases the number of charge carriers and defects. Key factors that influence the electrical conductivity of doped diamonds are the doping level, particle-particle connectivity and impurity. However, doping can induce instability of the diamond structures.

Doping diamond with boron improves the electric conductivity of the diamond. The microstructure of B-doped diamond does not change even when it is polarized to high electrode voltages. The B-doped diamond exhibits an extensive potential window with very low background current for electrochemical alcohol oxidation. Furthermore, catalyst nanoparticles all are deposited on its outer surface rather than inside micropores such as carbon black. All these unique features make the B-doped conductive diamonds as very attractive materials for highly durable catalyst support for DAFCs applications.

Mechanical and chemical stability of doped diamonds also allows the deposition of catalyst nanoparticles by using high-temperature processes such as thermal decomposition to produce new nanostructures and properties, without significant alternating their microstructures and properties [42]. A list of methods including electrochemical deposition, sol-gel and sputtering can be used to deposit nanoparticles of platinum and platinum alloy on doped diamonds. These doped diamonds supported catalysts have demonstrated excellent performance for electrochemical oxidation of methanol and ethanol [43–44]. Other bimetallic catalysts, such as Pt-Ru, Pt-Sn and tri-metallic catalysts, such as $Pt_{80}Ru_{10}Sn_{10}$, have also been supported on the doped diamond and were investigated for both methanol and ethanol electrooxidation [45]. Boron-doped diamond was also used by G. R. Salazar-Banda et al. [44] to support the $Pt-RuO_2-RhO_2$

composite catalyst. The multi-component catalyst shows better anti-CO poisoning effect for the methanol and the ethanol oxidation reactions when compared with Pt or Pt-RuO$_2$ supported on the same boron-doped diamonds.

Although doped diamonds have been demonstrated as good catalyst supports for DAFCs, there are still some challenges, such as relatively low conductivity and small surface area that need to be addressed. Also, it is still difficult to achieve uniformly and controllably boron doping in nanodiamond; therefore, improved synthesis methods and new ways of surface pretreatment are required. Catalyst nanoparticles are also hard to disperse uniformly on the doped diamond and the catalyst particles are typically too large.

2.5   Heteroatom-doped carbon

Doping of carbon nanomaterials with heteroatoms, such as N, S, B, P, and F, is an effective way to tailoring their electronic structures, chemical reactivity, as well as conductivity. Since the size and electronegativity of the heteroatoms are different from those of carbon atoms, the introduction of heteroatoms into carbon nanostructures could induce defects, as well as changes in charge distribution and electronic properties but maintain their intrinsic physical/chemical characteristics [46]. Since the atomic radius of the nitrogen atom is closest to that of carbon atom [47], it is, therefore, the most interesting heteroatom for doping the carbon framework. The most common methods used for doping nitrogen into carbon nanomaterials including thermal treatment with ammonia gas or nitrogen-containing compounds as well as nitrogen plasma bombardment. Nitrogen atoms were incorporated either at the edges or within the core structure of the carbon nanomaterials by replacing one of the sp^2 hybridized carbon atoms with the nitrogen atom. Nitrogen-doped carbon nanomaterials exhibit high electrical conductivity and stability as well as possess more anchored sites for catalyst nanoparticles in DAFCs, therefore, improve the performance of catalysts.

Chetty et al. [48] used nitrogen plasma to treat CNTs to introduce functional groups such as pyridinic and pyrrolic species on their surface, which can act as anchoring sites for the deposition of Pt-Ru particles with better dispersion. Pt-Ru catalyst nanoparticles supported on N-doped CNTs (PtRu/N-doped CNTs) exhibited enhanced catalytic activity for methanol oxidation in DMFCs as indicating by high oxidation current in the cyclic voltammetry. After a long operation time, Pt-Ru catalyst supported on N-doped CNTs maintained a high current of 18.9 mA mg^{-1} (for compassion, the value for Pt-Ru supported on acid-treated CNTs is 5.39 mA mg^{-1}, and on E-TEK carbon is 9.02 mA mg^{-1}, and on Vulcan black is 1.84 mA mg^{-1}). Nitrogen-doped carbon nanotubes were directly grown by H.-Y. Du et al. [49] on carbon cloth. Pt nanoparticles were then deposited on the N-doped CNTs/carbon cloth by chemical reduction of the Pt precursor salt in the ethylene glycol solution at an elevated temperature. The N-dopants in CNTs act as anchoring sites for the nucleation of Pt particles helping their dispersion and improving their stability. The Pt catalyst supported on N-doped CNTs/carbon cloth exhibited an enhanced catalytic activity due to the existence of N-doped CNTs that act as a fast electron-transfer path between the catalyst and the carbon cloth. Although carbon blacks possess relatively good electrical, their conductivity can be improved by nitrogen doping, G.Wu et al. [50] synthesized nitrogen-doped carbon black by carbonization of polyaniline in the presence of carbon black. The mass ratio of polyaniline to carbon black was used to adjust nitrogen content in the carbon black. N-doped carbon black increase the number of defects and electron density and appear to have more disordered structures. Pt-Ru catalyst supported on thus prepared N-doped carbon black was evaluated for CO stripping, and methanol oxidation and an enhanced catalytic activity was obtained owing to high accessible surface area for the electrochemical reactions and a lower charge transfer resistance at the catalyst/support and electrolyte interface. Well-aligned N-doped CNTs with open-ended were composited with graphene by a water-assisted CVD method [51]. The composite was used as the catalyst support for Pt-Ru nanoparticles in DMFCs which showed improved electrocatalytic performance and better long-term operation stability for methanol oxidation than that of Pt-Ru support on regular CNT or commercial carbon black. The authors assigned the performance improvement to the N-doping that helps immobilize the catalyst nanoparticles, and to the hierarchical CNTs-graphene structure that adds additional anchoring sites and forms a 3D electric conductive network.

2.6   Comparison of various carbon supports

Preparation, characterization and performance evaluation of various types of carbon catalyst supports are presented and discussed in the sub-section 2.1 to 2.5. In this sub-section, the advantages and disadvantages of these nancarbon catalyst supports are compared in table 1. Table 1 provides pro and cons of different nanocarbon supports which will facilitate the selection and application of them in DAFCs.The list of challenges of various carbon materials could also be used to guild the future research directions.

Table 1: Comparison of different carbon catalyst support materials

Carbon support	Advantages	Disadvantages
Carbon blacks	low cost, relatively good electrical conductivity, high surface areas	mixed microporous and macroporous structures, corrosion and impurities affects its stability, agglomeration reduce transport of reactants
Carbon nanotubes (CNTs)	unique and tunable structures, high surface areas, sufficient good electrical conductivity, high thermal stability, easy to create anchoring sites for catalyst particles	Inner surface is not accessible, high cost
Carbon nanofibers, nanohorns, nanocoils, nanowalls, and nanospheres	high surface area, high crystallinity, good electrical conductivity, entire surface is accessible, better durability and stability than CNTs	high cost, performance less known due to lack of investigation
Mesoporous carbons	three-dimensional structure, large number of mesopores, easy for the transport of reactants, easy to disperse catalyst uniformly	low stability, corrosion, low electrical conductivity
Graphene oxide and reduced graphene oxide	large surface area, good electric conductivity, high chemical stability, unique graphitized basal plane structure, more anchoring sites, better dispersion of catalyst particles	aggregation of graphene nanosheets, lack if high-quality graphene in mass production scales
Nanodiamond	mechanical and chemical stability, catalyst can be deposited at high temperatures, external surface	low surface area, insulator, elemental doping needed to improve conductivity, difficult to achieve uniformly and controllably doping, hard to disperse catalyst
Heteroatom-doped Carbon	high electrical conductivity, high stability, possess more anchored sites for catalyst nanoparticles	high cost due to doping process

3. Concluding outlook and future trends

The catalyst support materials play a critical role in determining the performance and durability of catalysts in DAFCs. Ideal carbon supports must possess high mesoporosity in the pore size range of 20–40 nm with a high accessible surface area. The most common catalyst supports used in DAFCs by far are porous carbon materials such as carbon blacks (CBs) due to their low cost, good electrical conductivity, and relatively high surface area. Although carbon blacks are still the most commonly used catalyst supports in DAFCs, the existence of a high amount of micropores in carbon black can hinder the flow of reactants/products and the accessibility by the polymer electrolyte.

To improve the performance of catalyst supports in DAFCs, new types of catalyst supports are, therefore, urgently required. Porous carbon nanomaterials, such as CNTs, mesoporous carbon, and graphene, have become more and more attractive and receiving increasing attention due to their unique features such as tunable structures and porosity, high surface areas and improved electrical conductivity and stability in fuel cell environments. CNTs are highly conductive due to their high crystallinity and also have high amounts of mesopores in the tubular structure that allows good transportation of reactant flux. Although higher catalytic activity and higher stability than carbon blacks were achieved from CNTs, their significantly high costs and complexity of synthesis methods compared to that of carbon blacks hinders their large-scale commercialization. Less studied carbon nanomaterials such as carbon nanohorns, carbon nanocoils, and carbon fibers have also shown some advantages due to their peculiar structure and have shown promising results as fuel cell catalyst supports. Ordered mesoporous carbons that possess high surface areas and high amounts of mesopores allow easy flow of reactants and well-disperse of catalyst nanoparticles; therefore, catalysts supported on the mesoporous carbons showed higher catalytic activity than the same catalysts supported on carbon blacks. Furthermore, synthesis methods of ordered mesoporous carbons are simple and not too expensive. The easy for tailoring the pore size and pore distribution makes the mesoporous carbons a good candidate to substitute carbon blacks as catalyst support in DAFCs. Graphene possesses a number of excellent properties such as a high surface area, high conductivity, high chemical stability, and unique graphitized basal plane structure. It becomes the most studied catalyst support for DAFCs. By using various methods for functionalization of graphene nanosheets, more anchoring sites can be created for higher and more homogeneous metal dispersion. This ultimately improves the durability and activity of catalyst nanoparticles. However, for graphene to be more cost-effective, novel methods for the synthesis of high-quality graphene nanostructures in a mass production scale with low production cost are required. Conductive nanodiamonds, formed by doping diamonds with a secondary element such as boron,

Nanomaterials for Alcohol Fuel Cells
Materials Research Foundations **49** (2019) 79-102

Materials Research Forum LLC
doi: https://doi.org/10.21741/9781644900192-2

have very attractive properties including high dimensional stability and chemical inert in fuel cell environments; therefore, conductive diamonds are intrinsically attractive for applications as durable catalyst support for DAFCs. However, low surface areas and difficulty anchoring the catalyst atoms on their surface have to be improved. Lastly, doping of CNTs, nanodiamond, and graphene with elements such as N have demonstrated better electronic conductivity in comparison to their pure forms. An enhanced catalytic activity for alcohol oxidation was obtained for catalyst supported on those doped carbon nanomaterials due to their more disordered nanostructures and defects, and the high electron density.

Although new types of carbon nanomaterials discussed above have very promising properties and have demonstrated great potential to replace carbon blacks as catalyst supports for DAFCs, cost-effective and straightforward methods for their large-scale production are required to facilitate their commercialization.

References

[1] S. Surampudi, S. R. Narayanan, E. Vamos, H. Frank, G. Halpert, A. LaConti, J. Kosek, G. K. S. Prakash, & G. A. Olah, Advances in direct oxidation methanol fuel cells. *Journal of Power Sources*, 47 (1994) 377–385. https://doi.org/10.1016/0378-7753(94)87016-0.

[2] S. Shahgaldi & J. Hamelin, Stability study of ultra-low Pt thin film on TiO 2 –C core-shell structure and TiO 2 encapsulated in carbon nanospheres as cathode catalyst in PEMFC. *Fuel*, 150 (2015) 645–655. https://doi.org/10.1016/j.fuel.2015.02.002.

[3] M. V. Martínez-Huerta & M. J. Lázaro, Electrocatalysts for low temperature fuel cells. *Catalysis Today*, 285 (2017) 3–12. https://doi.org/10.1016/j.cattod.2017.02.015.

[4] I. Katsounaros, S. Cherevko, A. R. Zeradjanin, & K. J. J. Mayrhofer, Oxygen Electrochemistry as a Cornerstone for Sustainable Energy Conversion. *Angewandte Chemie International Edition*, 53 (2014) 102–121. https://doi.org/10.1002/anie.201306588.

[5] H. Schulenburg, B. Schwanitz, N. Linse, G. G. Scherer, A. Wokaun, J. Krbanjevic, R. Grothausmann, & I. Manke, 3D Imaging of Catalyst Support Corrosion in Polymer Electrolyte Fuel Cells. *The Journal of Physical Chemistry C*, 115 (2011) 14236–14243. https://doi.org/10.1021/jp203016u.

[6] S. Shahgaldi & J. Hamelin, Improved carbon nanostructures as novel catalyst support in the cathode side of PEMFC: a critical review. *Carbon*, 94 (2015) 705–728. https://doi.org/10.1016/j.carbon.2015.07.055.

[7] Z. Liu, L. Hong, & S. W. Tay, Preparation and characterization of carbon-supported Pt, PtSnO2 and PtRu nanoparticles for direct methanol fuel cells. *Materials Chemistry and Physics*, 105 (2007) 222–228. https://doi.org/10.1016/j.matchemphys.2007.04.045.

[8] C.-H. Wang, H.-Y. Du, Y.-T. Tsai, C.-P. Chen, C.-J. Huang, L. C. Chen, K. H. Chen, & H.-C. Shih, High performance of low electrocatalysts loading on CNT directly grown on carbon cloth for DMFC. *Journal of Power Sources*, 171 (2007) 55–62. https://doi.org/10.1016/j.jpowsour.2006.12.028.

[9] Y. Zhao, L. Fan, H. Zhong, Y. Li, & S. Yang, Platinum Nanoparticle Clusters Immobilized on Multiwalled Carbon Nanotubes: Electrodeposition and Enhanced Electrocatalytic Activity for Methanol Oxidation. *Advanced Functional Materials*, 17 (2007) 1537–1541. https://doi.org/10.1002/adfm.200600416.

[10] L. V. Kumar, S. Addo Ntim, O. Sae-Khow, C. Janardhana, V. Lakshminarayanan, & S. Mitra, Electro-catalytic activity of multiwall carbon nanotube-metal (Pt or Pd) nanohybrid materials synthesized using microwave-induced reactions and their possible use in fuel cells. *Electrochimica Acta*, 83 (2012) 40–46. https://doi.org/10.1016/j.electacta.2012.07.098.

[11] X. Geng, J. Jing, Y. Cen, R. Datta, & J. Liang, In Situ Synthesis and Characterization of Polyethyleneimine-Modified Carbon Nanotubes Supported PtRu Electrocatalyst for Methanol Oxidation. *Journal of Nanomaterials*, 2015 (2015) 1–10. https://doi.org/10.1155/2015/296589.

[12] X. Hu, T. Wang, L. Wang, S. Guo, & S. Dong, A General Route to Prepare One- and Three-Dimensional Carbon Nanotube/Metal Nanoparticle Composite Nanostructures. *Langmuir*, 23 (2007) 6352–6357. https://doi.org/10.1021/la063246b.

[13] Y. Suda, Y. Shimizu, M. Ozaki, H. Tanoue, H. Takikawa, H. Ue, K. Shimizu, & Y. Umeda, Electrochemical properties of fuel cell catalysts loaded on carbon nanomaterials with different geometries. *Materials Today Communications*, 3 (2015) 96–103. https://doi.org/10.1016/j.mtcomm.2015.02.003.

[14] M. Okada, Y. Konta, & N. Nakagawa, Carbon nano-fiber interlayer that provides high catalyst utilization in a direct methanol fuel cell. *Journal of Power Sources*, 185 (2008) 711–716. https://doi.org/10.1016/j.jpowsour.2008.08.026.

[15] C. Zhou, Z. Liu, X. Du, D. Mitchell, Y.-W. Mai, Y. Yan, & S. Ringer, Hollow nitrogen-containing core/shell fibrous carbon nanomaterials as support to platinum nanocatalysts and their TEM tomography study. *Nanoscale Research Letters*, 7 (2012) 165. https://doi.org/10.1186/1556-276X-7-165.

[16] T. Yoshitake, Y. Shimakawa, S. Kuroshima, H. Kimura, T. Ichihashi, Y. Kubo, D. Kasuya, K. Takahashi, F. Kokai, M. Yudasaka, & S. Iijima, Preparation of fine platinum catalyst supported on single-wall carbon nanohorns for fuel cell application. *Physica B: Condensed Matter*, 323 (2002) 124–126. https://doi.org/10.1016/S0921-4526(02)00871-2.

[17] T. Hyeon, S. Han, Y.-E. Sung, K.-W. Park, & Y.-W. Kim, High-Performance Direct Methanol Fuel Cell Electrodes using Solid-Phase-Synthesized Carbon Nanocoils. *Angewandte Chemie International Edition*, 42 (2003) 4352–4356. https://doi.org/10.1002/anie.200250856.

[18] M. Sevilla, G. Lota, & A. B. Fuertes, Saccharide-based graphitic carbon nanocoils as supports for PtRu nanoparticles for methanol electrooxidation. *Journal of Power Sources*, 171 (2007) 546–551. https://doi.org/10.1016/j.jpowsour.2007.05.096.

[19] Z. Hamoudi, A. Brahim, M. A. El Khakani, & M. Mohamedi, Electroanalytical Study of Methanol Oxidation and Oxygen Reduction at Carbon Nanohorns-Pt Nanostructured Electrodes. *Electroanalysis*, 25 (2013) 538–545. https://doi.org/10.1002/elan.201200572.

[20] J. Wu, F. Hu, X. Hu, Z. Wei, & P. K. Shen, Improved kinetics of methanol oxidation on Pt/hollow carbon sphere catalysts. *Electrochimica Acta*, 53 (2008) 8341–8345. https://doi.org/10.1016/j.electacta.2008.06.051.

[21] G. Álvarez, F. Alcaide, O. Miguel, L. Calvillo, M. J. Lázaro, J. J. Quintana, J. C. Calderón, & E. Pastor, Technical electrodes catalyzed with PtRu on mesoporous ordered carbons for liquid direct methanol fuel cells. *Journal of Solid State Electrochemistry*, 14 (2010) 1027–1034. https://doi.org/10.1007/s10008-009-0913-3.

[22] L. Calvillo, M. J. Lázaro, E. García-Bordejé, R. Moliner, P. L. Cabot, I. Esparbé, E. Pastor, & J. J. Quintana, Platinum supported on functionalized ordered mesoporous carbon as an electrocatalyst for direct methanol fuel cells. *Journal of Power Sources*, 169 (2007) 59–64. https://doi.org/10.1016/j.jpowsour.2007.01.042.

[23] V. Celorrio, D. Sebastián, L. Calvillo, A. B. García, D. J. Fermin, & M. J. Lázaro, Influence of thermal treatments on the stability of Pd nanoparticles supported on graphitized ordered mesoporous carbons. *International Journal of Hydrogen Energy*, 41 (2016) 19570–19578. https://doi.org/10.1016/j.ijhydene.2016.05.271.

[24] L. Calvillo, M. Gangeri, S. Perathoner, G. Centi, R. Moliner, & M. J. Lázaro, Synthesis and performance of platinum supported on ordered mesoporous carbons as a catalyst for PEM fuel cells: Effect of the surface chemistry of the support. *International Journal of Hydrogen Energy*, 36 (2011) 9805–9814. https://doi.org/10.1016/j.ijhydene.2011.03.023.

[25] J. Zeng, C. Francia, C. Gerbaldi, V. Baglio, S. Specchia, A. S. Aricò, & P. Spinelli, Hybrid ordered mesoporous carbons doped with tungsten trioxide as supports for Pt electrocatalysts for the methanol oxidation reaction. *Electrochimica Acta*, 94 (2013) 80–91. https://doi.org/10.1016/j.electacta.2013.01.139.

[26] C. Alegre, L. Calvillo, R. Moliner, J. A. González-Expósito, O. Guillén-Villafuerte, M. V. M. Huerta, E. Pastor, & M. J. Lázaro, Pt and PtRu electrocatalysts supported on carbon xerogels for direct methanol fuel cells. *Journal of Power Sources*, 196 (2011) 4226–4235. https://doi.org/10.1016/j.jpowsour.2010.10.049.

[27] C. Alegre, D. Sebastián, M. E. Gálvez, R. Moliner, & M. J. Lázaro, Sulfurized carbon xerogels as Pt support with enhanced activity for fuel cell applications. *Applied Catalysis B: Environmental*, 192 (2016) 260–267. https://doi.org/10.1016/j.apcatb.2016.03.070.

[28] K. S. Novoselov, A. K. Geim, S. V Morozov, D. Jiang, Y. Zhang, S. V Dubonos, I. V Grigorieva, & A. A. Firsov, Electric Field Effect in Atomically Thin Carbon Films. *Science*, 306 (2004) 666–669. https://doi.org/10.1126/science.1102896.

[29] E. Antolini, Graphene as new carbon support for low-temperature fuel cell catalysts. *Applied Catalysis B: Environmental*, 123–124 (2012) 52–68. https://doi.org/10.1016/j.apcatb.2012.04.022.

[30] G. Wu, A. Santandreu, W. Kellogg, S. Gupta, O. Ogoke, H. Zhang, H.-L. Wang, & L. Dai, Carbon nanocomposite catalysts for oxygen reduction and evolution reactions: From nitrogen doping to transition-metal addition. *Nano Energy*, 29 (2016) 83–110. https://doi.org/10.1016/j.nanoen.2015.12.032.

[31] W. Gao, G. Wu, M. T. Janicke, D. A. Cullen, R. Mukundan, J. K. Baldwin, E. L. Brosha, C. Galande, P. M. Ajayan, K. L. More, A. M. Dattelbaum, & P. Zelenay, Ozonated Graphene Oxide Film as a Proton-Exchange-Membrane. *Angewandte Chemie International Edition*, 53 (2014) 3588–3593. https://doi.org/10.1002/anie.201310908.

[32] W. Zhao, X. Zhou, J. Chen, & X. Lu, Controllable Electrodeposition of Platinum Nanoparticles on Graphene Nanosheet for Methanol Oxidation Reaction. *Journal of Cluster Science*, 24 (2013) 739–748. https://doi.org/10.1007/s10876-013-0569-0.

[33] N. Soin, S. S. Roy, T. H. Lim, & J. A. D. McLaughlin, Microstructural and electrochemical properties of vertically aligned few layered graphene (FLG) nanoflakes and their application in methanol oxidation. *Materials Chemistry and Physics*, 129 (2011) 1051–1057. https://doi.org/10.1016/j.matchemphys.2011.05.063.

[34] S. Liu, L. Wang, J. Tian, W. Lu, Y. Zhang, X. Wang, & X. Sun, Microwave-assisted rapid synthesis of Pt/graphene nanosheet composites and their application for

methanol oxidation. *Journal of Nanoparticle Research*, 13 (2011) 4731–4737. https://doi.org/10.1007/s11051-011-0440-x.

[35] B. Luo, X. Yan, S. Xu, & Q. Xue, Polyelectrolyte functionalization of graphene nanosheets as support for platinum nanoparticles and their applications to methanol oxidation. *Electrochimica Acta*, 59 (2012) 429–434. https://doi.org/10.1016/j.electacta.2011.10.103.

[36] O. Akyıldırım, G. Kotan, M. L. Yola, T. Eren, & N. Atar, Fabrication of bimetallic Pt/Pd nanoparticles on 2-thiolbenzimidazole functionalized reduced graphene oxide for methanol oxidation. *Ionics*, 22 (2016) 593–600. https://doi.org/10.1007/s11581-015-1572-2.

[37] Y. Liu, Y. Xia, H. Yang, Y. Zhang, M. Zhao, & G. Pan, Facile preparation of high-quality Pt/reduced graphene oxide nanoscrolls for methanol oxidation. *Nanotechnology*, 24 (2013) 235401. https://doi.org/10.1088/0957-4484/24/23/235401.

[38] Z. Li, M. Ruan, L. Du, G. Wen, C. Dong, & H.-W. Li, Graphene nanomaterials supported palladium nanoparticles as nanocatalysts for electro-oxidation of methanol. *Journal of Electroanalytical Chemistry*, 805 (2017) 47–52. https://doi.org/10.1016/j.jelechem.2017.10.015.

[39] B. Su, G. Wu, C. J. Yang, Z. Zhuang, X. Wang, X. Chen, & X. Chen, Platinum nanoflowers supported on graphene oxide nanosheets: Their green synthesis, growth mechanism, and advanced electrocatalytic properties for methanol oxidation. *Journal of Materials Chemistry*, 22 (2012) 11284–11289. https://doi.org/10.1039/c2jm31133j.

[40] Y. Hu, H. Zhang, P. Wu, H. Zhang, B. Zhou, & C. Cai, Bimetallic Pt-Au nanocatalysts electrochemically deposited on graphene and their electrocatalytic characteristics towards oxygen reduction and methanol oxidation. *Physical Chemistry Chemical Physics*, 13 (2011) 4083–4094. https://doi.org/10.1039/c0cp01998d.

[41] S. Yang, F. Zhang, C. Gao, J. Xia, L. Lu, & Z. Wang, A sandwich-like PtCo-graphene/carbon dots/graphene catalyst for efficient methanol oxidation. *Journal of Electroanalytical Chemistry*, 802 (2017) 27–32. https://doi.org/10.1016/j.jelechem.2017.08.027.

[42] G. Siné, I. Duo, B. El Roustom, G. Fóti, & C. Comninellis, Deposition of clusters and nanoparticles onto boron-doped diamond electrodes for electrocatalysis. *Journal of Applied Electrochemistry*, 36 (2006) 847–862. https://doi.org/10.1007/s10800-006-9159-2.

[43] G. R. Salazar-Banda, K. I. B. Eguiluz, & L. A. Avaca, Boron-doped diamond powder as catalyst support for fuel cell applications. *Electrochemistry Communications*, 9 (2007) 59–64. https://doi.org/10.1016/j.elecom.2006.08.038.

[44] G. R. Salazar-Banda, H. B. Suffredini, M. L. Calegaro, S. T. Tanimoto, & L. A. Avaca, Sol–gel-modified boron-doped diamond surfaces for methanol and ethanol electro-oxidation in acid medium. *Journal of Power Sources*, 162 (2006) 9–20. https://doi.org/10.1016/j.jpowsour.2006.06.045.

[45] G. Siné, D. Smida, M. Limat, G. Fóti, & C. Comninellis, Microemulsion Synthesized Pt/Ru/Sn Nanoparticles on BDD for Alcohol Electro-oxidation. *Journal of The Electrochemical Society*, 154 (2007) B170–B174. https://doi.org/10.1149/1.2400602.

[46] H. Huang & X. Wang, Recent progress on carbon-based support materials for electrocatalysts of direct methanol fuel cells. *J. Mater. Chem. A*, 2 (2014) 6266–6291. https://doi.org/10.1039/C3TA14754A.

[47] T. Maiyalagan, Synthesis and electrocatalytic activity of methanol oxidation on nitrogen-containing carbon nanotubes supported Pt electrodes. *Applied Catalysis B: Environmental*, 80 (2008) 286–295. https://doi.org/10.1016/j.apcatb.2007.11.033.

[48] R. Chetty, S. Kundu, W. Xia, M. Bron, W. Schuhmann, V. Chirila, W. Brandl, T. Reinecke, & M. Muhler, PtRu nanoparticles supported on nitrogen-doped multiwalled carbon nanotubes as a catalyst for methanol electrooxidation. *Electrochimica Acta*, 54 (2009) 4208–4215. https://doi.org/10.1016/j.electacta.2009.02.073.

[49] H. Y. Du, C. H. Wang, H. C. Hsu, S. T. Chang, U. S. Chen, S. C. Yen, L. C. Chen, H. C. Shih, & K. H. Chen, Controlled platinum nanoparticles uniformly dispersed on nitrogen-doped carbon nanotubes for methanol oxidation. *Diamond and Related Materials*, 17 (2008)535-541. https://doi.org/10.1016/j.diamond.2008.01.116.

[50] G. Wu, R. Swaidan, D. Li, & N. Li, Enhanced methanol electro-oxidation activity of PtRu catalysts supported on heteroatom-doped carbon. *Electrochimica Acta*, 53 (2008) 7622–7629. https://doi.org/10.1016/j.electacta.2008.03.082.

[51] R. Lv, T. Cui, M.-S. Jun, Q. Zhang, A. Cao, D. S. Su, Z. Zhang, S.-H. Yoon, J. Miyawaki, I. Mochida, & F. Kang, Open-Ended, N-Doped Carbon Nanotube-Graphene Hybrid Nanostructures as High-Performance Catalyst Support. *Advanced Functional Materials*, 21 (2011) 999–1006. https://doi.org/10.1002/adfm.201001602.

Materials Research Forum LLC
doi: https://doi.org/10.21741/9781644900192-3

Chapter 3

Nanocatalysts for Direct 2-Propanol Fuel Cells

M. Nacef[1]*, M.L. Chelaghmia[1], A.M. Affoune[1], M. Pontié[2]

[1]Laboratoire d'Analyses Industrielles et Génie des Matériaux, Université 8 Mai 1945 Guelma, BP 401, Guelma 24000, Algérie

[2]Laboratoire GEIHP EA 3142, Institut de Biologie en Santé, PBH-IRIS, CHU, Université d'Angers, 4 Rue Larrey, 49933 Angers Cedex 9, Angers, France

nacef.mouna@univ-guelma.dz

Abstract

2-propanol is a new fuel source for fuel cells application. Its electrooxidation reaction has been developed in both acid and alkaline 2-propanol fuel cells. The Acid type fuel cell is the older one where platinum electrocatalysts have been widely investigated. More recently, alkaline medium and palladium electrocatalyst were considered as a suitable pair candidate for 2-propanol electrooxidation reaction. The present book chapter is designed to draw a clear picture of the state of the art of direct 2-propanol fuel cell. Thermodynamics aspects of direct 2-propanol fuel cell are presented and a review of catalysts used in 2-propanol oxidation reaction is exposed and especially nanocatalysts are emphasized. This chapter gives a starting point for future research on how 2-propanol would feed fuel cells.

Keywords

2-Propanol, Nanocatalyst, Electrooxidation, Alcohol Crossover, Platinum, Palladium, Direct Alcohol Fuel Cell

Contents

1.   Introduction

There is a global consensus on increasing energy demand around the world. This demand is in facing with the prospect of a shortage and even an energy crisis that are to be feared, given the progressive depletion of the most accessible fossil resources. International awareness of the need for non-fossil energy sources has led, for several decades of research, to the development of renewable energies. In this perspective, fuel cells, especially, direct liquid-feed fuel cells have emerged promising, in particular, those operating with alcohols.

Direct alcohol fuel cells (DAFCs) are the result of more than five decades of intense and progressive seeking for other fuels to replace hydrogen in H_2-fed fuel cells, such as proton exchange membrane fuel cells (PEMFCs). The research was mainly driven by the need to overcome hydrogen drawbacks. Indeed, liquid fuels, unlike gaseous fuels, have the advantage of high theoretical energy density, easy storage, handling, transportation, and distribution. Also, in a direct liquid fuel cell, the bulky reformer and humidifier are eliminated. Knowing that it is highly expected that the most likely commercial application for DAFCs is as a power source for portable electronic devices, such as mobile phone, laptop, digital camera, etc.; the size and weight of the energy supplier of such devices are significant criteria.

Methanol was the most investigated alcohol in DAFCs application because of its simple structure, availability and the ease of production. However, methanol is relatively toxic and more prone to the crossover phenomenon impeding the cathode performance. Thus, the quest for other green alcohols was conducted and the investigation has been oriented towards lightweight mono and bifunctional alcohols [1-4].

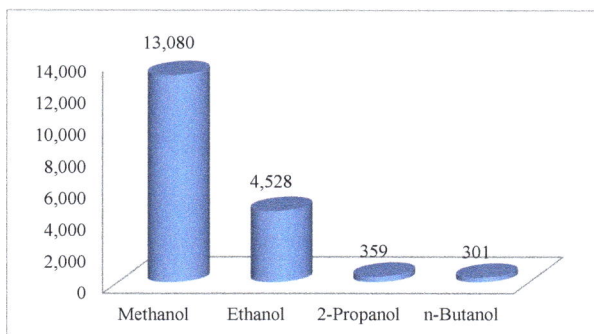

Figure 1.   A number of papers related to methanol, ethanol, 2-propanol, and n-butanol fuel cells. From SCOPUS database.

As depicted by the bibliometric analysis, the number of publications devoted to methanol, ethanol, 2-propanol, and n-butanol fuel cells decreases drastically as the number of carbons in the alcohol increases. Moreover, this decrease is significant between methanol and ethanol, and between ethanol and 2-propanol. This is not surprising as methanol has long been the first choice to feed direct alcohol fuel cells, fig. 1. The evolution of research interest for each of the alcohols, over the four past decades, seems to be guided by the first trend, fig. 2.

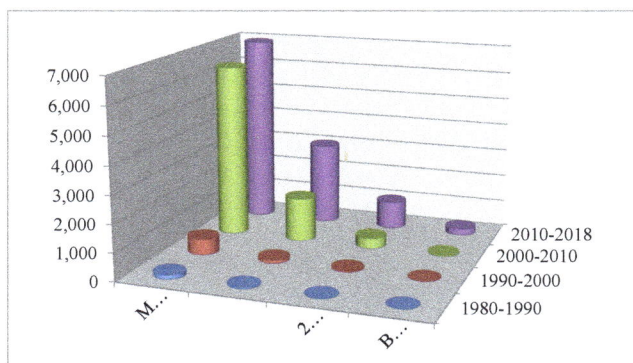

Figure 2.   The growth of publications devoted to direct methanol, ethanol, 2-propanol, and n-butanol fuel cells during 1980-2018. From SCOPUS database.

2-propanol which is commonly called isopropyl alcohol, isopropanol, propan-2-ol or rubbing alcohol, is C-3 type alcohol. It's the only isomer of n-propanol. This compound is a colorless liquid with an alcohol-like odor, volatile, and miscible in any proportion with water. 2-propanol is recognized as an irritant compound, highly flammable, and slightly polluting. It is toxic and affects seriously the ecosystem [5-7]. However, 2-propanol is not as toxic as methanol.

2-propanol is easily available and it is produced mainly by propylene hydration. The first commercial production of 2-propanol has been accomplished by the addition of water to propylene by Standard Oil of New Jersey, USA in 1930 [8]. This method consists of propylene of refinery grade indirectly hydration using the sulfuric acid. More recently, a new 2-propanol production method was employed which consists of the direct high purity propylene hydration. This method has the advantage to overcome the severe corrosion due to the use of concentrated sulfuric acid. In this case, propylene and water are heated at 130-150°C and 60-100 atm. After which, 2-propanol is recovered. Other processes, involving the gas phase or liquid phase mixtures, were also proposed [9].

In the above-mentioned methods, 2-propanol is produced from a petroleum-based compound. 2-propanol renewable production exists by means of converting acetone to 2-propanol, [10-12]. Acetone which is readily available from fermentation processes was hydrogenated to 2-propanol earlier. Nowadays, a very small amount of this alcohol is produced by the hydrogenation of acetone in the liquid phase.

2-propanol is widely used for multi purposes of industrial and household chemicals. It is essentially used as solvent, extractant and as a substitute for ethanol in cosmetic and pharmaceutical industries. It is also used as a gasoline additive to prevent carburetor icing.

The use of 2-propanol in fuel cells has long been restricted to the preparation of the Nafion® membrane in PEMFCs. 2-propanol utilization as fuel in direct alcohol fuel cells application was investigated first by Wang *et al.* in 1995 [13]. Since then, 2-propanol has rarely been considered alone for fuel cells application but frequently in comparison with other alcohol a the head of the list, such as methanol and ethanol. Quickly, researchers have realized that when increasing the number of carbons in the alcohol, the electrochemical activity was lowered, in particular onto platinum catalysts mainly due to the C-C bond cleavage difficulty [14-16]. Nevertheless, other catalysts have given better results towards alcohols with a higher number of carbons [17-19] which were sometimes superior to that of methanol [20].

2-propanol is the smallest secondary alcohol, so its catalytic oxidation is very interesting due to the particularity of the molecular structure. The complete electrooxidation of an alcohol yields to carbon dioxide and water. C3 primary alcohols electrooxidation mechanism was investigated in acidic media on Pt. The main products species were CO_2 and the corresponding aldehydes [13,21-23]. Unlike aliphatic alcohols, the electrooxidation of 2-propanol, as well as secondary alcohols produces the corresponding ketone and, at lesser amounts of carbon dioxide [13,24-28].

It is clear that gaining more knowledge in the field of direct 2-propanol fuel cells (D2PFC) and specifically with respect to the electrooxidation reaction is of paramount importance because it will allow understanding the reaction mechanisms of other secondary alcohols.

This chapter deals with an analysis of the history and current status of catalysts used for the anodic oxidation reaction as well as the cathodic reduction reaction in 2-propanol fuel cells. Despite the small number of papers dedicated to 2-propanol catalysts; <100, and even less to the nanocatalysts category, we tried to give an accurate background of 2-propanol electrooxidation following the chronological evolution, since there have been no reviews published on the topic till the present. A special focus on nanocatalysts for 2-propanol fuel cells was drawn.

2. Thermodynamics of direct 2-propanol fuel cell

D2PFC is an electrochemical device which converts the chemical energy of 2-propanol into heat and electrical energy through a catalytic reaction. As explained above, the direct feed of fuel cell with alcohol will avoid the use of a binding and a heavy reformer.

The complete electrooxidation of 2-propanol to carbon dioxide occurs at the anode, while the oxygen reduction reaction occurs at the cathode. The half reactions are as follows:

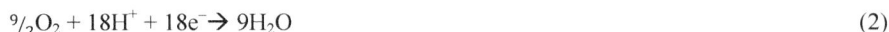

$$C_3H_7OH + 5H_2O \rightarrow 3CO_2 + 18H^+ + 18e^- \tag{1}$$

$$^9/_2O_2 + 18H^+ + 18e^- \rightarrow 9H_2O \tag{2}$$

The anodic reaction involves the transfer of 18 electrons. Hydrogen, methanol, and ethanol oxidation reactions involve 2, 6, and 12 electrons, respectively. D2PFC surpasses the pioneering PEMFC, DMFC, and DEFC. This is one of the advantages of feeding fuel cells with alcohol containing a higher number of carbons.

Nanomaterials for Alcohol Fuel Cells   Materials Research Forum LLC
Materials Research Foundations **49** (2019) 103-128   doi: https://doi.org/10.21741/9781644900192-3

The overall reaction is then:

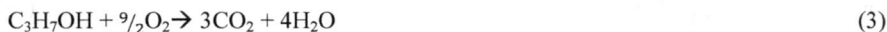

$$C_3H_7OH + \frac{9}{2}O_2 \rightarrow 3CO_2 + 4H_2O \tag{3}$$

From equation (3), it's easy to calculate the standard enthalpy change; ΔH^0 and the standard free Gibbs energy change; ΔG^0 at a given temperature. These values allow us to calculate thermodynamic parameters of fuel cell fed by 2-propanol and most generally for any other fuel.

The electromotive force is the electrical intensity developed by the fuel cell, it is directly proportional to the free Gibbs energy.

The specific energy, as well as the energy density, quantify the amount of stored energy in the system per unit mass and per unit volume, respectively.

The theoretical reversible efficiency is known as the ratio between the maximum work output of the fuel cell produced by the overall reaction and the input (total energy lost from the fuel) [28].

The pure compound capacity is calculated from Faraday's first law of electrochemistry. It states that 1 gram equivalent weight will deliver 96485.31 C (or 26.8 Ah).

All these parameters were calculated for a fuel cell fed with 2-propanol considering two cases: when the released water is in liquid or in gaseous form. Results are gathered in table 1, where the calculation method was briefly indicated [29].

Under standard conditions, the direct 2-propanol fuel cell produces specific energy of 9.02 kWh kg^{-1} which is 1.13 and 1.48 times greater than that of ethanol and methanol, respectively [29-31]. Thanks to its liquid state, 2-propanol could supply an energy density of 7080.7 kWh m^{-3} which is remarkably higher than that of hydrogen. The electromotive force at equilibrium is of 1.124 V, which is very close to that of lighter weight alcohols and H$_2$. The theoretical reversible efficiency is higher than that of alcohols with low carbon chain. This value passes from 97% when the released water is in liquid form to 105% when it is in gaseous form. These values denote the very good thermodynamic potentiality of 2-propanol when it is fed directly to the fuel cell [29].

The pure compound capacity of 2-propanol, which is not dependent on the free Gibbs energy, was the highest value when Demirci [30] compared it to other fuels used in direct liquid-feed fuel cells (note that pure compound value for 2-propanol is equal to that of n-propanol). As mentioned in the introduction section, 2-propanol electrooxidation is incomplete and commonly yields to acetone as the major reaction product. In such a case, the number of exchanged electrons in the reaction is only two. Even in this case, the pure compound capacity still higher than the benchmark NiMH and Li-ion batteries [32].

Table 1. Direct 2-propanol fuel cell's parameters calculations under standard conditions.

Fuel cell parameters	Form of the released water	
	Liquid	Gaseous
Standard free Gibbs energy change: $\Delta G^0 r$ [kJ mol^{-1}]	-1953.96	-1918.51
Standard enthalpy change: $\Delta H^0 r$ [kJ mol^{-1}]	-2008.48	-1831.25
Electromotive force: emf[V] $= -\Delta G_r^0/nF$	1.124	1.105
Specific energy: W[kWh kg^{-1}] $= -\Delta G_r^0/3600M$	9.02	8.87
Energy density: We[kWh m^{-3}] $= W \times d$	7080.70	/
Reversible efficiency: $\varepsilon[\%] = \Delta G_r^0/\Delta H_r^0$	97.2	104.76
Pure compound capacity:C[Ah kg^{-1}] $= 26.8\,n/M$	8027	8027

n is the number of electrons involved in the overall reaction (18 in the case of 2-propanol),

F is Faraday constant (96 485.31 C mol^{-1}),

d is the density of liquid 2-propanol (785.05 10^3 gm^{-3})

M is the molar weight of 2-propanol (60.09 g mol^{-1})

Several studies have been conducted aiming the investigation of temperature effect on the alcohols electrooxidation rate [33]. It was found that, except in the case of butanol [4], increasing temperature is of great interest in alcohols electrooxidation reaction [25,34].

From a thermodynamic point of view, investigation of 2-propanol oxidation reaction over a large temperature range was previously conducted by our group [29]. It was found that 2-propanol thermodynamic parameters evolution evolves in different ways. Some parameters increase with increasing temperature while others decrease.

In order to make the most accurate comparison, we have suggested in our previous work [29], to distinguish three ranges of temperature [298.15, 373.15], [373.15, 600], and [600, 1300] K. Other parameters should be considered such as the production way (renewable or not), the ease of transport and distribution, and the harmful effect of fuels taking into account the threshold limit value. From a concise comparison, we have found that 2-

propanol performances come after that of ethanol, methanol, hydrogen, and n-butanol, in this order, at low temperatures. However, for temperatures above 373.15, 2-propanol performances are inferior to that of methanol and ethanol but are superior to that of H_2. Thus, it could be concluded from this prospection that 2-propanol is a serious candidate as interesting as methanol or ethanol for fuel cells application at higher temperatures.

3.   Nanocatalysts for 2-propanol oxidation reaction

A fuel cell consists of an anode and a cathode separated by an electrolyte. 2-propanol may be electrooxidized in both acidic and alkaline environments where distinct mechanisms are involved. In acidic medium, where pH is commonly under 5, only noble metals could be used to oxidize the alcohol due to its corrosive properties. Since Pd and Au have shown poor reactivity towards alcohol oxidation reaction [35,36], Platinum-based electrodes exhibited better performances [37]. Thanks to the acidic environment, there is no salts formation, thus the electrode blockage is avoided. The electrolyte in acid type fuel cell is frequently a proton exchange membrane such as Nafion® type membranes. These membranes have a low operating temperature; around 90 °C. This essential component of the fuel cell system was a brake till polybenzimidazole (PBI) type electrolytes were tested. PBI has very good stability at higher temperatures.

In alkaline medium, the reaction kinetics oxidation is better. The availability of OH⁻ ions from the electrolyte enhances the oxidation reaction. The library of materials that could be used in alkaline fuel cell is larger in alkaline than in acid, avoiding by the same the way the costly platinum. Polymer electrolyte membranes are also expensive and are more prone to fuel crossover. In the following two sections, a review of electrocatalysts used in both acid and alkaline media will be presented, according to the scheme bellow (fig. 3).

3.1.   22-Propanol electrocatalysts in acidic media

For the last decades, research on the anode materials for direct 2-propanol fuel cells (D2PFCs) operating in acidic media has attracted a lot of attention as a result of the development of direct alcohol fuel cells [24-27,38-51]. Pt and Pt-based electrocatalysts could not be avoided if D2PFCs are expected to be widely used as energy suppliers for various devices.

On the other hand, the Pt-based, as well as noble metals electrocatalysts, should be well dispersed on the surfaces of supports, which will increase the catalytic active sites needed for the good charge mobility. Carbon materials have been traditionally used as electrodes support in fuel cells.

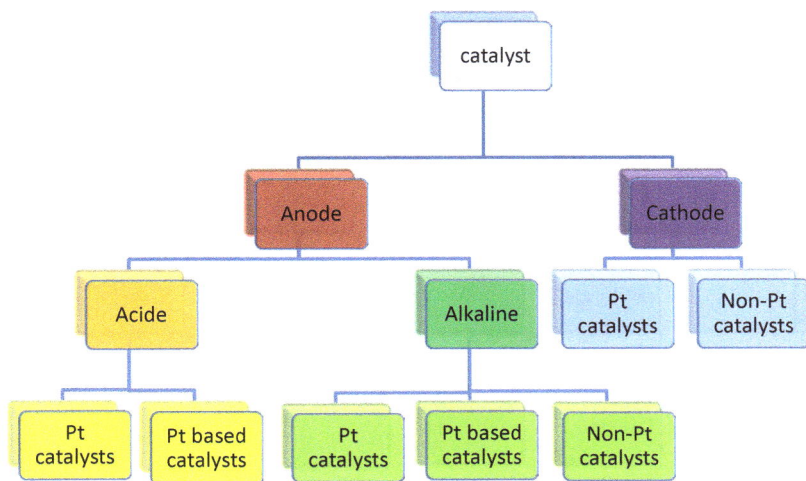

Figure 3. Classification of catalysts in D2PFC.

3.1.1 Platinum

The electrochemical behaviour of 2-propanol on platinum in acid solutions has been investigated by numerous groups [24,25,27,39,42-44,47,49,51]. Singularly, 2-propanol electrooxidation reaction pathway does not imply the formation of adsorbed CO on platinum surfaces, unlike other low carbon chain molecules. It is worth mentioning that the onset potential for the oxidation of 2-propanol was shifted to more negative values when compared to that of methanol, ethanol, and n-propanol, which is considered as a good performance [41].

In a series of papers, Sun *et al.* [24,27,42-44] have used electrochemical methods, Fourier transform infrared spectroscopy (FTIRS), as well as on-line mass spectroscopy to investigate the oxidation of 2-propanol at platinum single crystal electrodes. They found that the reaction pathways do not depend on the surface structure since the reaction products are not different on the basal planes surfaces as well as on the stepped surfaces. In all cases, acetone is the main product of 2-propanol electrooxidation and all adsorbates contained three C-atoms. Acetone acts as both product and intermediate. The formation of a further oxidized product, namely carbon dioxide was also detected. The reaction mechanism of 2-propanol on platinum single crystal electrodes was schematically explained by Sun *et al.* [24,27,42-44].

Pastor *et al.* and Reis *et al.* [25,45] examined the reactivity of 2-propanol on polycrystalline Pt in acid media using FTIRS and on-line differential electrochemical mass spectrometry (DEMS). They reported that for 2-propanol oxidation, the production of acetone at the polycrystalline platinum surface was faster than that of CO_2. On the other hand, according to Rodrigues *et al.* [46], a strong catalytic improvement towards the oxidation of 2-propanol to CO_2 compared to its oxidation to acetone was observed on carbon-supported Pt electrodes. Therefore proving the influence of the electrode roughness on the electrocatalytic properties.

Omoto *et al.* [47] compared between the electrooxidation reaction of 2-propanol and that of methanol on Pt/C catalysts using the electrochemical impedance spectroscopy. They found that the reaction rate of the intermediate for 2-propanol was lower than that of methanol under all the studied conditions.

3.1.2 Platinum-based catalysts

Due to the insufficient activity of pure Pt for 2-propanol oxidation, many studies have been focused on Pt modification using other metals [26,46,48-53]. The study of bimetallic catalysts is very interesting because the addition of a second metal usually enhances the activity and increases the selectivity of metallic catalysts for fuel cells, especially those operating at low temperature. These performances are explained by the bifunctional reaction mechanism [48,54].

Pt-Ru based electrocatalysts were the most widely explored binary catalysts for 2-propanol oxidation reaction. It was found that these catalysts are the most effective anode materials for direct 2-propanol fuel cells [26,46,48-51]. Similarly to the case of methanol and ethanol, Ru content in the Pt–Ru electrocatalysts is an important parameter that influences 2-propanol oxidation. Lee *et al.* [49] revealed that the electrocatalytic activity for the 2-propanol oxidation of Pt–Ru/C could be actually improved with increasing ruthenium amount in the catalyst Pt (35%)-Ru (65%). Cyclic voltammetry analysis revealed that Ru addition accordingly exhibited an evident enhancement in the oxidation currents of 2-propanol and reduced the onset potential. This implies that the Pt-Ru/C may be a promising anode catalyst for the direct 2-propanol fuel cells.

In the same way, Umeda *et al.* [50] observed that the highest current density for the 2-propanol oxidation is obtained at the Pt (35%)-Ru (65%). These authors as well as Lee *et al.* [51] employed the Tafel plots analysis and cyclic voltammetry in a temperature range between 25 °C and 100 °C to furnish additional information on the kinetics of 2-propanol oxidation.

Inversely, Rodrigues *et al.* [46] have pointed out that a small amount of Ru provided a poor activity to the Pt–Ru ($Pt_{0.89}Ru_{0.11}$) catalyst for 2-propanol oxidation at room temperature. They found that the morphology, as well as the nature of the surface, are crucial in the electrooxidation mechanism of 2-propanol. However, the incorporation of such a little amount of Ru in the catalyst does not bring any modification in the catalyst behaviour for 2-propanol oxidation.

On the other hand, Rodrigues and co-authors [26] used the DEMS technique to evaluate the electro-activity of electrodeposited Pt and a range of bimetallics: $Pt_{0.84}Rh_{0.16}$, $Pt_{0.70}Rh_{0.30}$, $Pt_{0.55}Rh_{0.45}$ towards 2-propanol oxidation reaction. They have found that the $Pt_{0.84}Rh_{0.16}$ catalyst composition gives the best performances towards 2-propanol electrooxidation.

Recently, Habibi and Dadashpour [52] used the platinum-nickel nanoparticles carbon-ceramic modified electrode (Pt-Ni/CCE) as an efficient electrocatalyst for 2-propanol electrooxidation reaction. They found that Pt-Ni/CCE electrocatalyst exhibited excellent activity and enhanced stability than either Pt/CCE or smooth platinum electrode. These features were ascribed to the role of Ni atoms in the electrocatalytic activity of Pt and its large electroactive surface area.

This results summary highlights the importance of platinum as an efficient catalyst towards 2-propanol oxidation reaction in acid media but suggests, also, to undertake more investigations in alloying Pt to other metals such as Ru. Studies have to be oriented preferentially to nanocatalysts for 2-propanol electrooxidation in acidic media, where a huge lack is to be filled.

3.2   22-Propanol electrocatalysts in alkaline media

The improvement of the alcohols electrooxidation kinetics could be affected by the switching from the acidic medium towards the alkaline medium [54-56]. Another aspect of the improvement of the performance of the fuel cell in the alkaline medium is the fact that in this type of devices, the conduction of the OH⁻ions allows the circulation of the ionic current in the opposite direction to that of the proton conducting membranes. Thereby, the electro-osmotic drag direction is reversed resulting in lower alcohol crossover [57].

Furthermore, in alkaline solutions different metals could be as active as platinum [58], opening the prospect of the use of non-Pt electrocatalysts [55,58]. It seems also that electrode poisoning from the reaction intermediates is less pronounced in alkaline media, knowing that hydroxide ions actually suppress the formation of the poisoning species[55,58]. In contrast to acidic media, alkaline media offer a wide range of possible options for cathodic and anodic catalysts material. Substitutes to platinum have been studied, showing good performances, sometimes comparable to that of Pt.

This part presents a summary of 2-propanol electrocatalysts in direct alkaline fuel cells.

3.2.1 Platinum

Several authors have explored the electrooxidation of 2-propanol on platinum in alkaline media. However, most of the research has been done with solid platinum electrodes and only very few studies investigated platinum nanoparticles. The electrooxidation of 2-propanol has been studied in alkaline media using, mainly, cyclic voltammetry and chronoamperometry [59,60]. Results indicated that the onset potential is about -0.7 V/ECS. 2-propanol typical voltammograms present a current peak during the anodic scan and another one on the cathodic scan. The first peak corresponds to the oxidation of freshly chemisorbed species which originate from 2-propanol adsorption. The carbonaceous species which have not completely reacted in the forward scan are oxidized during the reverse scan peak. 2-propanol electrooxidation on three platinum single crystals (Pt(100), Pt(110) and Pt(111)) indicated that Pt(111) yielded higher current densities [61].

The oxidation products as well as the adsorbates are, not identified by purely electrochemical methods. Hence, several spectroscopic methods, have been applied to analyze reaction intermediates and elucidate the action mechanism. Methanol electrochemical oxidation reaction yields to CO_2 which reacts with alkaline electrolyte. But in the case of 2-propanol, the only electrooxidation reaction product is acetone at low overpotential, independently of the catalyst [61,62]. Using attenuated total reflection infrared (ATR-IR), Okanishi and co-authors [63] indicated that 2-propanol electrooxidation produces, mainly, acetone; equation (4). However, acetone was catalytically oxidized to the enolate ion; equation (5). Furthermore, enolate ions accumulation decreases the oxidation current of 2-propanol. Neither CO_2 nor carbonate species were detected during 2-propanol oxidation suggesting that the complete oxidation of 2-propanol to CO_2 will be a minor reaction.

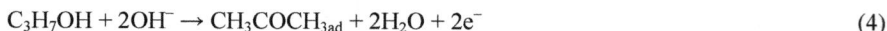

$$C_3H_7OH + 2OH^- \rightarrow CH_3COCH_{3ad} + 2H_2O + 2e^- \qquad (4)$$

$$CH_3COCH_{3ad} + OH^- \rightarrow CH_2COCH_3^-{}_{ad} + H_2O \qquad (5)$$

Prototype alkaline direct 2-propanol fuel cell (AD2PFC) using commercial Pt/C electrodes was tested [64]. The AD2PFC significantly surpassed the alkaline direct methanol fuel cells (ADMFC).

Results obtained on Pt nanoparticles electrocatalysts showed that the reversible 2-propanol/acetone redox couple is responsible for a low-potential current maximum [65].

3.2.2 Platinum-based catalysts

Since a long time, the platinum electrode has been considered as the best electrocatalyst for the electrooxidation reactions of small organic molecules. However, the C–C bond cleavage in 2-propanol molecule is very difficult to achieve onto Pt [66].

Pt-Ru is the main platinum-based catalyst tested for 2-propanol electrooxidation. Qi *et al.* [67] investigated direct oxidation of neat 2-propanol without adding water using Pt-Ru as anode. Results indicated that the achieved power density of the fuel cell was higher than that of the conventional ones supplied with 2-propanol aqueous solution. The Pt–Ru catalyst increased considerably the oxidation currents of 2-propanol while the onset potential was shifted to more negative values [65]. The currents measured at 15 min versus the applied potential for 2-propanol onto Pt, Ru, and Pt-Ru nanoparticle catalysts indicate that the current maximum amplitude was about three times higher on Pt-Ru than that on Pt and Ru single catalysts. Markiewicz [65] indicated also that the catalytic reduction of acetone at low potentials is enhanced when ruthenium was incorporated in platinum electrocatalysts, resulting on the reversible 2-propanol/acetone oxidation-reduction couple occurring onto Pt-Ru electrocatalyst. Yang *et al.* [68] prepared a composite PVA/TiO$_2$ membrane and used Pt-Ru for 2-propanol electrooxidation in KOH. Pt-Ru showed, however, a poor electrochemical activity towards 2-propanol.

In another side, performances of alkaline direct alcohol fuel cells fed with methanol, ethanol, and 2-propanol and using Pt-Ru on Vulcan as anode indicated that the highest power density was obtained with 2-propanol [69].

Electrochemical behaviour of intermetallic Pt-Pb/C nanoparticles for 2-propanol electrooxidation in alkaline medium was compared by various electrochemical techniques [70]. The results indicated that ordered intermetallic Pt-Pb/C are more electroactive for 2-propanol oxidation than either PtRu/C or Pt/C catalysts. Geometric and electronic structures are key parameters in the catalyst electroactivity enhancement and their modification should give significant improvements in 2-propanol electrooxidation.

3.2.3 Non-Pt based electrocatalysts

Pd and Au, are good candidates to replace Pt for 2-propanol oxidation in alkaline media. These materials are less expensive than platinum and their use in alkaline direct alcohol fuel cells could reduce the global costs of their production and allow a more extent of their widespread utilization.

Pd was investigated as an electrocatalyst for 2-propanol oxidation and compared with the traditional Pt catalyst in different alkaline media [59,63,71]. It was found that Pd electrocatalytic activity is better than that of Pt and Au in alkaline media. Furthermore, it was showed that 2-propanol oxidation reaction is more effective than that of ethanol onto Pd [60,72]. As highlighted above, acetone is the main product in 2-propanol electrooxidation reaction and its effect on the targeted reaction was not clarified. Hence, Pd/Ni-foam was used to clarify how much 2-propanol oxidation reaction is impeded by the presence of acetone [73] and clear evidence about the poisoning effect of acetone was provided. Tran *et al.* [74] constructed a Fuel-flexible alkaline direct liquid fuel cell using Pd as an anode. The maximum power density of 40 mW cm^{-2} was achieved at 60 °C.

An investigation conducted using CV and chronoamperometry indicated that Pd nanoparticles are effective catalysts for 2-propanol electrooxidation in alkaline medium [75]. Furthermore, gold nanoparticles incorporated to palladium have brought more stability to the catalyst. Also, Pd was less affected by poisoning species. The role of gold in PdAu/C electrocatalyst was not well explained and more investigations have to be conducted to clarify it. However, it seems that CO-like by-products are oxidized onto gold while Pd catalyzes alcohol dehydrogenation.

Electrochemically elaborated high-quality poly(p-phenylene) (PPP) film and its composite catalyst (Pd-Au/PPP) have been utilized for 2-propanol electrooxidation in alkaline medium [76]. Results indicated that Pd-Au nanoparticles homogenously distributed onto PPP, have enhanced catalytic activity and are less poisoned than Pd-Au deposited on the bare glassy carbon electrode. Two-dimensional assemblies of Pd nanoparticles on functionalised ITO are effective electrocatalyst for 2-propanol oxidation [77]. Serov *etal.* [78] investigated Pd-Cu nanoparticles catalysts for 2-propanol electrooxidation and found that Pd$_3$Cu exhibits the highest mass activity. They found that, effectively, Pd based materials are seriously poisoned by acetone. Nevertheless, Pd-Cu are more acetone tolerant compared to Pd-Ni. Yi *et al.* [79] used membrane-less direct alcohol fuel cell with PdSnNi/MWCNT as well as Pd/MWCNT nanoparticles. They found that the cell exhibited improved cell potential and cell power density with PdSnNi/MWCNT anode when compared to that of Pd/MWCNT anode. Furthermore, the maximum cell current density displayed with PdSnNi/MWCNT anode is 2×higher than that of Pd/MWCNT anode.

While, Au is not regarded as a good electrocatalyst for alcohol oxidation in acid media, it is highly active for 2-propanol oxidation in alkaline media [80]. Liu *et al.* [62] studied the poisoning of acetone to Pt and Au electrodes for electrooxidation of 2-propanol and found that around 90% of the electrocatalytic activity of 2-propanol oxidation was lost for Pt and Au electrode, when the acetone concentration reached 1.0 and 0.3 mol L^{-1}, respectively. Gold electrode was less influenced by acetone than platinum in the presence of low acetone amounts. Also, it seems that increasing the operating temperature emphasizes the poisoning effect of acetone. Zhang *et al.* [81] found that the peak current density of 2-propanol oxidation on gold is higher than those of methanol, ethanol, and n-propanol.

Experiments conducted by Markiewiez group, [65] showed that ruthenium is far less active than platinum for 2-propanol oxidation reaction. Nevertheless, Ru seems to have a little activity in the hydride region of the forward scan; at more negative potentials than the onset potential on Pt.

Transition metals based electrocatalysts such as nickel or copper have shown good electrocatalytic activity towards a wide range of organic molecules in alkaline media [18,82-85]. However, in the best of our knowledge, transition metals based catalyst utilization in 2-propanol electrooxidation reaction was reported only by Jafarian *et al.*[86] and Wang *et al.* [87]. Complexes of Ni^{II}–(N, N'-bis(2-hydroxy, 3-methoxy benzaldehyde)-1,2-propandiimine) electropolymerized onto GC electrode in concentrated sodium hydroxide solution were tested versus 2-propanol oxidation [86]. Authors reported that the obtained films represent a potent and persistent electrocatalytic activity towards 2-propanol oxidation. More recently, a novel N-doping carbon nanotubes anchored Pd_3Fe alloy nanoparticles (Pd_3Fe/CN) was prepared and its reactivity towards 2-propanol electrooxidation in 0.1 M KOH media was investigated using cyclic voltammetry, chronoamperometry, and CO stripping. Pd_3Fe/CN nanoparticles with an average size of 5.1±0.8 nm exhibited larger electrochemical surface area as well as higher activity than Pd/C against 2-propanol oxidation reaction in potassium chloride solution [87].

It is complex to make an accurate comparison between the differently employed electrocatalysts in direct 2-propanol electrooxidation because experiments were not performed in the same conditions. However, it seems that palladium and Pd-Au are the best choices for the electrocatalysis of 2-propanol fuel cells in alkaline media, as indicated in Table 2.

Table 2. Peak current density ratios of best 2-propanol catalysts in alkaline media.

Catalyst 1	Catalyst 2	Peak current densityratio = i_p(catalyst 2)/i_p(catalyst 1)	Reference
Pt	Pt-Ru	3	[65]
Pt	Pd	13	[59]
Pd	Pd-Au	2.5	[76]

4. Cathode electrocatalysts for 2-propanol fuel cell

The most recurrent problem in DMFC is methanol crossover through the membrane from the anodic to the cathodic compartment. This alcohol permeation is driven by the concentration gradient and electroosmosis. This phenomenon has a double effect; reducing the fuel efficiency and lowering cathode potential by an alcohol oxidation reaction at the cathode. Furthermore, it has been reported that crossover causes cathode flooding [38].

Thus, the fundamental challenges, in DMFCs and more widely in DAFCs, are how to overcome the sluggish in oxygen reduction reaction (ORR) kinetics and alcohol crossover.

As in the case of the anode, D2PFCs adequate electrocatalyst for the cathode electrode is still missing. So, the development of a new cathode electrocatalyst that can be used to efficiently electro-catalyze ORR in the presence of 2-propanol is of utmost importance.

D2PFC tests were conducted using Pt black [20,38,48] or Pt on Vulcan [20,48] as a cathode catalyst for oxygen reduction reaction and Nafion® membranes such as Nafion 115® and the thicker Nafion 117®.

It was shown that increasing cathode loading with Pt had no impact on the improvement of the cell voltage at low current densities. Also, cell operating on 2-propanol require lower cathode loading than methanol fed cell [38]. 2-propanol crossover is 3×lower than that of methanol in the same conditions when Nafion 115® membrane is used [20,48].

The open circuit voltage (OCP) is the difference of electrical potential between the anode and the cathode when the device is disconnected from any electrical circuit. OCP is a key parameter in a fuel cell operation and it is largely related to the fuel permeation. Higher OCP implies lower fuel permeation. Hence, it was found that 2-propanol OCP is 0.8 V [40]. It is higher than that of fuel cells fed with methanol, ethanol, n-propanol, formaldehyde and ethylene glycol. Nevertheless, 2-propanol crossover rate was superior to that of methanol even when the commercial fumasep® FAA-2 membrane was used [69].

The lower solubility of 2-propanol, as well as ethanol in PBI membrane, gave lower crossover through the membrane. The permeation is higher by a factor of ten compared to that through Nafion® membrane [13].

As introduced above, cathode flooding is a major problem in DAFCs which could be overcome by feeding the anodic compartment by neat 2-propanol and using sulphonated polyetheretherketone (SPEEK) membrane [67]. The absence of water in fuel cell feeding has a direct effect on SPEEK membrane properties. This later is much less subject to swelling and yields to better blocking of the 2-propanol crossover.

Pt electrocatalysts were used as a cathode in alkaline direct 2-propanol fuel cells [64,74]. Authors have found that increasing cathode as well as anode loading on Pt, increases the current densities provided by the fuel cell, in the potential range from the open circuit voltage and cell voltage limit of 0.5 V. Also, they found that the polarization phenomenon was ascribed mainly to the cathode when the cell operates reversibly at relatively high voltage. Alcohol permeation from the anodic compartment to the cathodic compartment will cause an increase in the cathode polarization due to mixed potential as a result of fuel oxidation reaction at the cathode.

The effect of gas composition on cell performance was investigated by Kariya *et al.* [28]. They have found that the OCP of 2-propanol increased accordingly by raising the partial pressure of O_2 in the cathode gas.

Cheaper than platinum, MnO_2/C could be effectively used as cathode electrocatalyst together with PVA/TiO_2 composite membrane. In this case, the alcohol crossover is much lower than the reported values for Nafion® membranes [68].

Another way to prevent the mixed potential at the cathode is to design a cathode catalyst with high selectivity which will not be affected by fuel crossover. This goal seems achieved by the work of Yi *et al.* [79], where authors used Fe/PANI electrocatalyst as cathode in a membrane-less alcohol fuel cell. Even though 2-propanol and the other tested alcohols can reach the cathode surface, their electrooxidation rate remains negligible.

In the more recent work of Yi group, they have shown that rather than having an oxygen reaction at the cathode compartment by feeding it with pure oxygen or an oxygen-nitrogen mixture, Fe^{3+} could be reduced at the cathode in a liquid catholyte. However, 2-propanol supplies the worst maximum power density compared to methanol, ethanol, and n-propanol [88].

At last, it is obvious that the number of studies devoted to 2-propanol fuel cell cathode materials is insufficient and seems clear that more research has to be conducted in improving the cathode selectivity and stability.

Conclusion

2-propanol promotion as the best choice for direct alcohol fuel cell depends on two addressed challenges:

(i) Developing an efficient electrocatalyst for 2-propanol electrooxidation which is able to break the C-C bond and leads to a complete oxidation reaction. The catalyst has to be somewhat cheap. These criteria could be achieved by using transition metals and/or nanoparticle materials. The control of particle size and particle distribution is of paramount significance and depends on the preparation method. In this field, reports focused on 2-propanol are rather limited and a lot of research is to be undertaken. The benchmark methanol and ethanol background could be helpful.

Platinum remains very efficient towards 2-propanol electrooxidation. But alloying platinum to other metals by performing binary or ternary alloys would certainly improve the anodic reaction in particular if metals are dispersed in very small sizes.

Suitable materials for 2-propanol electrooxidation is still under research. It seems that palladium and related materials are the best choices for the electrocatalysis of 2-propanol fuel cells in alkaline media.

(ii) Enhancing the oxygen oxidation reaction by designing an efficient, nonexpensive electrocatalyst and lowering the alcohol crossover. This goal would be achieved by employing of non-noble metals and alcohol blocking membrane.

As fossil fuels price is expected to increase due to their gradual decrease, 2-propanol production from petroleum derivative will impede their wide extent. So, 2-propanol renewable production routes development is a key parameter for the widespread use of D2PFC.

References

[1] F. Fathirad, A. Mostafavi, D. Afzali, Bimetallic Pd–Mo nanoalloys supported on vulcan XC-72R carbon as anode catalysts for direct alcohol fuel cell, Int. J. Hydrogen Energy. 42 (2017) 3215-3221. https://doi.org/10.1016/j.ijhydene.2016.09.138

[2] L. An, R. Chen, Recent progress in alkaline direct ethylene glycol fuel cells for sustainable energy production, J. Power Sources. 329 (2016) 484-501. https://doi.org/10.1016/j.jpowsour.2016.08.105

[3] B. Hasa, E. Kalamaras, E.I. Papaioannou, J. Vakros, L. Sygellou, A. Katsaounis, Effect of TiO_2 loading on Pt-Ru catalysts during alcohol electrooxidation, Electrochem. Acta. 179 (2015) 578-587. https://doi.org/10.1016/j.electacta.2015.04.104

[4] D. Takky, B. Beden, J.M. Leger, C. Lamy, Evidence for the effect of molecular structure on the electrochemical reactivity of alcohols: Part III. Electro-oxidation of the butanol isomers on platinum single crystals in an alkaline medium, J. Electroanal. Chem. Interfacial Electrochem. 256 (1998) 127-136. https://doi.org/10.1016/0022-0728(88)85012-5

[5] www.sigmaaldrich.com

[6] U.B. Demirci, How green are chemicals used as liquid fuels in direct liquid-feed fuel cells, Environ. Int. 35 (2009) 626-631. https://doi.org/10.1016/j.envint.2008.09.007

[7] CLIP, Chemical Laboratory Information Profile, 2-propanol, J. Chem. Educ.85 (2008) 1186. https://doi.org/10.1021/ed085p1186

[8] K. Weissermel, H.J. Harp. Industrial organic chemistry. VCH Publishers, New York, 1997. https://doi.org/10.1002/9783527616688

[9] Y. Onoue, Y.Mizutani, S. Akiyama, Isopropyl alcohol by direct hydration of propylene, Bull. Japan Petroleum Institut. 15 (1973) 50-55. https://doi.org/10.1627/jpi1959.15.50

[10] Z. Dai, H. Dong, Y. Zhu, Y. Zhang, Y. Li, Y. Ma, Introducing a single secondary alcohol dehydrogenase into butanol tolerant clostridium acetobutylicum Rh8 switches ABE fermentation to high level IBE fermentation, Biotechnol. Biofuels. 5 (2012) 44-53. https://doi.org/10.1186/1754-6834-5-44

[11] A. Rahman, S. S-Al-Deyab, A review on reduction of acetone to isopropanol with Ni nano superactive, heterogeneous catalysts as an environmentally benevolent approach, Appl. Catal. A. 469 (2014) 517-253. https://doi.org/10.1016/j.apcata.2013.10.015

[12] E. Grousseau, J. Lu, N. Gorret, S.E. Guillouet, A.J. Sinskey, Isopropanol production with engineered Cupriavidus necator as bioproduction platform, Appl. Microbiol. Biotechnol. 98 (2014) 4277-4290. https://doi.org/10.1007/s00253-014-5591-0

[13] J. Wang, S. Wasmus, R.F. Savinell, Evaluation of ethanol, 1-propanol, and 2-propanol in a direct oxidation polymer electrolyte fuel cell, J. Electrochem. Soc. 142 (1995) 4218-4224. https://doi.org/10.1149/1.2048487

[14] M. Nacef, A.M. Affoune, M. Umeda, Kinetic of the anodic reaction in direct alcohol fuel cells, Algerian J. Adv. Mater. 3 (2006) 95-98.

[15] J.P. I. de Souza, S.L. Queiroz, K. Bergamaski, E.R. Gonzalez, F.C. Nart, Electro-oxidation of ethanol on Pt, Rh, and Pt-Rh electrodes. A study using DEMS and in-situ

FTIR techniques, J. Phys. Chem. B. 106 (2002) 9825-9830.
https://doi.org/10.1021/jp014645c

[16] S.S. Gupta, J. Datta, A comparative study on ethanol oxidation behavior at Pt and PtRh electrodeposits, J. Electroanal. Chem., 594 (2006) 65-72.
https://doi.org/10.1016/j.jelechem.2006.05.022

[17] M. Zhiani, S. Majidi, H. Rostami, M.M. Taghiabadi, Comparative study of aliphatic alcohols electrooxidation on zero-valent palladium complex for direct alcohol fuel cells, Int. J. Hydrogen Energy. 40 (2015) 568-576.
https://doi.org/10.1016/j.ijhydene.2014.10.144

[18] M.L. Chelaghmia, M. Nacef, A.M. Affoune, Ethanol electrooxidation on activated graphite supported platinum-nickel in alkaline medium, J. Appl. Electrochem. 42 (2012) 819-826. https://doi.org/10.1007/s10800-012-0440-2

[19] M.L. Chelaghmia, M. Nacef, A.M. Affoune, Elaboration d'électocatalyseur platine-nickel pour l'oxydation anodique du méthanol, Algerian J. Adv. Mater. 5 (2008) 439-442.

[20] Z. Qi, M. Hollett, A. Attia, A. Kaufman, Low temperature direct 2-propanol fuel cells, Electrochem. Solid-State Lett. 5 (2002)129–130.
https://doi.org/10.1149/1.1475197

[21] E. Pastor, S. Wasmus, T. Iwasita, Spectroscopic investigations of C_3 primary alcohols on platinum electrodes in acid solutions. Part I. n-propanol, J. Electroanal. Chem. 350 (1993) 97-116. https://doi.org/10.1016/0022-0728(93)80199-R

[22] E. Pastor, S. Wasmus, T. Iwasita, DEMS and in-situ FTIR investigations of C_3 primary alcohols on platinum electrodes in acid solutions. Part II. Allyl alcohol, J. Electroanal. Chem. 353 (1993) 81-100. https://doi.org/10.1016/0022-0728(93)80288-S

[23] E. Pastor, S. Wasmus, T. Iwasita, Spectroscopic investigations of C_3 primary alcohols on platinum electrodes in acid solutions. Part III. Propargyl alcohol, J. Electroanal. Chem. 371 (1994) 167-177. https://doi.org/10.1016/0022-0728(93)03250-S

[24] P. Gao, S.C. Chang, Z. Zhou, M.J. Weaver, Electrooxidation pathways of simple alcohols at platinum in pure nonaqueous and concentrated aqueous environments as studied by real-time FTIR spectroscopy, J. Electroanal. Chem. 212 (1989) 161–178.
https://doi.org/10.1016/0022-0728(89)87077-9

[25] E. Pastor, S. Gonzalez, A. J. Arvia, Electroreactivity of isopropanol on platinum in acids studied by DEMS and FTIRS, J. Electroanal. Chem. 395 (1995) 233–242.
https://doi.org/10.1016/0022-0728(95)04129-C

[26] I. A. Rodrigues, F. C. Nart, 2-Propanol oxidation on platinum and platinum–rhodium electrodeposits, J. Electroanal. Chem. 590 (2006) 145–151. https://doi.org/10.1016/j.jelechem.2006.02.030

[27] S.G. Sun, D.F. Yang, Z.W. Tian, In-situ FTIR studies on the adsorption and oxidation of n-propanol and ispropanol at a plaint electrode in sulphuric acid solutions, J. Electroanal. Chem. 289 (1990) 177–187. https://doi.org/10.1016/0022-0728(90)87215-6

[28] J. Larminie, A.L. Dicks, Fuel Cell Systems Explained, New York, Wiley, 2000.

[29] M. Nacef, A.M. Affoune, Comparison between direct small molecular weight alcohols fuel cell's and hydrogen fuel cell's parameters at low and high temperature. Thermodynamic study, Int. J. Hydrogen Energy. 36 (2011) 4208-4219. https://doi.org/10.1016/j.ijhydene.2010.06.075

[30] U.B. Demirci, Direct liquid-feed fuel cells: thermodynamic and environmental concerns, J. Power Sources. 157 (2006) 164 -168.

[31] M. Nacef, M.L. Chelaghmia A.M. Affoune, Etude comparative des propriétés thermodynamiques des piles à combustible à hydrogène et aux alcools, Algerian J. Adv. Mater. 5 (2008) 447-450.

[32] N. Kariya, A. Fukuoka, M. Ichikawa, Direct PEM fuel cell using "organic chemical hydrides" with zero-CO_2 emission and low crossover, Phys. Chem. Chem. Phys. 8 (2006) 1724-1730. https://doi.org/10.1039/b518369c

[33] J. Otomo, I. Shimada, F. Kosaka, K. Ishiyama, Y. Oshima, Reaction analysis of alcohol electro-oxidation at intermediate temperatures, ECS Trans., 50 (2013) 2009-2017. https://doi.org/10.1149/05002.2009ecst

[34] H. Wang, Z. Jusys, R.J. Behm, Ethanol electrooxidation on a carbon-supported Pt catalyst: Reaction kinetics and product yields, J. Phys.Chem. B. 108 (2004) 19413-19424. https://doi.org/10.1021/jp046561k

[35] F. Kardigan, B. Beden, J.M. Leger, C. Lamy, Synergistic effect in the electrocatalytic oxidation of methanol on platinum+palladium alloy electrodes, J . Electroanal. Chem. 125 (1981) 89-103. https://doi.org/10.1016/S0022-0728(81)80326-9

[36] T. Lopes, E. Antolini, E.R. Gonzalez, Carbon supported Pt–Pd alloy as an ethanol tolerant oxygen reduction electrocatalyst for direct ethanol fuel cells, Int. J . Hydrogen Energy. 33 (2008) 5563-5570. https://doi.org/10.1016/j.ijhydene.2008.05.030

[37] E. Antolini, Catalysts for ethanol fuel cells, J. Power Sources. 170 (2007)1-12. https://doi.org/10.1016/j.jpowsour.2007.04.009

[38] D. Cao, S.H. Bergens, A direct 2-propanol polymer electrolyte fuel cell, J. Power Sources. 124 (2003) 12–17. https://doi.org/10.1016/S0378-7753(03)00613-X

[39] S.S. Gupta, J. Datta, An investigation into the electro-oxidation of ethanol and 2-propanol for application in direct alcohol fuel cells (DAFCs), J. Chem. Sci. 117 (2005) 337–344. https://doi.org/10.1007/BF02708448

[40] A. Santasalo, T. Kallio, K. Kontturi, Performance of liquid fuels in a platinum-ruthenium-catalysed polymer electrolyte fuel cell, Platinum Metals Rev. 53 (2009) 58–66. https://doi.org/10.1595/147106709X416040

[41] S.L.J. Gojković, A.V. Tripković, R.M. Stevanović, Mixtures of methanol and 2-propanol as a potential fuel for direct alcohol fuel cells, J. Serb. Chem. Soc. 72 (2007) 1419–1425. https://doi.org/10.2298/JSC0712419G

[42] S.G. Sun, Y. Lin, Kinetic aspects of oxidation of isopropanol on Pt electrodes investigated by in situ time-resolved FTIR spectroscopy, J. Electroanal. Chem. 375 (1994) 401–404. https://doi.org/10.1016/0022-0728(94)03536-9

[43] S.G. Sun, Y. Lin, Kinetics of isopropanol oxidation on Pt(111), Pt(110), Pt(100), Pt(610) and Pt(211) single crystal electrodes-studies of in situ time-resolved FTIR spectroscopy, Electrochim. Acta. 44 (1998) 1153–1162. https://doi.org/10.1016/S0013-4686(98)00218-7

[44] S.G. Sun, Y. Lin, In situ FTIR spectroscopic investigations of reaction mechanism of isopropanol oxidation on platinum single crystal electrodes, Electrochim. Acta. 41 (1996) 693–700. https://doi.org/10.1016/0013-4686(95)00358-4

[45] R.G.C.S. Reis, C.A. Martens, G.A. Camara, The electrooxidation of 2-propanol: An example of an alternative way to look at in situ FTIR data, Electrocatal. 1 (2010) 116–121. https://doi.org/10.1007/s12678-010-0018-x

[46] I.de A. Rodrigues, J.P.I. De Souza, E. Pastor, F.C. Nart, Cleavage of the C–C bond during the electrooxidation of 1-propanol and 2-propanol: Effect of the Pt morphology and of codeposited Ru,Langmuir. 13 (1997) 6829–6835. https://doi.org/10.1021/la9704415

[47] J. Otomo, X. Li, T. Kobayashi, C.J. Wen, H. Nagamoto, H. Takahashi, AC-impedance spectroscopy of anodic reactions with adsorbed intermediates: electro-oxidations of 2-propanol and methanol on carbon-supported Pt catalyst, J. Electroanal. Chem. 573 (2004) 99–109.

[48] Z. Qi, A. Kaufman, Performance of 2-propanol in direct-oxidation fuel cells, J. Power Sources. 112 (2002) 121–129. https://doi.org/10.1016/S0378-7753(02)00357-9

[49] C.G. Lee, H. Ojima, M. Umeda, Electrooxidation of C1 to C3 alcohols with Pt and Pt–Ru sputter deposited interdigitated array electrodes, Electrochim. Acta. 53 (2008) 3029–3035. https://doi.org/10.1016/j.electacta.2007.11.019

[50] M. Umeda, H. Sugii, I. Uchida, Alcohol electrooxidation at Pt and Pt–Ru sputtered electrodes under elevated temperature and pressurized conditions, J. Power Sources. 179 (2008) 489–496. https://doi.org/10.1016/j.jpowsour.2008.01.011

[51] C.G. Lee, M. Umeda, I. Uchida, Cyclic voltammetric analysis of C1–C4 alcohol electrooxidations with Pt/C and Pt–Ru/C microporous electrodes, J. Power Sources. 160 (2006) 78–89. https://doi.org/10.1016/j.jpowsour.2006.01.068

[52] B. Habibi, E. Dadashpour,Electrooxidation of 2-propanol and 2-butanol on the Pt–Ni alloy nanoparticles in acidic media, Electrochim. Acta. 88 (2013) 157–164. https://doi.org/10.1016/j.electacta.2012.10.020

[53] Y. H. Chu, Y. G. Shul, Combinatorial investigation of Pt–Ru–Sn alloys as an anode electrocatalysts for direct alcohol fuel cells, Int. J. Hydrogen. Energy.35 (2010) 11261–11270. https://doi.org/10.1016/j.ijhydene.2010.07.062

[54] L. An, T. S. Zhao, R. Chen, Q. X. Wu, A novel direct ethanol fuel cell with high power density, J. Power Sources. 47 (2002) 3707-3714.

[55] E. Antolini, E. R. Gonzalez, Alkaline direct alcohol fuel cells, J. Power Sources. 195(2010) 3431-3450. https://doi.org/10.1016/j.jpowsour.2009.11.145

[56] A. Tripković, K. Popović, B. Grgur, B. Blizanac, P. Ross, N. Marković, Methanol electrooxidation on supported Pt and PtRu catalysts in acid and alkaline solutions, Electrochim. Acta. 47 (2002) 3707–3714. https://doi.org/10.1016/S0013-4686(02)00340-7

[57] E.H. Yu, K. Scott, Development of direct methanol alkaline fuel cells using anion exchange membrane, J. Power Sources. 137 (2004) 248-256. https://doi.org/10.1016/j.jpowsour.2004.06.004

[58] R. Parsons, T. VanderNoot, The oxidation of small organic molecules. A survey of recent fuel cell related research, J. Electroanal. Chem. 257 (1988) 9-45. https://doi.org/10.1016/0022-0728(88)87028-1

[59] J. Liu, J. Ye, C. Xu, S.P. Jiang, Y. Tong, Electro-oxidation of methanol, 1-propanol and 2-propanol on Pt and Pd in alkaline medium, J. Power Sources. 177 (2008) 67–70. https://doi.org/10.1016/j.jpowsour.2007.11.015

[60] A. Santasalo-Aarnio, Y. Kwon, E. Ahlberg, K. Kontturi, T. Kallio, M. T.M. Koper, Comparison of methanol, ethanol and iso-propanol oxidation on Pt and Pd electrodes in alkaline media studied by HPLC, Electrochem. Commun. 13 (2011) 466–469. https://doi.org/10.1016/j.elecom.2011.02.022

[61] A. Santasalo, F.J. Vidal-Iglesias, J. Solla-Gullón, A. Berná, T. Kallio, J.M. Feliu, Electrooxidation of methanol and 2-propanol mixtures at platinum single crystal electrodes, Electrochim. Acta. 54 (2009) 6576-6583. https://doi.org/10.1016/j.electacta.2009.06.033

[62] Y. Liu, Y. Zeng, R. Liu, H. Wu, G. Wang, D. Cao, Poisoning of acetone to Pt and Au electrodes for electrooxidation of 2-propanol in alkaline medium, Electrochim. Acta. 76 (2012) 174–178. https://doi.org/10.1016/j.electacta.2012.04.130

[63] T. Okanishi, Y. Katayama, R. Ito, H. Muroyama, T. Matsui, K. Eguchi, Electrochemical oxidation of 2-propanol over platinum and palladium electrodes in alkaline media studied by in situ attenuated total reflection infrared spectroscopy, Phys. Chem. Chem. Phys. 18 (2016) 10109-10115. https://doi.org/10.1039/C5CP07518A

[64] M. E. P. Markiewicz, S. H. Bergens, A liquid electrolyte alkaline direct 2-propanol fuel cell, J. Power Sources. 195 (2010) 7196–7201. https://doi.org/10.1016/j.jpowsour.2010.05.017

[65] M.E.P. Markiewicz, S.H. Bergens, Electro-oxidation of 2-propanol and acetone over platinum, platinum–ruthenium, and ruthenium nanoparticles in alkaline electrolytes, J. Power Sources. 185 (2008) 222–225. https://doi.org/10.1016/j.jpowsour.2008.06.023

[66] H. Lin, G.L. Chen, Z.S. Zheng, J.Z. Zhou, S.P. Chen, Z.H. Lin, The adsorption and oxidation of isopropanol at platinum electrode in alkaline media, Acta Phys.Chim. Sin. 21 (2005) 1280-1284.

[67] Z. Qi, A. Kaufman, Liquid-feed direct oxidation fuel cells using neat 2-propanol as fuel, J. Power Sources. 118 (2003) 54-60. https://doi.org/10.1016/S0378-7753(03)00061-2

[68] C.C. Yang, S.J Chiu, K.T. Lee, W.C. Chien, C.T. Lin, C.A. Huang, Study of poly(vinyl alcohol)/titanium oxide composite polymer membranes and their application on alkaline direct alcohol fuel cell, J. Power Sources. 184 (2008) 44-51. https://doi.org/10.1016/j.jpowsour.2008.06.011

[69] A. Santasalo-Aarnio, P. Peljo, E. Aspberg, K. Kontturi, T. Kallio, Methanol, ethanol and iso-propanol performance in alkaline direct alcohol fuel cell (ADAFC), ECS Trans. 33 (2010) 1701-1714. https://doi.org/10.1149/1.3484660

[70] Y. Feng, W. Yin, Z. Li, C. Huang, Y. Wang, Ethylene glycol, 2-propanol electrooxidation in alkaline medium on the ordered intermetallic Pt-Pb surface, Electrochim. Acta. 55 (2010) 6991–6999. https://doi.org/10.1016/j.electacta.2010.06.080

[71] J. Ye, J. Liu, C. Xu, S.P. Jiang, Y. Tong, Electrooxidation of 2-propanol on Pt, Pd and Au in alkaline medium, Electrochem. Commun. 9 (2007) 2760–2763. https://doi.org/10.1016/j.elecom.2007.09.016

[72] Y. Su, C. Xu, J. Liu, Z. Liu, Electrooxidation of 2-propanol compared ethanol on Pd electrode in alkaline medium, J. Power Sources. 194 (2009) 295-297. https://doi.org/10.1016/j.jpowsour.2009.04.076

[73] Y. Cheng, Y. Liu, D. Cao, G. Wang, Y. Gao, Effects of acetone on electrooxidation of 2-propanol in alkaline medium on the Pd/Ni-foam electrode, J. Power Sources. 196 (2011) 3124-3128. https://doi.org/10.1016/j.jpowsour.2010.12.008

[74] K. Tran, T.Q. Nguyen, A.M. Bartrom, A. Sadiki, J.L. Haan, A fuel-flexible alkaline direct liquid fuel cell, Fuel Cells. 14(2014) 834-841. https://doi.org/10.1002/fuce.201300291

[75] C. Xu, Z. Tian, Z. Chen, S.P. Jiang, Pd/C promoted by Au for 2-propanol electrooxidation in alkaline media, Electrochem. Commun. 10 (2008) 246–249. https://doi.org/10.1016/j.elecom.2007.11.036

[76] W. Zhou, C. Wang, J. Xu, Y. Du, P. Yang, Enhanced electrocatalytic performance for isopropanol oxidation on Pd–Au nanoparticles dispersed on poly(p-phenylene) prepared from biphenyl, Mater. Chem. Phys. 123 (2010) 390-395. https://doi.org/10.1016/j.matchemphys.2010.04.027

[77] D. Renard, C. Mc Cain, B. Baidoun, A. Bondy, K. Bandyopadhyay, Electrocatalytic properties of in situ-generated palladium nanoparticle assemblies towards oxidation of multi-carbon alcohols and polyalcohols, Colloids Surf. A: Physicochem. Eng. Aspects. 436 (2014) 44-54. https://doi.org/10.1016/j.colsurfa.2014.09.027

[78] A. Serov, U. Martinez, A. Falase, P. Atanassov, Highly active PdCu catalysts for electrooxidation of 2-propanol, Electrochem. Commun. 22 (2012) 193-196. https://doi.org/10.1016/j.elecom.2012.06.023

[79] Q. Yi, Q. Chen, Z. Yang, A novel membrane-less direct alcohol fuel cell, J. Power Sources. 298(2015) 171-176. https://doi.org/10.1016/j.jpowsour.2015.08.050

[80] T.R. Vidakovic, M.L. Avramov-Ivic, B.Z. Nikolic, The influence of 2-propanol on the reaction of formaldehyde electrooxidation or vice versa on gold (100) and (111) single crystal planes in alkaline medium, J. Serb. Chem. Soc. 65 (2000) 915-922. https://doi.org/10.2298/JSC0012915V

[81] J.H. Zhang, Y.J. Liang, N. Li, Z.Y. Li, C.W. Xu, S.P. Jiang, A remarkable activity of glycerol electrooxidation on gold in alkaline medium, Electrochim. Acta. 59 (2012) 156–159. https://doi.org/10.1016/j.electacta.2011.10.048

[82] M.L. Chelaghmia, M. Nacef, A.M. Affoune, M. Pontié, Facile synthesis of Ni(OH)$_2$ modified disposable pencil graphite electrode and its application for highly sensitive non-enzymatic glucose sensor, Electroanalysis. 30 (2018) 1117-1124. https://doi.org/10.1002/elan.201800002

[83] M. Nacef, M.L. Chelaghmia A.M. Affoune, M. Pontié, Electrochemical investigation of glucose on a highly sensitive nickel-copper modified pencil graphite electrode, Electroanalysis. in press. https://doi.org/10.1002/elan.201800622

[84] Q. Lin, Y. Wei, W. Liu, Y. Yu, J. Hu, Electrocatalytic oxidation of ethylene glycol and glycerol on nickel ion implanted-modified indium tin oxide electrode, Int. J. Hydrogen Energy. 42 (2017) 1403-1411. https://doi.org/10.1016/j.ijhydene.2016.10.011

[85] K. Ye, G. Wang, D. Cao, G. Wang, Recent advances in the electro-oxidation of urea for direct urea fuel cell and urea electrolysis, Top. Curr. Chem. 376 (2018) 42. https://doi.org/10.1007/s41061-018-0219-y

[86] M. Jafarian, M. Rashvandavei, F. Gobal, S. Rayati, M.G. Mahjani, Electrocatalytic oxidation of 1-propanol and 2-propanol on electro-active films derived from NiII-(N,N′-bis(2-hydroxy, 3-methoxy benzaldehyde)-1,2-propandiimine) modified glassy carbon electrode, Electrocatal. 2 (2011) 163–171. https://doi.org/10.1007/s12678-011-0049-y

[87] W. Wang, S. Liu, Y. Wang, W. Jing, X. Niu, Z. Lei, Achieving high electrocatalytic performance towards isopropanol electrooxidation based on a novel N-doping carbon anchored Pd$_3$Fe alloy, Int. J. Hydrogen Energy. 43 (2018) 15952-15961. https://doi.org/10.1016/j.ijhydene.2018.06.159

[88] Q. Yi, T. Zou, Y. Zhang, X. Liu, G. Xu, H. Nie, X. Zhou, A novel alcohol/iron (III) fuel cell, J. Power Sources. 321 (2016) 219-225. https://doi.org/10.1016/j.jpowsour.2016.04.134

Nanomaterials for Alcohol Fuel Cells   Materials Research Forum LLC
Materials Research Foundations **49** (2019) 129-158   doi: https://doi.org/10.21741/9781644900192-4

Chapter 4

Polymer Electrolyte Membranes for Direct Methanol Fuel Cells

S. Mehdipour-Ataei*, M. Mohammadi

Iran Polymer and Petrochemical Institute, Tehran, Iran

s.mehdipour@ippi.ac.ir*, ma.mohammadi@ippi.ac.ir

Abstract

Fuel cells as a new source of energy generation are under the focus. According to research results, not only are polymer electrolyte fuel cells superior but also the use of methanol solution instead of hydrogen as a fuel will eliminate problems associated with the use of high purity hydrogen. So development and design of a polymeric membrane as a vital component which determines the performance is the priorities of research in this area. Although Nafion membranes are the most common, high permeability of these membranes to methanol lead to a significant loss of performance. To overcome this obstacle, modifying existing polymers or preparing new structures have been proposed. In this regard, this chapter introduces the polymeric membranes used in methanol fuel cells. Therefore, applied polymers in methanol fuel cells including perfluorinated ionomers; partially- and non-fluorinated polymers including polypropylene-based membranes, and the other polymer structures are described. Also, modified membranes based on these polymers with improved properties are presented. Finally, organic-inorganic composites, newly developed structures with high performance, and some common polymer electrolyte membranes are explained.

Keywords

Direct Methanol Fuel Cell, Polymeric Membranes, Nafion, Hydrocarbon-Based Polymers, Organic-Inorganic Composites, Multilayer, Pore-Filled Structures

List of abbreviations

AMPS	2-acrylamido-2-methyl propane sulfonic acid
DMFC	direct methanol fuel cell
ETFE-SA	poly(ethylene-alt-tetrafluoroethylene)
GO	graphene oxide

LDPE	low-density polyethylene
MCO	methanol crossover
MMT	Montmorillonite
PAN	polyacrylonitrile
PBI	polybenzimidazoles
PEM	polymer electrolyte membranes
PES	poly(ether sulfone)
PFA	polyfurfuryl alcohol
PI	polyimides
PPZ	polyphosphazene
PS	poly (styrene)
PSSA	poly (styrene sulfonic acid)
PSU	polysulfones
PTFE	polytetrafluoroethylene
PVA	polyvinyl(alcohol)
PVDF	polyvinylidene fluoride
SPEEK	sulfonated poly (ether ether ketone)
SPEEKK	sulfonated poly(ether ether ketone ketone)
SPS	sulfonated poly (styrene)
SPSU	sulfonated poly(sulfone)
TCPB	three component blend
ZrP	zirconium phosphate

Contents

1. Introduction

Polymer electrolyte membranes (PEM) as a kind of ion exchange materials class that is able to transport ions and separate them are ion exchange polymeric sheets or thin films. They have been applied in various industries and applications such as dialyzers, electrolyzer, and water desalination prior to the polymer electrolyte membrane fuel cell

(PEMFC). DMFC as a kind of PEMFC with methanol and O_2/air fuels (as shown in Fig. 1) contains a proton conducting polymeric membrane between two electrodes that avoids the electrons flow and acts as a barrier between the fuel and the oxidant. In fact, the critical component of DMFC i.e., a thin PEM, makes ionic pathways accessible for transport of protons, acts like a gas and electronic barrier, separates the anode and the cathode from each other, and enables the current via an external circuit. The membrane thickness is generally in the range of 30-200 μm [1-3].

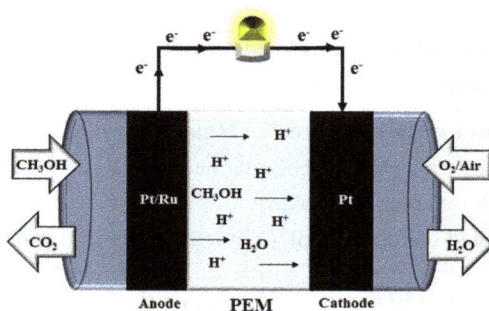

Figure 1. Schematic illustration of the polymeric membrane role.

In the case of DMFCs, high methanol permeability in the PEM causes a competitive reaction at the cathode which hinders the ability to reach high efficiency and decreases its lifetime. So the membrane is a critical component which one of its tasks is to block the methanol permeability. As an applicable illustration, the U.S. Department of Energy determined some specifications for DMFCs as follows: 100 W/l power density, 1000 Wh/l energy density, 5000 h lifetime, and 3 \$/W cost. The characteristics for a polymeric membrane in DMFC to provide these requirements include [3-5]:

1) low methanol crossover (MCO) (less than 10^{-6} mol/min/cm or less than 5.6×10^{-6} cm^2/s methanol diffusion coefficient at temperature of 25 °C) to prevent direct oxidation on the cathode, 2) high proton conductivity (more than 80 mS/cm) and thin to minimize membrane resistance, 3) high dimensional stability in aqueous media to minimize any membrane/electrode interface resistance, 4) high mechanical and chemical sustainability

in particular at the temperatures above 80 °C; 5) low cost; 6) low ruthenium crossover; and 7) electrically insulator [5,6].

In a general view, five classes of applied polymers in DMFC are perfluorinated ionomers, partially-fluorinated polymers, acid-base complexes, non-fluorinated ionomers, and hydrocarbon aromatic polymers. Up to now, Nafion[®] (DuPont) is the most common polymer electrolyte in PEMFCs. Other perfluorinated polymers such as XUS (Dow Chemicals), Aciplex (Asahi Kasei), Aquivion (Solvay), Fumion (Fumatech), and Flemion (Asahi Glass Engineering) are being evaluated. In spite of all its advantages, Nafion membrane does not meet all the mentioned requirements and suffer from obstacles including high MCO and its humidity dependent proton conductivity. Methanol is entirely soluble in water, so water and fuel (methanol) permeate through the membrane simultaneously. On the other hand, reducing the water uptake reduces MCO, but it sacrifices the proton conductivity. Other disadvantages include a low-temperature limit (lee than 80 °C); high cost; high ruthenium crossover; and degradation under dry conditions during start-stop cycles. Thus as a replace for Nafion membrane, the non-fluorinated hydrocarbon polymers are promising candidates due to their properties which will be explained in the next sections. These PEMs comprise a versatility of approaches to design, for instance, random and block copolymers synthesis, to graft ionic polymers onto hydrophobic membranes, to blend ionic and nonionic polymers, synthesis of ionic and nonionic interpenetrating networks, incorporating functionalized or original fillers into polymer membranes, and coating ionic membranes with one or several thin barriers [3-5,7].

2.   Perfluorosulfonic acid ionomers and Nafion

The perfluorinated polymers are commercialized for PEMFC and alcoholic fuel cells. These membranes have demonstrated acceptable performance because of low polarization capability and high electronegative F-atom. More attention has been paid to these polymers because hydrocarbon ion exchange membranes deteriorate in the presence of the oxidizing agent, especially at high temperatures. The most significant shortcoming of these ionomers in DMFC is their high MCO. Perfluorinated polymers have been produced widely by Asahi Glass (Flemion), Asahi Chemical (Aciplex), and in the 1970s DuPont developed Nafion. Among these, Nafion as a perfluorinated sulfonic acid ionomer is superior due to its high proton conductivity at optimal water contents, chemical and oxidative stability, and relative durability. Other perfluorinated ionomers with similar chemistry have a variety of equivalent weights by means of different ratios of backbone to side chain monomer. Other companies, such as Dow Chemical (USA) (Dow membrane), Asahi Glass Engineering (Japan) (Flemion[®]), Asahi Kasei (Japan)

(Aciplex-S®), W.L. Gore (USA) (Gore-Tex®, Gore-Select®), 3P Energy (Germany), Asahi Chemical, and Solvay Solexis are also developed their own Nafion-like membranes. These membranes have been extensively tested in hydrogen fuel cells, however, data on DMFC performance of these materials are not available. Nafion was developed explicitly for hydrogen fuel cells nonetheless today it is one of the main membranes for DMFC. The chemical structure of Nafion as shown in Fig. 2 comprises of a hydrophobic fluorocarbon backbone with pendant vinyl ether (perfluoroether) side chain terminating with strong acidic and a hydrophilic group, i.e., sulfonic acid ($-SO_3H$). The thermal and chemical stability of Nafion is due to the hydrophobic polytetrafluoroethylene (PTFE) backbone, and the hydrophilic side chains provide the channels for conduction of proton, but unfortunately, these channels also allow to pass methanol in DMFC. So, the main preventing factor for using Nafion membrane in DMFCs is its high MCO. To overcome this problem, the thicker membrane, Nafion 117 is more common. The commercialized Nafion 1110 is even thicker than Nafion 117 and have slower rates of MCO, but the latter is still the standard membrane for DMFC. Thicker membrane consumes more fuel due to the higher thickness that is, MCO reduction by increasing mass transfer resistance. Moreover, it is proven that increasing the equivalent weight is the other solution to high MCO [2,3,5,8-11].

Other membranes may be better than Nafion in some properties, but so far no suitable alternative with comparable properties to Nafion is introduced. Hyflon as a type of perfluorinated ionomers family with short side chain was developed in 2007. This membrane produces 300 mW/cm^2 at 140 °C by supplying methanol (1 M) and air as fuel and oxidant respectively [3-5,9].

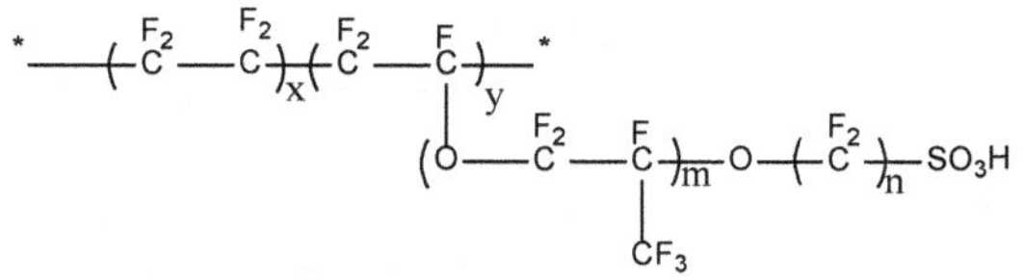

Figure 2. Chemical structure of Nafion-like ionomers.

2.1   Modified perfluorosulfonic acid membranes

Perfluorosulfonic acid membranes are currently used as electrolytes in DMFCs. Thus, developing composites based on these ionomers or synthesizing alternative new structures capable of limiting MCO is necessary. The polymer electrolyte membranes should have high ion conduction, chemical, and electrochemical stability in fuel cell media, and low MCO. As mentioned above the most critical challenge in these types of membranes specially Nafion membrane is limiting their high MCO. To reduce MCO, some modifications including composites development such as Nafion-silica, Nafion–cesium ions, Nafion–zirconium phosphate, and Nafion–poly (furfuryl alcohol) membranes have been studied. Sandwiching Pd thin film membranes with Nafion, or depositing nanoparticles of Pd on the surface of Nafion can also reduce the MCO [7,12]. Some of these modified membranes are presented in the next sections.

2.2   Modified Nafion-based membranes

The main Nafion modification strategies are as follows:

One of the proposed routes to decrease MCO in the Nafion membrane is preparing Nafion-polyfurfuryl alcohol (PFA) nanocomposite membranes. In situ polymerization of furfuryl alcohol with Nafion membrane, turn it to a hydrophobic component. In this way, the chemically stable PFA induces low MCO by developing Nafion-PFA nanocomposites. The Nafion-PFA demonstrates approximately three times lower MCO which compensate its lower conductivity [5].

Nafion-silica membranes, another solution to MCO problem, can be prepared by several methods including casting, the sol-gel reaction followed by solution casting of the Nafion solution, composite mixture casting and silica oxide casting with Nafion ionomer. Heat-treating to achieve a stable structure is required at the final stage.

In another strategy, the combination of polyaniline and silica is applied. The conductive polyaniline reduces the MCO by changing the structure. On the other hand, silica enhances conductivity. So sol-gel reaction to prepare polyaniline-Nafion-silica nanocomposite membrane, at first provide Nafion having silica within its hydrophilic clusters, and subsequently, redox polymerization results to deposition of polyaniline on the silica-Nafion membranes [5].

As another modifying route, Nafion-zirconium membrane can be impregnated with insoluble ZrP. The ZrP additive enhances water retention capability so raises the operating temperature to 150 °C. Increasing the membrane dry weight and its thickness is also obtained in these structures [5].

The final example is Nafion/polypyrrole membranes. These membranes can be developed by in-situ polymerization or polymerization in hydrogen peroxide in the presence of Fe (III) as the oxidizing agent. The membranes prepared via the latter method have low MCO but high resistance and poor performance in comparison with Nafion. This is attributed to the weak bonding in the electrode [5,13]. Modified Nafion-based membranes by applying polyvinyl(alcohol) (PVA) is explained in the next sections [14].

3.   Partially fluorinated hydrocarbon polymers

In hydrocarbon polymers containing polar groups, the water absorption is limited due to the existence of these groups. These polymers reveal high amounts of water uptake in a wide temperature range. These membranes are commercially available, less expensive, and easily reused by ordinary techniques. The chemical structure of these polymers allows the pendant groups to be attached to the polar site [3,5,15]. Chemical structure of this type of membranes is presented in Fig. 3.

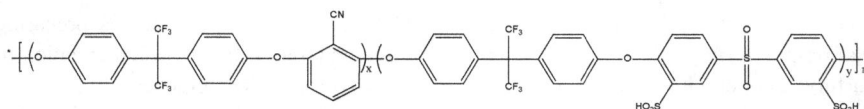

Figure 3. Chemical structure of an example of partially fluorinated hydrocarbon polymers.

4.   Non-fluorinated aromatic hydrocarbon membranes

When the electrolyte is Nafion membrane, less than 10 wt% of the methanol solution, is fed as fuel to reduce MCO and to get high fuel cell performance. Hydrocarbon electrolyte membranes with specific structures have been synthesized to lower the MCO effect. These membranes show about 10 times lower MCO than a Nafion membrane, and high DMFC performance by using approximately 30 wt% methanol concentration. However, achieving high power densities need much more MCO reduction [16].

The structure of hydrocarbon polymers can be aliphatic or aromatic while, aromatic polymers are under intense focus owing to their readily availability, capable of processing, structurally diverse and susceptible in the fuel cell environment. Poly(arylene ether) polymers just like poly(ether ketone)s, poly(ether sulfone)s (PES), polysulfones (PSU), polyimides (PI), polybenzimidazoles (PBI) are examples of the polyarylenes. Polyesters are not applicable because the ester bond does not have sufficient stability in watery media. The existence of bulky and inflexible aromatic groups has turned these polymers to high-temperature resistant material with a glass transition temperature of above 200 °C. Due to the presence of aromatic rings, both electrophilic and nucleophilic substitution reactions are possible. Concentrated sulfuric acid, fuming sulfuric acid, chlorosulfonic acid, or sulfur trioxide can be used as the sulfonating agent of these polymers. The post-modification via electrophilic aromatic sulfonation and/or synthesis of a polymer from sulfonated monomer have been used to prepare functionalized aromatic polymers for DMFC and PEMFC. However, in post-modification, sulfonation degree and position of sulfonate groups cannot be controlled precisely, and also a wide range of properties are caused by side reactions or polymer chain degradation. On the other hand in polymer synthesis from the functionalized monomer by controlling position, number, and distribution of these functional groups along the polymer chain more thermo-hydrolytically stable functionalized polymers are achievable and high conductivity, and low swelling is obtainable by changing the microstructure. These membranes may cost more than post-sulfonated membranes [3,5,7].

At a high degree of sulfonation, the performance of this group of polymers may be reduced because of the high MCO. In order to improve high-temperature performance, the introduction of bulky groups in the backbone of the aromatic polymer and incorporation of aromatic hydrocarbon into the polymer backbone can be applied [3]. Preparation of composite membranes is another solution to improve the properties. Changing the polymer structure by using different monomer structures and/or functional groups can also change the MCO, for example in Cs^+-doped membranes, cesium cations alter the microstructure of the membrane and conductivity is slight decreases in comparison with H^+-filled membranes. However, Cs^+ has smaller hydration energy which means a reduced affinity for water. In these types of membranes, an optimum 2M concentration of methanol is applied [17]. Another example of altering microstructure in order to decrease MCO may be done by changing isomerism of the monomer structure, i.e., using para and/or meta substitutions and like this [18]. A more detailed introduction to this group of PEMs and modified membranes thereof is explained in the following sections.

4.1 Poly(styrene)-based membranes

The first PEMFCs which were used in the Gemini space program consisted of crosslinked poly (styrene) and poly (styrene sulfonic acid) (PSSA) membranes as the electrolyte. Today, sulfonated poly (styrene) (SPS), poly (styrene) (PS) and PSSA random copolymers from this family are studied extensively in DMFCs. Moreover, sulfonated crosslinked polystyrene can be applied with other polymers such as partially-fluorinated polymers or with chemically crosslinked blends, etc. Chemical structure of these polymers is shown in Fig. 4. Copolymerization of styrene with monomers of styrene sulfonic acid or direct sulfonation of PS by acetyl sulfate, sulfur trioxide, etc. as sulfonating agents can be used to prepare SPS [7]. These copolymers have advantages like acceptable mechanical properties, suitable thermal stability, and good conductivity. Selectivity of this type of PEMs is higher than that of sulfonated poly (ether ether ketone) (SPEEK) and sulfonated poly (ether ether ketone ketone) (SPEEKK) but higher than that of post-sulfonated membranes. However, the DMFC performance of the post-sulfonated type is shown to be better than copolymers which are synthesized by functionalized monomers. The concern which has made these membranes to be put aside is their high oxidative degradation that is limited to the operating temperature below 60 °C [4,8].

Grafting of SPS or PSSA on hydrophobic porous frameworks such as polyethylene-tetrafluoroethylene, polyvinylidene fluoride (PVDF) or low-density polyethylene (LDPE) which are inert leading to a partially-fluorinated polymer can be done by activation of framework via radiation or electron beam irradiation source and subsequently in situ graft copolymerization of styrene or styrene sulfonic acid monomers onto it. In the styrene case, post-sulfonation should be done after copolymerization in order to functionalize the polystyrene backbone. To reduce water swelling of these membranes, crosslinking of grafts by divinylbenzene would be effective. Radiation treatment causes polymer chain degradation which may lead to durability problems. The lifetime of these materials is about 2000 h with a negligible DMFC performance loss. 20 times cost reduction per area by using graft copolymer membranes, compared with Nafion is notable [4,8,19].

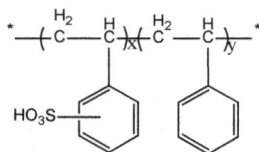

Figure 4. Chemical structure of SPS-based membranes for DMFC.

4.2   Polyarylene-type membranes

These thermoplastic materials that are engineering polymers show high thermal and mechanical properties, suitable durability under the oxidative condition, and processability. The proton conductivity and methanol permeability of these polymers are the other characteristics that have made them be a Nafion alternative candidate. The lifetimes of about 3000 h is reported for these membranes. These polymers have relatively high water uptake compared to perfluorinated ionomers. This characteristic is related to membrane-electrode interfacial problems and needs to challenge for reduction.

Proposed approaches to this point include copolymerization using sulfonated monomer, synthesis of block copolymers, applying fluorine and polar groups in the polymer structure, and blending [8]. A number of most important polymers from this family are introduced in the following sections.

4.2.1 Sulfonated poly(aryl ether ketone)-based membranes

This group of PEMs general structure is shown in Fig. 5 and they are strongly investigated due to their interesting properties for DMFC. PEEKs and PEEKKs as two kinds of this group and are semi-crystalline thermally stable polymers with chemical stability. These polymers have low MCO, acceptable hydrolytic susceptibility, good mechanical strength, and proton conductivity, thermo-oxidative stability, low cost, and a lifetime of more than several thousand hours under fuel cell conditions. Due to the different microstructure of Nafion and SPEEK membranes, these polymers have lower MCO than Nafion in the temperatures between 25 $^\circ$C and 80 $^\circ$C [7]. These SPEEK features results in a significantly reduced hydrodynamic flow of water, i.e., electro-osmotic drag and water permeation. On the other hand, it leads to disadvantageous swelling behavior and a stronger decrease in water and proton transport coefficients with decreasing water content. At higher sulfonation degrees, the MCO of these membranes increases, so their DMFC performance decreases. One useful solution to this problem is cross-linking. Suitable compounds for this purpose are aliphatic and cyclic diamines. To minimize MCO, the optimization of diamines chain length can be applicable [3,10,20].

The FUMA-Tech GmbH has introduced sulfonated poly (aryl ether ketone) membranes. These membranes have higher mechanical strength and significantly lower MCO than Nafion. The operation temperature of these membranes is in the range of 100–160 $^\circ$C. Inorganic or polymeric materials have been suggested for blending with these membranes to increase stability [5].

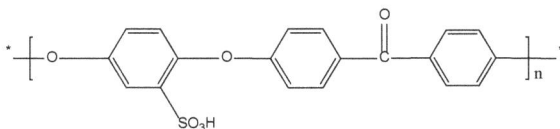

Figure 5. Chemical structure of SPEEK.

4.2.2 Sulfonated poly(aryl ether sulfone)-based membranes

Such membranes with proposed structure in Fig. 6 have not only comparable but also higher performance especially in DMFC compared to Nafion 117 membranes. High mechanical, thermal, and chemical stability are the characteristics of this type of polymers. These membranes have good proton conductivities (more than 100 mS/cm under the temperature of 65 °C and 85% RH) and approximately one order of magnitude lower MCO in comparison with Nafion at room temperature so; they have comparable performance with lower cost in DMFC applications [4,7]. Los Alamos National Laboratory and Virginia Polytechnic Institute created the bi-phenol based poly(arylene ether sulfone) membrane and a poly(arylene ether benzonitrile) membrane. Beside this, Air Force Research Laboratory synthesized sulfonated poly(arylene thioether sulfone). The water uptake is the major problem of these groups of polymers leading to flooding, dimensional instability and so catalyst layer delamination which results in performance loss. This obstacle can be modified by applying fluorine groups into the polymer backbone [5,21].

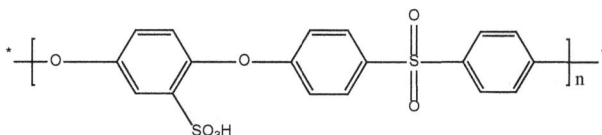

Figure 6. Chemical structure of SPAES.

4.2.3 Poly(imide)s

Chemical structure of sulfonated PI (Fig. 7) is similar to the polyarylenes additionally containing imide bonds along the polymer backbone. Sulfonated PIs due to their lower MCO, water uptake, and also higher thermal and oxidative stability are interesting for DMFC applications compared to polyarylenes. The chemical interaction of polymer backbone with the methanol solution in DMFC plasticizes polyimide membrane, which improves the MCO problem [7]. The hydrolytic instability is the main shortcoming of these materials. Over time by operating in fuel cell conditions, due to the chemical degradation of the PI backbone, conductivity decreases and loss of mechanical properties results in decreasing lifetime. Six-member ring poly(imide)s are preferred in comparison with five-member ring types since the electron density of nitrogen atom increases, and the hydrolytic stability of the imide bond improves leading to lowering ring strain and higher hydrolytic stability [4,8,19].

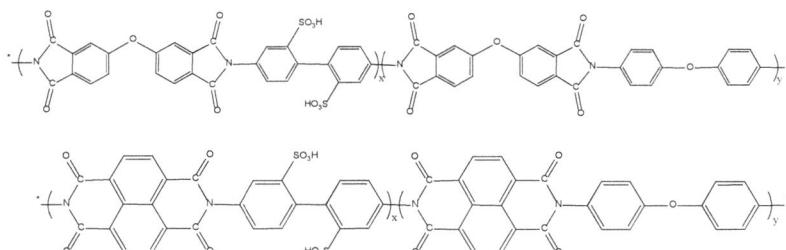

Figure 7. Chemical structure of SPI (top: five-member ring, bottom: six-member ring.

4.2.4 Poly(phosphazene)s

Another group of polymer electrolyte membranes is inorganic polyphosphazene (PPZ) polymers which contain phosphorus and nitrogen atoms alternatively in the polymer backbone, and two aliphatics, aromatic, inorganic or also organometallic pendant groups are attached to each atom of phosphorus (Fig. 8). The synthesis of the sulfonated polymer by using functionalized PPZ is preferred in comparison with post-sulfonation because the properties and side reactions which cause to polymer degradation are controllable in this way. Sulfonated PPZ has high thermo-chemical stability and low MCO in DMFC performance test. These polymers have good chemical and thermal stability. Reasonable mechanical properties and good proton conductivity at higher sulfonation degrees with extraordinary stability in the Fenton test, and low cost are other properties of this kind of PEMs [8]. In addition, various chemical structures can be synthesized. Low water and methanol diffusivities in the crosslinked PPZ (1.2×10^{-7} cm^2/s) with high mechanical strength, without losing rigidity up to 173 °C under pressure (800 kPa); and high durability in Fenton condition is also notable [5,7]. Contact resistance with Nafion binder in the catalyst layer is a problem in these membranes, leads to weak stability under test condition. A proposed route to improve these membranes mechanical and dimensional stability is cross-linking and blending with polyacrylonitrile (PAN) and PBI as a kind of hydrophobic polymers. In this way, selectivity and MCO will also be improved compared to Nafion [4,8].

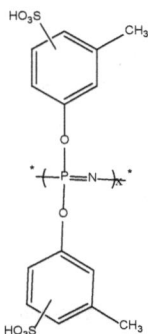

Figure 8. Chemical structure of Sulfonated PPZ.

4.2.5 Polybenzimidazole-based membranes

The basic PBI (pK_a=5.5) under the trademark of Celazol® with benzimidazole units that have hydrogen bonding sites for both donor and acceptors, is from the aromatic heterocyclic polymers group and have high thermal stability, chemical resistance, and mechanical strength. The acid-doped PBI also have desirable mechanical, thermal and oxidative stability at temperatures from 100 °C to 200 °C and dry condition. Lower MCO, higher proton conductivity that is also nearly humidity independent, and a little lower cost in comparison with Nafion membranes are other characteristics of this type of membranes. The blend membranes comprised from this polymer with PEEK and PSU and also its N-substitution form with methyl and ethyl groups and phosphoric acid-doped PBI have been reported [4,22]. The general route for membrane preparation from these polymers as other types of polymers is dissolving the polymer in a suitable solvent, mostly dimethylacetamide, then activation of prepared membranes by phosphoric acid 11 M (the most common concentration) solution. In Fig. 9 chemical structure of this polymer is shown. The major disadvantage of these membranes is swelling in water and methanol at high temperature and leaching of H_3PO_4. One solution to this problem is the addition of phosphotungstic acid with higher molecular weight instead of H_3PO_4 [7]. Recently, PEMEAS (USA) has commercialized Celtec® V from these polymers family for DMFC application. This membrane can operate in the temperatures between 60 °C to 160 °C under dry condition. The lower MCO, good mechanical properties, and much lower price than Nafion are the other benefits of these membranes [5].

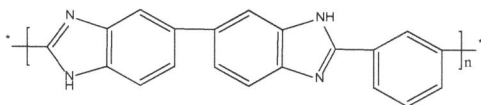

Figure 9. Chemical structure of PBI-based membranes.

5. Modification techniques and modified membranes

In a DMFC, due to the excessive swelling, more than 40 percent of the methanol permeates the membrane and is lost. MCO can be suppressed by adjusting the surface or bulk of the membrane via coating a methanol-impermeable layer (polymeric or metal), forming a hybrid membrane by impregnating inorganic filler, blending with an alcohol barrier polymer such as polyvinylidene alcohol, and such. [5,23,24]. In the following section, some kinds of these approaches are presented.

5.1 Block copolymers

Changing chemical microstructure via different monomer structures (as explained before) or synthesizing polymers with diverse arrangements of monomers are other ways to modify membranes. In block copolymers, it is possible to arrange a unique template in a well-ordered arrangement where ionic as well as non-ionic monomers chemically connect on one backbone simultaneously providing hydrophobic–hydrophilic blocks. In this morphology, microphase separation occurs on a nanometric scale. Since the different blocks are thermodynamically incompatible, different kinds of morphologies like spheres on a cubic lattice, hexagonally packed cylinders, interpenetrating gyroids, and alternating lamellae form. These membranes at higher IECs (1–2 meq/g) show similar conductivities and selectivities compared to Nafion with IEC=0.9 meq/g. The ability to use different solvent-casting systems has a profound effect on morphology and proton conductivity that result in methanol transport in this type of polymer structure [3,4,8,25,26].

5.2 Cross-linked, blend, and graft-type membranes

Polymer-polymer blending and covalent cross-linking via activation of sulfonic acid groups are two techniques for reducing swelling and MCO. But reduced conductivity due to the immobilization of protons and brittleness in the dry state emerge new problems. Blends as a kind of composite membranes with acid-base interactions or covalent crosslinking are more favorable than physically blended polymers; however the latter with PVDF and PVA as highly selective polymers might show a positive effect on lowering MCO. The most recent compatible blends have been reported between SPEEK–

polyetherimide, SPEEK–PBI, PVA and Nafion, PVA and PSSA, PVA and poly(styrene sulfonic acid-co-maleic acid), sulfonated poly(sulfone) (SPSU) and acid-doped PBI, SPSU/PES, and SPEEK/PES [27]. Blending of PVA with PAN, PVDF, and PBI applies in ethanol dehydration by separating ethanol and water during pervaporation process. So, water selectivity over ethanol can increase proton/methanol selectivity in these polymers [3,4,20,28,29].

Two commercial grafted copolymer membranes include an IonClad® R-1010 membrane, composed of PSSA side chains grafted to a perfluorinated polymeric backbone, and IonClad® R-4010 membrane, composed of PSSA grafted to a tetrafluoroethylene/perfluoropropylene copolymer. The proton conductivity of these two membranes is comparable to Nafion, whereas the MCO is four times lower. The point about these membranes is that surface incompatibility of binder and the polymer leading to delamination of catalyst layer [4,7,19,30].

5.3 Electrospun membranes

Electrospinning process as a technique for membrane fabrication modifies the electrospun nanofibers morphology via controlling the specific surface area and structural alignment of ionic domains. The orientation of ionic domains along the nanofiber axis is reached due to shear force applies during the electrospinning. The aligned ionic structure results in higher conductivity and this orientation can be extended if the shear force increases as well which decreases the fiber diameter. However, it should be highlighted that if this process is run at its optimum variables, i.e., polymer solution concentration, voltage, feed rate, needle to collector distance, etc. will reach the desired result. Even at a small amount of additive, this technique provides a fully exfoliated structure, which produces a high-performance PEM with low cost. Other than the electrospinning, nanofibers can be provided by template synthesis, drawing, self-assembly, and phase separation methods. In the case of Nafion, due to the high molecular weight and insolubility in most common solvents, the electrospun membrane is hard to produce [3,31].

5.4 Organic-inorganic composite membranes

Polymeric composites that are applied to improve the physical and electrochemical properties of polymer membranes are materials comprising of a polymeric matrix and another polymer, ceramic, or metal as a filler or reinforcer. The filler upgrades one or more properties of the polymer and/or decreases the cost. The advantage of using composites over blend membranes is the preserving of the properties of both filler and matrix as well as providing synergistic properties that are not achieved with blend membranes. The organic fillers are polymeric fibers or particles. In the most common organic composites, PTFE is used as the hydrophobic and ionic insulator filler or matrix. PTFE has no influence on improving the proton conductivity or water holding characteristic at high temperatures. It only functions as a fuel crossover barrier. Organic composites comprising a porous membrane which is filled with a conductive polymer solution are also helpful. These membranes are considered as an intermediate between composite and multilayer membranes [2]. Inorganic composite membranes consist of a micrometer to nanometer size inorganic fillers including silica, zirconium phosphate, phosphotungstic acid, molybdophosphoric acid, aerosil, ormosil, silane-type fillers, titanium oxide, hydroxyapatite, laponite, montmorillonite, zeolites, palladium, and so on in a number of different polymers. Generally, the inorganic materials have been used with the purpose of improving proton conductivity and/or lowering MCO. For highly conductive membranes, inorganic fillers with exfoliated structure are applied to decrease MCO, whereas highly conductive inorganic fillers are used for non-conductive membranes [8]. Inorganic fillers on the basis of the ingredient materials are classified into metal oxides and carbon-based materials. The second one can be further sub-divided into carbon nanotubes and graphene oxide which are different from the aspect of carbon nanostructure. These fillers can be used in their original as well as modified forms. In another classification, inorganic fillers can be divided into three groups [32]: inert hygroscopic, proton conductive, and hydrophilic and proton conductive. The first category is used for decreasing the fuel crossover and retaining water which is useful under high temperature and dry conditions. These groups may not necessarily improve conductivity. On the other hand, the second category guaranties higher proton conductivity. In the third category, fillers are proton conductors and hygroscopic and typically, are sulfonated form of the first group. In general, the purpose of adding all types of fillers in DMFCs membranes is to increase the operating temperature and reduce MCO by increasing the tortuosity in the methanol permeation path in comparison with unfilled polymer membranes. This concept is illustrated in Fig. 10. In fact, in the presence of particles, the fuel crossover path across the membrane is longer. Therefore in DMFC the longer path of methanol permeation reduces the rate of MCO and thus improves the membrane durability. So, membrane morphology is a promising parameter in altering the path of methanol permeation. Among three types of composite morphologies (ordinary composites, intercalated, and exfoliated) the exfoliated structure is the best which create a tortuous pathway for methanol. The exfoliated structure is obtainable by incorporation of clay material into the polymer. One way to reach the exfoliated structure is by developing the functionalized polymer membrane by electrospinning at its optimum parameters. Nanoceramic fillers not only induce these two advantages but also improve both hydrolytic and mechanical stability of the membrane in concentrated methanol solutions. The improvements induced by this type of modification are applicable for many types of polymer electrolytes such as perfluorinated ionomers, PS, poly(arylene ether ketone)s, PBI, PSU, PVA and other polyarylene-based membranes [3,7,33].

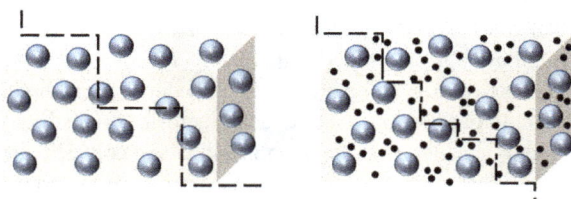

Figure 10. Schematic illustration of increasing path tortuosity by applying fillers in composite membranes.

The metal oxides as a family of inorganic fillers are inert hygroscopic materials and most widely used fillers in fuel cell membranes. Adding the metal oxides is with the purpose of increasing the water uptake of PEM, and they include TiO_2, ZrO_2, Fe_3O_4, and SiO_2 of which the latter is the most common one. Silica-containing membranes have excellent thermal and mechanical properties, water retention at high temperature, sustainability to MCO, and proton conductivity. However, several problems including low water transport results to lower proton conductivities and brittleness when the loaded inorganic phase reaches to its critical level may be met. The most common functional groups for functionalization of oxides to overcome the problem of proton conductivity are sulfonic acid groups. This causes strong interaction between functionalized nanomaterial and polymer and forms a cross-linked ionic structure leading to the swelling to decrease and thermal stability to increase. Besides that, the homogeneous dispersion of nanoparticles in the polymer matrix and hydrogen bonding between the functional groups is feasible [34]. In the case of PBI incorporated sulfonated silica nanoparticles, the ionic linkage between PBI and silica help to increase in tensile strengths and decrease in elongations at break. Despite the advantages of metal oxides usage as filler, they have some problems such as difficult homogenous dispersion in the polymer membrane. Additionally, the metal oxides may accelerate the oxidation and degradation rate of the membrane. Carbon nanomaterials have received much attention due to their high surface area, inertness and the capability of functionalization in different ways. These nanomaterials induce outstanding properties to the membranes such as increased chemical stability, reducing the fuel crossover, improving the thermal and mechanical stability, and increasing proton conductivity. Moreover, the amount of carbon nanomaterials to achieve good performance is much less than that of metal nanoparticle and this also allows easier uniform dispersion. Carbon nanotubes, histidine, and imidazole amino acid doped multi-walled carbon nanotubes are examples of this type of materials. The histidine makes the membrane for absorbing the water through hydrogen bonding, thus facilitating the transport of protons. Graphene offers many advantages like high mechanical strength, the barrier to gases, and large surface area and have been used in initial studies of fuel cell membranes but showed no promising results due to the electrical conductivity that facilitates the electron transport and reduces the fuel cell voltage. On the other hand, graphene oxide (GO) an electronically insulator filler with different oxygen groups (hydroxides, epoxides, carboxyls, and carbonyls) makes GO to a highly hydrophilic and electrically insulator but maintaining high mechanical strength and thermal properties and impermeability of gases because of the higher tortuosity. GO could also facilitate the transport of protons and show higher water uptake due to its high surface area [2]. Heteropolyacids are made up of heteropolyanions and protons as the counter-ions in which the heteropolyanion consists of one or more heteroatoms like phosphorus surrounded by different metal-oxygen polyhedrons (metals include zirconium, tungsten, molybdenum) are strong acids and show good proton conductivity in the crystalline state. This strategy may reduce MCO because of the existence of impermeable fillers which increase tortuosity, while heteropolyacids enhance proton mobility and increasing proton conductivity. An important issue for the heteropolyacids is their leaching in methanol solution, which results in performance loss. A technique to trap the acid inside the membrane needs to be developed for this problem. The stability is usually improved by modification of silica because of the covalent bonds strength or columbic interactions increases. Leaching of heteropolyacid can also be reduced by silane and surface-modified fumed silica. The inorganic silane phase reduces the absorption of water and MCO and fixes the heteropolyacid on the membrane [5]. Adding layered silicate nanoparticles such as montmorillonite (in its original or modified form) and laponite into polymers is another strategy to make nanocomposite membranes for DMFC. The silicates make the membrane tortuous and thus decrease MCO. Though, the silicates also act as physical barriers to water. Montmorillonite (MMT) and its derivatives improve the mechanical and barrier properties of the membrane, prevent excessive swelling, and conduct ions. By adding sulfonated–MMT with large surface area to modify Nafion, blocking the methanol passage by reducing the size of the ion channel in the composite membrane occurs while by the presence of sulfonic group proton conductivity is preserved. The most promising route to prepare this type of membranes is solvent casting. The hot-pressed membranes exhibit amazingly low conductivities because of the lack of interactions between the polymer and the filler [4]. Using a surfactant made up of a hydrophobic core, and a hydrophilic shell with hydrophilic or hydrophobic surface-treated nanoparticles is another way to improve conductivity and hydrolytic stability due to the strong physical interaction between polymer and nanoparticles [35]. This is as a result of the better miscibility comprises the structural similarity between polymer and nanoparticle.

5.5   Multilayer membranes

The choice of a suitable polymer structure and the membrane preparation method has an essential role in determining the membrane performance. The concept of multilayer membranes, as a subset of composite concept, is new to fuel cells. These structures were not used extensively in the field of PEMFCs; however, nowadays there is much focus on applying them in DMFC. These structures are able to bring together multiple layers of different materials which are designed in a manner each keeping its unique characteristic (high conductivity, low permeability, mechanical strength, water retention and compatibility between layers and with the electrodes) intact and simultaneously overcoming the drawbacks of the other layer. Various kinds of these structures including a varying number of layers, and combining organic and/or inorganic layers can be prepared. Multilayer membranes can be addressed as a solution to the sulfonated polymers solubility in water, and the resultant decline of proton conductivity. In these structures usually sulfonated polymers combined with mechanically strong polymers in one or all layers. Other than polymers, multilayer structures the same as composite membranes may also be developed through materials instead of polymers. Initially, this approach was designed for Nafion membranes with high MCO. Some barrier layers of methanol with Nafion consist of palladium, polypyrrole, PBI and PVA [2, 8]. In these structures, the inner layer can be a "reaction" layer such as Pt/Ru and SiO_2. In this way, the water generated in the structure would help to keep the membrane humidified and to reduce the internal resistance with improving the proton conductivity. In these types of composite materials, conductivity decreases with increasing more layers on the surface [4].

Classification of multilayer membranes based on preparation methods categorizes them into three main groups including hot-pressed, solution cast, and dip coating. There are also new methods these do not seem relevant to any of the three common categories. Hot-pressing is the simplest approach that includes pressing of two or more layers under high pressure and a temperature close to the polymer glass-rubber transition temperature, which makes them stick together through a mechanical bond. In this method the only bond formed to keep the layers is mechanical, so under the fuel cell condition where the membrane is in contact with continuous tension and compression, the delamination of the layers is possible. The commercial extruded membranes or solution cast ones can be used for hot-pressing. Together with hot-pressing, casting as the oldest method is also common for multilayer membranes preparation. In other words, the combination of two processes is common, and it should be pointed out that hot pressing without cast membranes cannot be applied. But the casting method induces the mechanical as well as chemical interaction between the layers, which forms the stronger bond between the layers, thus, decreases delamination. In this method after preparing the first layer, the second polymer solution is poured on that and is dried. The critical point in this approach is the choice of a solvent so that the solvent of the polymer solution in one layer does not dissolve other layers. Besides, using extruded membranes is omitted in this method, which is a disadvantage in comparison with hot-pressing because generally, extruded membranes get higher mechanical properties. Instead, dip coating provides a membrane with a different number of bilayers, even up to 20 or 50, which are extremely thin. The solution cast and hot-pressed membranes usually contain 3-5 layers. The advantage of this method due to the presence of a large number of layers is to minimize dissolving of the highly sulfonated and hydrophilic polymers in water and suppression of dimensional changes in the inner layers. The main point in this approach is the high mechanical strength of the base membrane in order to stand repeated dipping. This approach involves providing a base membrane by any of the two methods mentioned above or using a commercial membrane, then dipping this membrane in a solution of the polymer that will form the next layer, and finally, the prepared membrane with three-layer is then allowed to dry. After the membrane drying, the process of dipping may be repeated as many times needed to form the desired membrane using another or the same polymer solution. As mentioned above the major advantage of this method is the formation of membranes with 10-20 thin layers so that the final membrane thickness is still in the range of 15-20 mm. In order to improve the mechanical strength and lifetime, heating to achieve a crosslinked structure can be done which reduces the MCO by higher selectivity. However, crosslinking causes a loss of some hydrophilic sites. Multilayer membranes made by layer-by-layer (LbL) method and dip coating are similar. In LbL self-assembly technique, an aqueous solution of two oppositely charged polyelectrolytes are prepared separately and deposited on a support surface alternately. After each dipping cycle, the surface charge is reversed, and this enables the deposition of a subsequent layer. Finally, a multilayer structure that is stabilized by strong electrostatic attraction is formed [12]. To obtain the optimum performance, the number and thickness of layers must be controlled [2].

It is notable that a balanced structure of both multilayer and composite/blend membranes is needed for MCO reduction. A general scheme of these membranes is shown in Fig. 11.

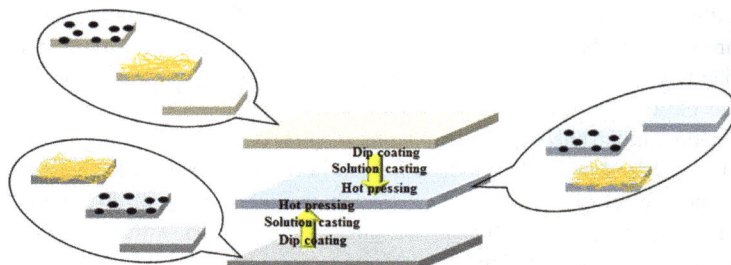

Figure 11. Schematic representation of composite/blend multilayer membranes concept.

5.6   Impregnated membranes and pore-filled structures

The concept of the pore-filled membrane is developed to control membrane swelling and liquid permeation. This structure is composed of a porous substrate with sub-micrometer or fewer pore sizes and a graft or gel type polymer electrolyte that fills the pores of the substrate and exhibits proton conductivity (Fig. 12). The porous substrate is entirely inert for liquid fuels or to gas and is mechanically stable to prevent excessive swelling of filling polymer, so decreases MCO and provides dimensional stability. Small amounts of water can be contained in the filling polymer electrolyte for proton migration. It is notable that these membranes don't show high enough conductivity that results from an inert substrate. The most important reason for using these structures is that polymer-filled microporous membranes without proton conduction suppression can reduce the same amount of swelling which a highly crosslinked polymer can do, in other words increasing selectivity. In brief, restricting a polymer electrolyte by a microporous host membrane suppresses swelling without significant compromising of proton conductivity. It is notable that different host membranes do not show the same performance [4-6, 28]. Ionomers with high conductivity such as poly(acrylic acid), poly(acrylic acid-co-vinyl sulfonic   acid),   poly(acrylamide-tert-butylsulfonic   acid),   and   also   acid-doped nanoparticles can be used as filling material. Porous silica, alumina, PVDF, PTFE, crosslinked high-density polyethylene and PI are some examples which can be used as a substrate. Cross-linked PVA nanofibers porous thin film can also be used as a substrate which by increasing tortuosity making the methanol transport difficult i.e., lower MCO. Different manners including spin coating, dipping, and screen printing can be applied to fill a porous substrate. In these structures, optimizing the substrate strength and the filling ratio to achieve a balanced MCO and proton conductivity is necessary [2,6,8,36].

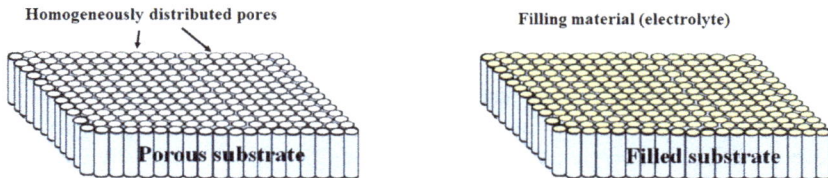

Figure 12. Schematic illustration of the pore-filled structure concept.

5.7 Nano-porous proton conducting membranes

To independent proton conductivity from the concentration of the ionic group and water content, and control the MCO a similar approach, based on the proton conducting membranes with nano-porous structure is also proposed. These membranes include high surface area inorganic, electronically insulating nanoparticles with the size of 15 nm or less with good aqueous acids retention (suitable materials are silica, zirconia, titania, or alumina); polymer binders such as PVDF; and aqueous acid or mixtures of acids to fill the nano-pores. An issue for these types of membranes, during the operation in DMFC, is the leakage of acids that must be cared [7]. The membranes, with PVDF binder to embed SiO_2 and acid into the polymeric matrix, have a very high surface area and thus high ionic conductivity. Meanwhile, the MCO of these membranes is low, due to small pore sizes of 1.5–3 nm in comparison with 3 nm for Nafion. Further MCO reduction is obtained by impregnating the pores with Na_2SiO_3 solution and hydrolyzing the silicate in sulfuric acid. These membranes can operate in a temperature range of 0 °C to over 90 °C which is an advantage in comparison with Nafion. They are also less sensitive to iron impurities and can tolerate more than 500 ppm of Fe, which allows the use of Pt–Fe catalysts or stainless steel fuel cell hardware. These types of modified PVDF membranes have excellent mechanical properties and ductility at membrane thicknesses of 30–1000 μm. They have a high dimensional and thermal stability (below 0 °C to over 100 °C) [5].

6. Other types of polymer electrolyte membranes

6.1 Semi-interpenetrating polymer network

The main characteristic of these membranes is high performance due to suppression of MCO. Proton-conducting membranes based on chemically cross-linked [poly(vinyl alcohol) and poly(2-acrylamido-2-methyl-1-propane sulfonic acid)] composites are as a

Nanomaterials for Alcohol Fuel Cells   Materials Research Forum LLC
Materials Research Foundations 49 (2019) 129-158   doi: https://doi.org/10.21741/9781644900192-4

kind of these membranes in which the poly(ethylene glycol) bis(carboxymethyl) ether act as plasticizer [7].

6.2   PVA-based membranes

The primary influence of membrane modification by PVA is the MCO reduction. PVA-based membranes have been studied for DMFCs because of their good selectivity. Nevertheless, it must be noted that these polymers are almost chemically unstable. To improve proton conductivity and/or decrease MCO, several solutions such as cross-linking, blending with conductive inorganic fillers or highly conductive polymers have been tested. In sum, they do not show a promising feature for DMFC electrolytes [8].

6.3   Irradiated sulfonated poly(ethylene-alt-tetrafluoroethylene) membranes

Radiation techniques and low-dose electron beam mainly affect the hydrophilic side chains resulting in a thin barrier layer which can provide some advantages for the resulted PEMs [4]. Nowofol GmbH Co. manufactures poly(ethylene-alt-tetrafluoroethylene) polymer known as ETFE-SA. The ETFE-SA is prepared by irradiation followed by sulfonation. These membranes are typically semi-crystalline, but by sulfonation, water adsorbs into the amorphous phase and decreases the initial crystallinity (3–12% of Nafion is crystalline). ETFE-SA has a large number of chemical cross-links made within sulfonation and irradiation. ETFE-SA membranes are cheaper than Nafion 115 (due to the thinner thickness; 35 μm compared to 127 μm) and show lower water and methanol uptake at 25 $^{\circ}$C. The lifetime of these membranes in a DMFC is over 2000 h without loss of performance. The main drawback of ETFE-SA is low power density (4.5 mW/cm^2 in comparison to 10.5 mW/cm^2 for Nafion 115 at 30 $^{\circ}$C and 1M methanol) [5,37].

6.4   Acrylic membranes with an asymmetric structure

The University of Singapore has designed a three component blend (TCPB) membrane based on the acrylic polymer with low solubility in methanol [5]. In this structure, the hydrophilic 4-vinyl phenol segments in 4-vinyl phenol-co-methyl methacrylate form the ionic channels within the TCPB. The hydrophilic components, 2-acrylamido-2-methyl propane sulfonic acid (AMPS), 2-hydroxyethyl methacrylate and poly(ethylene glycol) dimethylacrylate are polymerized after embedding them in the TCPB matrix to gain a homogeneous membrane structure. In fact, in this new structurally asymmetric membrane the hydrophilic part is placed between two matrixes containing high amounts of TCPB. The proton conductivity of AMPS-based membranes is lower than Nafion due to lower sulfonic acid group content, but the MCO is lower which is, due to the two exterior layers. Association of functional groups, due to the ionic interaction leads to be stable up to 270 °C and show good mechanical properties [5].

Another type of PEMs developed by Grandfield University is based on the copolymers of PVDF and LDPE with styrene. The copolymers prepared via three-step process; radiation with a cobalt source in the air and proxy radicals formation on the backbone of the polymer, then polymerizing with styrene, and subsequently sulfonation and hydrolyzing in hot water. These membranes are cheaper and have lower methanol diffusion coefficients in 2M methanol at 20 ˚C in comparison with Nafion [5].

7.   Some commercially available membranes

7.1   Polycarbon membranes of Polyfuel

In 2005, Polyfuel, Inc. introduced novel polycarbon membranes with the thicknesses of 62 μm and 45 μm which demonstrate power densities of 60 mW/cm^2 and 80 mW/cm^2 respectively. The lifetime of 5000 h for these membranes in comparison with the required lifetime of 2000–3000 h for a commercially portable fuel cell is their important character. The water water back diffusion has been improved in these membranes, which helps to the MCO issue [5].

7.2   Trifluorostyrene composite membranes of Ballard

Ballard power system Inc. has tested these composite membranes in DMFC cells. However, originally these membranes were developed for PEMFC applications. The DMFC test conditions for a Ballard Mark IV single cell with para-methyl-α,β,β-trifluorostyrene grafted poly(ethylene-co-tetrafluoroethylene) (Tefzel.RTM) as membrane is; temperature of 110 °C, methanol concentration of 0.5 M, oxygen pressure of 3 bar, and oxidant and methanol stoichiometry of 2.0 and 3.0 respectively [5].

7.3   Pall Ionclad membranes

Pall IonClad$^®$ membranes from Pall Gelman Sciences Inc. are tetra-fluoroethylene/perflouropropylene-based membranes. IonClad$^®$ R-1010 and IonClad R-4010 with 36 μm and 63 μm thickness respectively, show 2.5–3 times lower MCO compared to Nafion 117 with a thickness of 180 μm. The selectivity of Pall R4010 membranes at 80 °C is 130, compared to 37 for Nafion 117 [5]

7.4   XUS membranes of Dow chemical

The Dow membrane with its short side chain in spite of similar structure to Nafion functions very differently. It has higher MCO than Nafion 117 which is due to its thinner thickness. Testing a Dow membrane in a six-cell Ballard power system in 1988 showed higher performance in comparison with Nafion [5].

7.5   3P energy membranes

The German 3P-energy PFSA membrane which has 20 times lower MCO show higher performance compared to Nafion that is due to the capability of using higher concentrations of the methanol solution. Unfortunately, data on the durability and lifetime of these membranes are not available [5].

Conclusion and future remarks

In brief, with regard to the problems associated with the use of fossil fuels, fuel cells and, more precisely, polymer electrolyte fuel cells are a good source of energy generation. Direct methanol fuel cells with methanol solution as fuel are preferred upon hydrogen fuel cells provided that the increased efficiency and power output. Therefore, in order to improve the generated power of these cells, the development and improvement of the most critical component, namely, the polymer electrolyte membrane, which determines the performance of the system, is necessary. As explained in this chapter, the major problem with current membranes is the high permeability of methanol, which causes a dramatic loss of cell performance. Thus, the synthesis of novel and controlled structures using monomers with specific microstructures, designed copolymers, and such has gained a lot of attention. Also, development of membranes by new methods, including organic-inorganic composites, electrospun mats of polymeric and/or non-polymeric nanofibers, multi-layer, and pore-filled membranes, are some of the prospects as alternative candidates for available commercial membranes.

References

[1] M. Sajgure, B. Kachare, P. Gawhale, S. Waghmare, G. Jagadale, Direct methanol fuel cell: A review, I.J.C.E.T. 6 (2016) 8-11.

[2] C.M. Branco, S. Sharma, M.M. de Camargo Forte, R. Steinberger-Wilckens, New approaches towards novel composite and multilayer membranes for intermediate temperature-polymer electrolyte fuel cells and direct methanol fuel cells, J. Power Sources. 316 (2016) 139-159. https://doi.org/10.1016/j.jpowsour.2016.03.052

[3] N. Awang, A. Ismail, J. Jaafar, T. Matsuura, H. Junoh, M. Othman, M. Rahman, Functionalization of polymeric materials as a high-performance membrane for direct methanol fuel cell: A review, React. Funct. Polym. 86 (2015) 248-258. https://doi.org/10.1016/j.reactfunctpolym.2014.09.019

[4] N.W. Deluca, Y.A. Elabd, Polymer electrolyte membranes for the direct methanol fuel cell: a review, J. Polymer Sci. B Polymer Phys. 44 (2006) 2201-2225. https://doi.org/10.1002/polb.20861

[5] V. Neburchilov, J. Martin, H. Wang, J. Zhang, A review of polymer electrolyte membranes for direct methanol fuel cells, J. Power Sources. 169 (2007) 221-238. https://doi.org/10.1016/j.jpowsour.2007.03.044

[6] T. Yamaguchi, F. Miyata, S.I. Nakao, Pore-filling type polymer electrolyte membranes for a direct methanol fuel cell, J. Membr. Sci. 214 (2003) 283-292. https://doi.org/10.1016/s0376-7388(02)00579-3

[7] F. Lufrano, V. Baglio, P. Staiti, V. Antonucci, Performance analysis of polymer electrolyte membranes for direct methanol fuel cells, J. Power Sources. 243 (2013) 519-534. https://doi.org/10.1016/j.jpowsour.2013.05.180

[8] Y.S. Kim, B.S. Pivovar, Polymer electrolyte membranes for direct methanol fuel cells, advances in fuel cells, Elsevier, 2007, pp. 187-234. https://doi.org/10.1016/s1752-301x(07)80009-3

[9] V. Baglio, A. Di Blasi, V. Antonucci, L. Cirillo, A. Ghielmi, V. Arcella, Proton exchange membranes based on the short-side-chain perfluorinated ionomer for high temperature direct methanol fuel cells, Desalination. 199 (2006) 271-273. https://doi.org/10.1016/j.desal.2006.03.065

[10] T. Xu, Ion exchange membranes: state of their development and perspective, J. Membr. Sci. 263 (2005) 1-29.

[11] N. Radenahmad, A. Afif, P.I. Petra, S.M. Rahman, S.G. Eriksson, A.K. Azad, Proton-conducting electrolytes for direct methanol and direct urea fuel cells–a state-of-the-art review, Renew. Sustain. Energ. Rev. 57 (2016) 1347-1358. https://doi.org/10.1016/j.rser.2015.12.103

[12] S.P. Jiang, Z. Liu, Z.Q. Tian, Layer-by-layer self-assembly of composite polyelectrolyte–nafion membranes for direct methanol fuel cells, Adv. Mater. 18 (2006) 1068-1072. https://doi.org/10.1002/adma.200502462

[13] M. Smit, A. Ocampo, M. Espinosa-Medina, P. Sebastian, A modified nafion membrane with in situ polymerized polypyrrole for the direct methanol fuel cell, J. Power Sources. 124 (2003) 59-64. https://doi.org/10.1016/s0378-7753(03)00730-4

[14] Z.G. Shao, X. Wang, I.M. Hsing, Composite Nafion/polyvinyl alcohol membranes for the direct methanol fuel cell, J. Membr. Sci. 210 (2002) 147-153. https://doi.org/10.1016/s0376-7388(02)00386-1

[15] N.Y. Arnett, W.L. Harrison, A.S. Badami, A. Roy, O. Lane, F. Cromer, L. Dong, J.E. McGrath, Hydrocarbon and partially fluorinated sulfonated copolymer blends as functional membranes for proton exchange membrane fuel cells, J. Power Sources. 172 (2007) 20-29. https://doi.org/10.1016/j.jpowsour.2007.04.051

[16] T. Yamaguchi, H. Zhou, S. Nakazawa, N. Hara, An extremely low methanol crossover and highly durable aromatic pore-filling electrolyte membrane for direct methanol fuel cells, Adv. Mater. 19 (2007) 592-596. https://doi.org/10.1002/adma.200601086

[17] L. Carrette, K.A. Friedrich, U. Stimming, Fuel cells: principles, types, fuels, and applications, ChemPhysChem 1 (2000) 162-193. https://doi.org/10.1002/1439-7641(20001215)1:4<162::aid-cphc162>3.3.co;2-q

[18] M. Oroujzadeh, S. Mehdipour-Ataei, M. Esfandeh, Preparation and properties of novel sulfonated poly (arylene ether ketone) random copolymers for polymer electrolyte membrane fuel cells, Eur. Polymer J. 49 (2013) 1673-1681. https://doi.org/10.1016/j.eurpolymj.2013.03.008

[19] V. Tricoli, N. Carretta, M. Bartolozzi, A comparative investigation of proton and methanol transport in fluorinated ionomeric membranes, J. Electrochem. Soc. 147 (2000) 1286-1290. https://doi.org/10.1149/1.1393351

[20] K. Kreuer, On the development of proton conducting polymer membranes for hydrogen and methanol fuel cells, J. Membr. Sci. 185 (2001) 29-39. https://doi.org/10.1016/s0376-7388(00)00632-3

[21] X. Ren, P. Zelenay, S. Thomas, J. Davey, S. Gottesfeld, Recent advances in direct methanol fuel cells at Los Alamos National Laboratory, J. Power Sources. 86 (2000) 111-116. https://doi.org/10.1016/s0378-7753(99)00407-3

[22] J.T. Wang, J. Wainright, R. Savinell, M. Litt, A direct methanol fuel cell using acid-doped polybenzimidazole as polymer electrolyte, J. Appl. Electrochem. 26 (1996) 751-756. https://doi.org/10.1007/bf00241516

[23] C. Karthikeyan, S. Nunes, L. Prado, M. Ponce, H. Silva, B. Ruffmann, K. Schulte, Polymer nanocomposite membranes for DMFC application, J. Membr. Sci. 254 (2005) 139-146. https://doi.org/10.1016/j.memsci.2004.12.048

[24] W.C. Choi, J.D. Kim, S.I. Woo, Modification of proton conducting membrane for reducing methanol crossover in a direct-methanol fuel cell, J. Power Sources. 96 (2001) 411-414. https://doi.org/10.1016/s0378-7753(00)00602-9

[25] W. Harrison, M. Hickner, Y. Kim, J. McGrath, Poly (arylene ether sulfone) copolymers and related systems from disulfonated monomer building blocks: synthesis, characterization, and performance–a topical review, Fuel cell. 5 (2005) 201-212. https://doi.org/10.1002/fuce.200400084

[26] Y.A. Elabd, C.W. Walker, F.L. Beyer, Triblock copolymer ionomer membranes: part II. Structure characterization and its effects on transport properties and direct methanol fuel cell performance, J. Membr. Sci. 231 (2004) 181-188. https://doi.org/10.1016/j.memsci.2003.11.019

[27] C. Manea, M. Mulder, Characterization of polymer blends of polyethersulfone/sulfonated polysulfone and polyethersulfone/sulfonated polyetheretherketone for direct methanol fuel cell applications, J. Membr. Sci. 206 (2002) 443-453. https://doi.org/10.1016/s0376-7388(01)00787-6

[28] B.S. Pivovar, Y. Wang, E. Cussler, Pervaporation membranes in direct methanol fuel cells, J. Membr. Sci. 154 (1999) 155-162. https://doi.org/10.1016/s0376-7388(98)00264-6

[29] M. Shen, S. Roy, J. Kuhlmann, K. Scott, K. Lovell, J. Horsfall, Grafted polymer electrolyte membrane for direct methanol fuel cells, J. Membr. Sci. 251 (2005) 121-130. https://doi.org/10.1016/j.memsci.2004.11.006

[30] K. Scott, W. Taama, P. Argyropoulos, Performance of the direct methanol fuel cell with radiation-grafted polymer membranes, J. Membr. Sci. 171 (2000) 119-130. https://doi.org/10.1016/s0376-7388(99)00382-8

[31] T. Tamura, H. Kawakami, Aligned electrospun nanofiber composite membranes for fuel cell electrolytes, Nano Lett. 10 (2010) 1324-1328. https://doi.org/10.1021/nl1007079

[32] H. Zhang, P.K. Shen, Recent development of polymer electrolyte membranes for fuel cells, Chem. Rev. 112 (2012) 2780-2832. https://doi.org/10.1021/cr200035s

[33] S. Nunes, B. Ruffmann, E. Rikowski, S. Vetter, K. Richau, Inorganic modification of proton conductive polymer membranes for direct methanol fuel cells, J. Membr. Sci. 203 (2002) 215-225. https://doi.org/10.1016/s0376-7388(02)00009-1

[34] A. Arico, V. Baglio, A. Di Blasi, P. Creti, P. Antonucci, V. Antonucci, Influence of the acid–base characteristics of inorganic fillers on the high temperature performance of composite membranes in direct methanol fuel cells, Solid State Ionics. 161 (2003) 251-265. https://doi.org/10.1016/s0167-2738(03)00283-2

[35]   H. Thiam, W.R.W. Daud, S.K. Kamarudin, A. Mohammad, A.A.H. Kadhum, K.S. Loh, E. Majlan, Overview on nanostructured membrane in fuel cell applications, Int. J. Hydrogen Energ. 36 (2011) 3187-3205. https://doi.org/10.1016/j.ijhydene.2010.11.062

[36]   M. Nasef, N.A. Zubir, A. Ismail, K. Dahlan, H. Saidi, M. Khayet, Preparation of radiochemically pore-filled polymer electrolyte membranes for direct methanol fuel cells, J. Power Sources. 156 (2006) 200-210. https://doi.org/10.1016/j.jpowsour.2005.05.053

[37]   A. Arico, V. Baglio, P. Cretı, A. Di Blasi, V. Antonucci, J. Brunea, A. Chapotot, A. Bozzi, J. Schoemans, Investigation of grafted ETFE-based polymer membranes as alternative electrolyte for direct methanol fuel cells, J. Power Sources. 123 (2003) 107-115. https://doi.org/10.1016/s0378-7753(03)00528-7

Materials Research Forum LLC
doi: https://doi.org/10.21741/9781644900192-5

Chapter 5

Fabrication and Properties of the Polymer Electrolyte Membrane (PEM) for Direct Methanol Fuel Cell Applications

C. Moganapriya[1], R. Rajasekar[1,*], V. K. Gobinath[1], A. Mohankumar[1]

[1]Department of Mechanical Engineering, Kongu Engineering College, Tamilnadu, India

rajasekar.cr@gmail.com

Abstract

This chapter provides an overview of recent research on the fabrication of polymer electrolyte membrane and their properties for direct methanol fuel cell applications. Precise importance has been focused towards the basic principles involved in fabricating polymer electrolyte membrane and the characterization of different properties. Importance of the chapter is the detailed disclosure regarding fabrication techniques such as sol-gel technique, spray coating technique, dip coating and several novel methods with their procedure. Nafion PEMs are widely used for direct methanol fuel cell applications. It has high methanol permeability and expensive material. Similar to Nafion, various thermoplastic polymers namely poly (ether ether ketone) (PEEK), polysulfone (PSF) and polybenzimidazole (PBI) have been used as substitute PEMs owing to their nominal cost, excellent thermal and mechanical stability at elevated temperature. In recent times, there has been ample research on fabricating PEM that can satisfy all the required characteristics for the desired performance of direct methanol fuel cells. Recent research achievements and their application in DMFC also been reported in this chapter.

Keywords

PEM, DMFC, Nafion, Sol-Gel, Dip Coating, Spray Coating

Contents

1.   Introduction

Fuel cells are extensively recognized as a clean and effective technology for producing electricity as they possess high energy density, zero pollutants and nominal charging time [1]. Direct Methanol fuel cells has gained interest due to its simple design, better efficiency, improved functionality and cost reduction [2]. The transfer of methanol is more expedient as compared to hydrogen since methanol holds high energy density in a liquid state at standard atmospheric temperature and pressure. The liquid energy carrier permits the energy generation for a higher operating time without refilling of fuel as collated to the pressurized hydrogen storage [3, 4]. Furthermore, methanol with simple and basic molecular structure and higher electrochemical affinity enable DMFCs as the promising proton exchange membrane in fuel cell applications [1,2].

The essential part in the design of polymer-electrolyte membrane (PEM) fuel cell is the selection of polymer membrane, which acts a medium for proton conduction and a barrier to prohibit the direct contact of oxidant and fuel [2]. It also prevents the entry of fuel into fuel cell during operation [2,5]. However, methanol can penetrate through the PEMs from the anode to cathode, and it results in the methanol crossover which in turn reduces the function of DMFCs. Hence, the polymer veil should possess the key properties like low methanol permeability, better resistance to chemical, thermal effects, good proton conductivity with mechanical stability [6]. Several sorts of sulfonated polymers have been broadly used as PEMs likeNafion, poly(ether ketone), chitosan, sulfonate and sulfonated poly(arylene ether sulfone) [7-11]. Many researchers studied the performance of different membranes for DMFC application [2, 12]. Ahmed et al. proposed Nafion as

Nanomaterials for Alcohol Fuel Cells   Materials Research Forum LLC
Materials Research Foundations **49** (2019) 159-176   doi: https://doi.org/10.21741/9781644900192-5

the best-suited PEM owing to its higher chemical, thermal, mechanical stability with better proton conductivity [13]. However, Nafion shows high alcoholic permeability which reduces the performance of DFMC [14]. Several approaches have been developed to enhance the alcoholic resistance of PEM by adopting alternative PEM and membrane modification techniques [15,16]. Some researchers proposed non-hybrid thermoplastic polymer membrane such as poly ether ether ketone (PEEK) [17,18], polysulfones (PSF) [17,19] and polybenzimidazole (PBI) [6,17,20] as an alternate to replace Nafion PEM for DMFC application [1]. Therefore, substantial effort is focused on augmenting the efficiency and miniaturizing of the DMFC. The evolution of new generation proton conductive medium for enhanced performance of DMFC is of crucial research interest. The following highlights are some of the characteristics of recently developed technologies which narrowed technical gaps significantly

- Inexpensive and tough membranes (hydrocarbon membranes made of poly fuel)
- High performance non-platinum anode catalysts (<0.2 mg cm^{-2}) or low platinum
- Non-platinum cathode catalysts with minimal metal loading
- Non-carbon cathode affirms highly unaffected by oxidation (porous titanium).

1.1   Basic working of a fuel cell

In recent times, several electronic companies including Sanyo, Toshiba, Hitachi,and Fujitsu developed DMFC powered prototype models for laptops, mobile phones and personal assistant digital devices. Fig.1 depicts the basic working of DMFC. In general, chemical energy is converted into electrical energy and work is extracted for the functioning of the fuel cell from the produced energy. When methanol in its liquid state is catalytically oxidized at the anode, the energy is converted which results in the ejection of electrons and protons as shown in fig.1. The liberated proton then diffuses through PEM and to the cathode, where the proton combines with electrons to generate electricity. The byproducts of this chemical reaction are water and carbon dioxide. As depicted in fig.1, the combination of the anode, PEM and cathode catalyst are termed as a membrane electrode assembly (MEA). The catalyst is generally bounded to the cloth made of carbon fiber and membrane or gas diffusion layer (GDL). It is positioned on either side of MEA while conducting performance tests for PEM fuel cells [21].

2.   Fabrication of MEA for DMFCs

There are various methods adopted by different researchers for the fabrication of DMFC. Some of them are

Nanomaterials for Alcohol Fuel Cells   Materials Research Forum LLC
Materials Research Foundations **49** (2019) 159-176   doi: https://doi.org/10.21741/9781644900192-5

- Sol-gel method

- Spray coating

- Dip coating

- Hot pressing and Hydrolysis

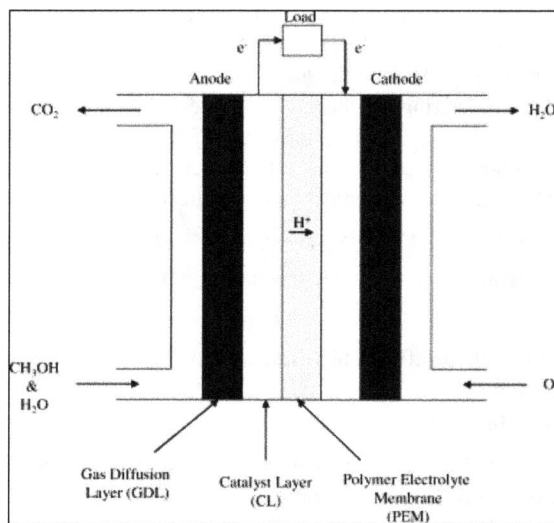

Figure 1. Schematic working of DMFC [21].

2.1   Sol-gel method

The sol-gel method is widely employed to prepare various DMFC Nafion/ silica hybrid composites. Tetraethoxysilane (TEOS) is used to produce silica in PEM [22-24]. Deng et al. [25] adopted a sol-gel method to develop Nafion/silica hybrid membrane for PEM fuel cell application. The result from his studies showed higher water uptake with nominal methanol permeability and higher mechanical strength as compared to unmodified PEM. By employing the same process, Jalani et al. [26] fabricated Nafion –metal oxide (Metal oxides - Zr, Si, Ti) nanocomposite PEM membranes through in situ sol-gel method. The synthesized layer showed better water uptake properties than unmodified PEM. Moreover, physical, thermal, mechanical properties were enhanced by adding inorganic nanosized add-ons which exhibit higher acidity along with properties to uptake water.

In the investigation of Yen et al., the Nafion based silica– silane composite PEMs were synthesized through in-situ sol-gel technique. The procedure adopted for the fabrication of nanocomposite membrane is depicted in fig. 2. A diluted solution of methanol/ Silane was prepared by introducing (3-Mercaptopropyl) trimethoxysilane into the beaker along with 99.9% pure methanol. The variation in the content of SiO2 based on the weight percentage of Nafion was obtained by varying the quantity of SH-silane present in that membrane. The obtained solutions were blended for 1 hour at room temperature to initiate the sol-gel reaction. The sol-gel prepared mixture was poured slowly into the glass dish for further process. The thickness of the developed composite membrane depends on the quantity of poured solution. Then the full container was positioned on the surface of a dry oven maintained at a vacuum, and the mixture was desiccated by raising the temperature from 300 to 323 K [13].

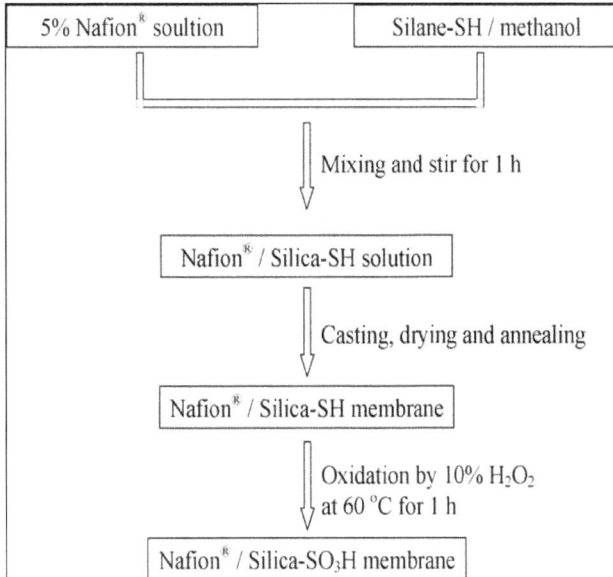

Figure 2. The process for preparing the silica–SO₃ H/Nafion ® composite membranes [13]

The temperature was raised slowly to prevent crevice in the composite membrane. The residuals present in the composite membrane were separated through evacuation technique at a temperature of 373 K for 12 h. The synthesized membrane was then

Nanomaterials for Alcohol Fuel Cells   Materials Research Forum LLC
Materials Research Foundations **49** (2019) 159-176   doi: https://doi.org/10.21741/9781644900192-5

oxidized by immersing in the solution of 10 wt% water at 337 K for 1 h and dehydrated in an oven. The Nafion based silica– silane composite PEM was synthesized by sol-gel method [13]. Yang et al. attempted to combine a tetraethoxysilane solution (TEOS) with polymer Poly(vinyl alcohol) (PVA) to produce a three-dimensional network of composite PEMs by employing a sol-gel process. The technique involves hydrolysis and condensation. The cross-linking reaction was initiated by directly adding the 5wt-% glutaraldehyde (GA) solution into the adherent solution of PVA mixture. The silica nanoparticles have the potential to intensify the thermal and chemical phenomenon and increase the dimensional stability of prepared composite PEMs. It also acts as a solid plasticizer. The lab made Pt- Ru anode Ti screen was developed for DMFC to facilitate the mass transfer rate of methanol [27].

Fig. 3(a)–(b) depicts the SEM images of prepared PEMs by sol-gel and blend method. From fig. 3 a, the presence of uniformly distributed round shaped white colored SiO_2 particles in PVA polymer matrix was evident. As seen from fig. 3 b, the dispersion of silica nanoparticles was poor in PVA/SiO_2 nanocomposite PEM, and some of the clusters were exposed on the higher surface of synthesized PEM through blend process [27].

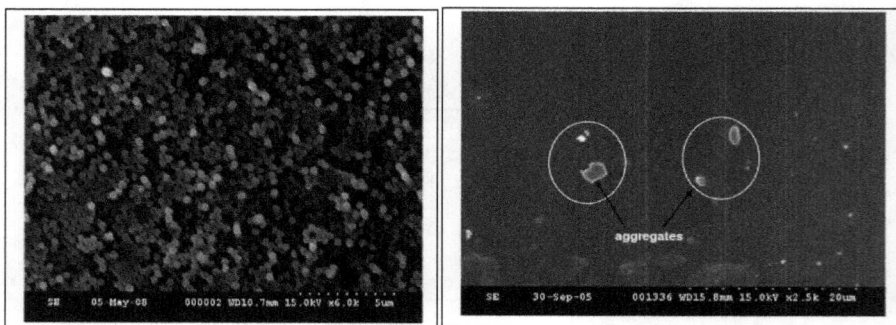

Figure 3. SEM photographs: (a) SiO_2 fillers by sol–gel process (b) PVA/15 wt. % SiO_2 SPE by a blend process [27]

2.2   Spray coating method

Fig. 4 portrays the general procedure of spray coating Sun et al. developed a spray coating method for fabricating composite PEMs [28]. Many researchers adopted the same technique for their studies [29]. Fig. 5 depicts the schematic representation of the spray coating technique. The equipment comprises of the spray gun, stainless steel template, temperature controller, heater and a gas line for nitrogen supply as shown in fig. 4. Tricoli and Nannetti [30] presented a clear procedure for the fabrication composite membrane through this technique. The mixture was synthesized by dissolving 20 wt -% of Nafion with methanol, ethanol and DMF solution. Then the functionalized mordenite powders were slowly added to the mixture through ultrasonication technique. During ultrasonication, a magnetic stirrer was used for 2 h with an interval of 15 min for different weight percentages of mordenite. During spraying, nitrogen gas maintained at 2 bar pressure was supplied to a spray gun, and the spraying temperature was kept at 100 n°C for steel template utilizing temperature sensor and controller. The prepared blend was slowly discharged into a spray container and is sprinkled onto the surface of a stainless steel template fixed at an interval of 5 cm top of the template. Then the blend was subjected to evaporation which resulted in the formation of composite PEMs. The developed membrane was annealed at 150 °C for a time period of 4 h. Full hydration was achieved by treating the prepared membrane in boiled deionized water for 10 min. The organic residue present in PEM was removed by heating the membrane in 5% H at 80 °C for 30 min. The acidic Nafion membrane was obtained by boiling SO_2 for roughly around 30 min. The prepared membrane was finally rinsed thoroughly in deionized boiling water for 10 min [1]. The prepared membranes were tested for solubility and water uptake. The mechanical strength of the membrane plays a vital role in determining the life span of DMFC. For all the performance testing, MEA was incorporated in the fuel cell with the specified electrode area.

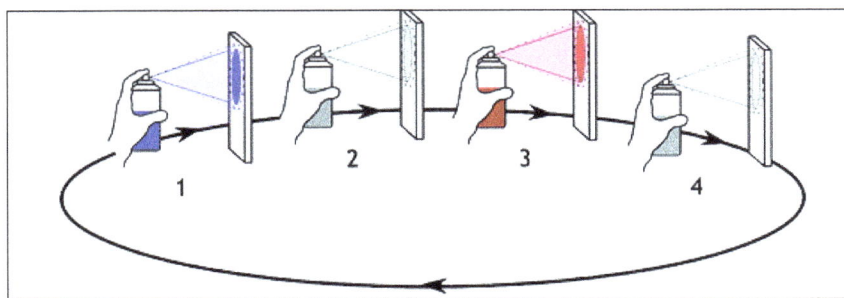

Figure 4. Spray coating.

Andre Wolz et al. [31] fabricated membrane electrode assembly with a multilayered electrode through spray coating technique. Two different inks were utilized for preparation. Sonication was applied to produce good dispersion of Polyaniline (PANI) in ethanol catalyst, and the mixture is used for first ink. Second ink constituted carbon nanotubes dissolved in deionized water and ethanol. The PEM assembly was fabricated

Nanomaterials for Alcohol Fuel Cells   Materials Research Forum LLC
Materials Research Foundations **49** (2019) 159-176   doi: https://doi.org/10.21741/9781644900192-5

by using eco-spray containers. PANI was sprayed for 4 s as the first layer on to the surface of the Nafion membrane. Dwell time of 2 s was allowed for drying of the first layer. Sequentially, the next layer of carbon nanotubes was sprayed for 4 s. The procedure was alternatively repeated until the completion of the multilayer membrane. It is possible to deposit 40 bilayers by a spray coating technique. Fig. 6 portrays the TEM micro images of as-prepared samples. It shows the homogeneous and uniform distribution of nanoparticles through the entire length of the PANI membrane [31].

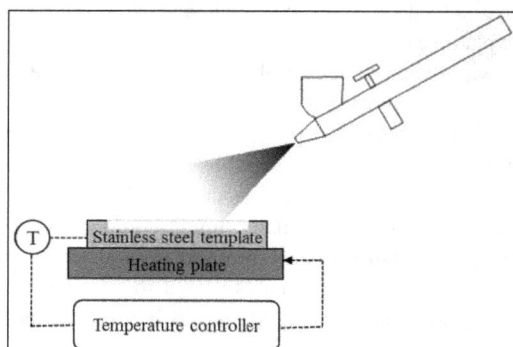

Figure 5. Schematic diagram for fabrication of sprayed membranes [1].

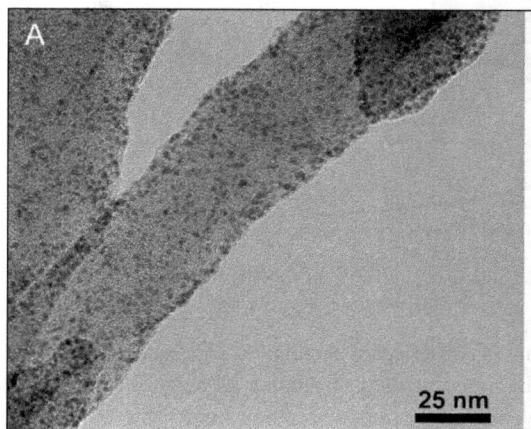

Figure 6. TEM images of prepared sample[31].

2.3   Dip coating

Jaeyoung Lee employed a dip coating technique for developing Co- B catalyst on nickel foam. Cobalt chloride was used as a precursor, and sodium boride was used as reducing agent. The catalysts were prepared through chemical reaction of the reactants. Varying concentrations of cobalt and different weight proportions of sodium hydroxide were used to prepare aqueous precursor solutions. Nickel foam acts a support for cobalt catalyst. It consists of the open pore with 0.55-0.7 mm cell size and 600 g/m^2. First, the cleaned nickel foam was immersed for 10 s in the precursor solution, and the coated foam was dipped in reducing solution to enable reduction reaction. Then the catalyst was rinsed in distilled water for the duration of 30 s. Dip coating was consequentially performed until the expected quantity of catalyst was coated on the surface of nickel foam [32]. Sudipta Mondal et al. studied the effect of dip coating of the blended sulfonated membrane. The neutrality of the membrane was obtained by washing with deionized water. Then the membrane was dipped in the coating mixture for 10- 15 min. The coated membrane was then air dried, and the procedure was repeated 5 times. The coated membranes then thoroughly cleaned with deionized water [33].

2.4   Hot pressing and hydrolysis

Many researchers developed several novel methods for fabricating membrane assembly in PEM for DMFC application. Perfluoro sulfonyl fluoride copolymer resin and porous Polytetrafluoroethylenewere utilized for the synthesis of perfluoro sulfonyl fluoride composite polymer membrane. Fig. 7 portrays the step by step procedure for the fabrication of PEM [34]. Copolymer resin was filled in a cubic frame made of stainless steel, and it was performed by hot pressing into steel shape. The membrane was hot pressed with poly tetrafluoroethylene. The mixture was hydrolysis in a sodium chloride solution. The mixture was stirred by using a magnetic stirrer. Then the catalyst was deposited on film by employing a decal method. The printed catalyst pattern was transferred to PEM from the film through hot pressing. Finally, the membrane assembly was washed repeatedly for several times with deionized water.

3.   Properties of PEM for DMFC

This section discusses the previous works regarding the properties of PEMs like water uptake, proton conductivity, electro-osmotic coefficient, and swelling ratio when these membranes are unprotected from liquid water.

Nanomaterials for Alcohol Fuel Cells   Materials Research Forum LLC
Materials Research Foundations **49** (2019) 159-176   doi: https://doi.org/10.21741/9781644900192-5

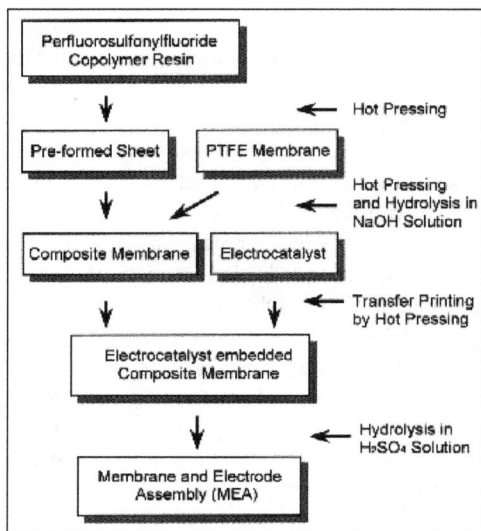

Figure 7. MEA preparation [34].

3.1 Water uptake and swelling

Water uptake of PEM can be expressed in two ways namely water (w) weight percentage and content of water (λ)[35]. The water uptake is calculated by using the weight of the wet sample w_{wet}, dry sample w_{dry}

$W\% = (w_{wet} - w_{dry}/ w_{dry}$

The water uptake denoted as λ constitutes the quantity of water molecules present in the unit mass of sulfonic acid. An expression relating λ and w is given by

$$\lambda = \frac{\omega * EW}{M_{H_2O}}$$

$$M_{H_2O}$$

where EWis the equivalent weight in 1100 g/mol of Nafion and misconsidered as the molar weight of water. The PEMs is equilibrated through water vapor, and water is carried into the cell by the stream of humidified gas. During equilibration, the property of water uptake decreases with increase in temperature [36-38]. Moreover, the water uptake differs during electrolysis from liquid water and water vapor. Fig. 8 depicts the temperature dependency of the Nafion membrane from liquid water vapor [37, 39-41]. λ mainly depends on the pretreatment technique of membrane during equilibration with liquid water. As seen from the figure, Zawodzinski et al. [39]discussed the water uptake λ of PEM at room temperature and elevated the high temperature. He reported that the λ is approximately twice for the membrane dehydrated at room temperature than the membrane dried at elevated temperature (105 °C).

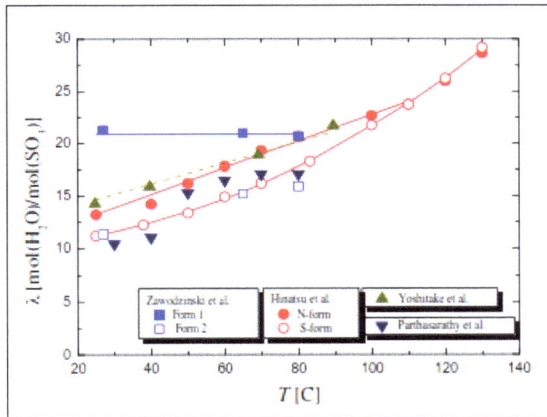

Figure 8. Water uptake λ of a Nafion membrane immersed in liquid water at different temperatures.

The water uptake λ was evaluated to be independent of immersion temperature when dried at ambient atmosphere. Hinatsu et al. [37] reported the effect of pretreatment on the water uptake properties of fabricated PEM. He found that the λ was remarkably higher without adopting pretreatment using vacuum drying and it remains constant up till 100 °C. He utilized two different types of pretreated membranes; N form (pretreated at 80 °C) and S form (pretreated at 105 °C). The water uptake of N form PEM was better compared to the S form at 110 °C immersion temperature. It was noted that the value of water uptake was found to be the same for both N and S form when immersed at a

Nanomaterials for Alcohol Fuel Cells   Materials Research Forum LLC
Materials Research Foundations 49 (2019) 159-176   doi: https://doi.org/10.21741/9781644900192-5

temperature larger when compared to the glass transition temperature. The results of Hinatsu et al. [37] was found to be similar to that of Zawodzinski et al. [39]. Broka and Ekdunge [38] elaborated the difference in the value of water uptake λ between N and S form. The water uptake λ studied by other researchers [37, 41] is in line with the results of the N and S form.

The water uptake by PEM was higher in liquid water than saturated water vapor [37, 39,41]. The variation in water uptake properties of PEM can be precisely described through Schroeder's paradox [39, 42]. According to the paradox, the minimal water uptake in vapor form was owing to complexity in condensation of water vapor between the minute holes in PEM. Moreover, Weber and Newman [43] reported the formation of bulk-like PEM when counterpoised with liquid water while discrete liquid water was observed in the saturated vapor membrane.

3.2   Proton conductivity

The proton conductivity of synthesized PEM depends on the percentage content of water and temperature of operation. Zawodzinski et al. [39] calculated the conductivity of as developed Nafion PEM in terms of the percentage of water content at ambient temperature. He found that the conductivity of the membrane linearly increases with an increase in water content (2-22). The effect of temperature during immersion in liquid water on the conductivity of the Nafion membrane was also measured for a temperature range of 25- 90 °C. During the experiments, the water content was kept constant for the entire range of measurements, and the tests were performed at ambient temperature. From the results obtained by Zawodzinski et al. [39], the proton conductivity can be expressed by the following relation [44]

$$k[S\ cm^{-1}] = (0.005139\lambda - 0.00326)\exp[1268\left(\frac{1}{303} - \frac{1}{T}\right)]$$

$$\kappa[S\ cm^{-1}] = (0.005139\lambda - 0.00326)\exp\left[1268\left(\frac{1}{303} - \frac{1}{T}\right)\right]$$

where T is the membrane temperature. Kopitzke et al. [45] and Parthasarathy et al. also studied the proton conductivity as a ramification of temperature for a Nafion membrane submerged in liquid water. The details reported by Kopitzke et al. for pretreated PEM are in line with the results of Zawodzinski et al.

Fig. 9 depicts the comparative data reported by various researchers [39,40,45,46]. The conductivity was found to be similar for the pretreated membrane at high temperature.

Hence, the proton conductivity of PEM also decreases when equilibrated with water through drying at elevated temperatures similar to the water content.

3.3 Electro-osmotic drag coefficient

The water molecules are accompanied while the proton crosses through PEM by applying the electric field. It is considered a electroosmosis process. Electroosmotic coefficient of drag is defined as the number of water molecules that are dragged along H+ ions. Zawodzinski et al. also described the effect of osmotic drag and medium of hydration with liquid water. MEA was fabricated through chemical plating, and it depends on the membrane temperature and electrolysis condition. Onda et al. [47] described experimentally the correlation between drag and temperature.

$$n_{drag} = 0.0134 \times T + 0.03$$

Marangio et al. [48]reported the drag coefficient using pressure differential PEM that can generate H+ ions at the highest elevated pressure. He found that the number of water molecules was approximately half of the value of membrane compared to membrane without considering the effect of current efficiency.

3.4 Methanol permeability

Methanol permeability through PEM was determined by adopting the diffusion technique. It is used to find out the rate of permeability of alcohol through PEM and from which methanol permeability can be determined. Many researchers reported the experimental setup and step by step procedure for determining methanol permeability [29,49,50]. The temperature for measurement was varied between 30-70 $^\circ$C. For measuring the permeability of methanol at varying temperature, diffusion cell and containers were submerged in temperature managed water bath. Finally, the concentration of methanol was measured by gas chromatography.

Conclusion

This chapter addresses the different types of fabrication methods used for developing PEM by using different materials. In addition, the properties influencing the performance of DMFC and related applications are also discussed. In overview, it was observed that DMFC has potential characteristics to implement as a future alternative source of power. Membrane plays a crucial part in the development of DMFC. The selection of membrane has potential effects on the performance of DFMC. It was proven that membrane possessing high proton conductivity, low methanol permeability, high thermal and mechanical stability can certainly enhance the performance of DMFCs.

References

[1] P. Prapainainar, S. Maliwan, K. Sarakham, Z. Du, C. Prapainainar, S.M. Holmes, P. Kongkachuichay, Homogeneous polymer/filler composite membrane by spraying method for enhanced direct methanol fuel cell performance, Int. J. Hydrogen Energ. 43 (2018) 14675-14690. https://doi.org/10.1016/j.ijhydene.2018.05.173

[2] H. Ahmad, S.K. Kamarudin, U.A. Hasran, W.R.W. Daud, Overview of hybrid membranes for direct-methanol fuel-cell applications, Int. J. Hydrogen Energ. 35 (2010) 2160-2175. https://doi.org/10.1016/j.ijhydene.2009.12.054

[3] M.U. Guruz, V.P. Dravid, Y.W. Chung, Synthesis and characterization of single and multilayer boron nitride and boron carbide thin films grown by magnetron sputtering of boron carbide, Thin Solid Films. 414 (2002) 129-135. https://doi.org/10.1016/s0040-6090(02)00422-4

[4] A. Ainla, D. Brandell, Nafion®–polybenzimidazole (PBI) composite membranes for DMFC applications, Solid State Ion. 178 (2007) 581-585. https://doi.org/10.1016/j.ssi.2007.01.014

[5] C. Li, Z. Yang, X. Liu, Y. Zhang, J. Dong, Q. Zhang, H. Cheng, Enhanced performance of sulfonated poly (ether ether ketone) membranes by blending fully aromatic polyamide for practical application in direct methanol fuel cells (DMFCs), Int. J. Hydrogen Energ. 42 (2017) 28567-28577. https://doi.org/10.1016/j.ijhydene.2017.09.166

[6] R.K. Abdul Rasheed, Q. Liao, Z. Caizhi, S.H. Chan, A review on modelling of high temperature proton exchange membrane fuel cells (HT-PEMFCs), Int. J. Hydrogen Energ. 42 (2017) 3142-3165. https://doi.org/10.1016/j.ijhydene.2016.10.078

[7] N.N. Krishnan, H.-J. Lee, H.-J. Kim, J.-Y. Kim, I. Hwang, J.H. Jang, E.A. Cho, S.-K. Kim, D. Henkensmeier, S.A. Hong, T.H. Lim, Sulfonated poly(ether sulfone)/sulfonated polybenzimidazole blend membrane for fuel cell applications, Eur. Polym. J.46 (2010) 1633-1641. https://doi.org/10.1016/j.eurpolymj.2010.03.005

[8] C. Li, G. Sun, S. Ren, J. Liu, Q. Wang, Z. Wu, H. Sun, W. Jin, Casting Nafion–sulfonated organosilica nano-composite membranes used in direct methanol fuel cells, J.Membr. Sci.272 (2006) 50-57. https://doi.org/10.1016/j.memsci.2005.07.032

[9] J. Wang, X. Zheng, H. Wu, B. Zheng, Z. Jiang, X. Hao, B. Wang, Effect of zeolites on chitosan/zeolite hybrid membranes for direct methanol fuel cell, J. Power Sourc.178 (2008) 9-19. https://doi.org/10.1016/j.jpowsour.2007.12.063

[10] Q. Luo, H. Zhang, J. Chen, D. You, C. Sun, Y. Zhang, Preparation and characterization of Nafion/SPEEK layered composite membrane and its application in

vanadium redox flow battery, J.Membr. Sci.325 (2008) 553-558.
https://doi.org/10.1016/j.memsci.2008.08.025

[11] C. Li, L. Xiao, Z. Jiang, X. Tian, L. Luo, W. Liu, Z.L. Xu, H. Yang, Z.J. Jiang, Sulfonic acid functionalized graphene oxide paper sandwiched in sulfonated poly(ether ether ketone): A proton exchange membrane with high performance for semi-passive direct methanol fuel cells, Int. J. Hydrogen Energ. 42 (2017) 16731-16740. https://doi.org/10.1016/j.ijhydene.2017.05.126

[12] F.A. Zakil, S.K. Kamarudin, S. Basri, Modified Nafion membranes for direct alcohol fuel cells: An overview, Renew. Sustain. Energ. Rev. 65 (2016) 841-852.
https://doi.org/10.1016/j.rser.2016.07.040

[13] C.Y. Yen, C.H. Lee, Y.F. Lin, H.L. Lin, Y.H. Hsiao, S.H. Liao, C.Y. Chuang, C.C.M. Ma, Sol–gel derived sulfonated-silica/Nafion® composite membrane for direct methanol fuel cell, J. Power Sour.173 (2007) 36-44.
https://doi.org/10.1016/j.jpowsour.2007.08.017

[14] C. Lamy, A. Lima, V. LeRhun, F. Delime, C. Coutanceau, J.-M. Léger, Recent advances in the development of direct alcohol fuel cells (DAFC), J. Power Sour.105 (2002) 283-296. https://doi.org/10.1016/s0378-7753(01)00954-5

[15] D.J. Kim, H.J. Lee, S.Y. Nam, Sulfonated poly(arylene ether sulfone) membranes blended with hydrophobic polymers for direct methanol fuel cell applications, Int. J. Hydrogen Energ. 39 (2014) 17524-17532.
https://doi.org/10.1016/j.ijhydene.2013.09.030

[16] S. Zhong, X. Cui, Y. Gao, W. Liu, S. Dou, Fabrication and properties of poly(vinyl alcohol)-based polymer electrolyte membranes for direct methanol fuel cell applications, Int. J. Hydrogen Energ. 39 (2014) 17857-17864.
https://doi.org/10.1016/j.ijhydene.2014.08.040

[17] M. Nacef, A.M. Affoune, Comparison between direct small molecular weight alcohols fuel cells' and hydrogen fuel cell's parameters at low and high temperature. Thermodynamic study, Int. J. Hydrogen Energ. 36 (2011) 4208-4219.
https://doi.org/10.1016/j.ijhydene.2010.06.075

[18] E. Kjeang, J. Goldak, M. R. Golriz, J. Gu, D. James, K. Kordesch, Modeling methanol crossover by diffusion and electro-osmosis in a flowing electrolyte direct methanol fuel cell, Fuel Cell. 5 (2005) 486-498.
https://doi.org/10.1002/fuce.200400087

[19] K. Kang, S. Park, S.O. Cho, K. Choi, H. Ju, Development of lightweight 200-W direct methanol fuel cell system for unmanned aerial vehicle applications and flight demonstration, Fuel Cell. 14(2014) 694-700. https://doi.org/10.1002/fuce.201300244

[20] M. Müller, N. Kimiaie, A. Glüsen, Direct methanol fuel cell systems for backup power – Influence of the standby procedure on the lifetime,Int. J. Hydrogen Energ. 39(2014) 21739-21745. https://doi.org/10.1016/j.ijhydene.2014.08.132

[21] N. W. DeLuca, Y. A. Elabd, Polymer electrolyte membranes for the direct methanol fuel cell: A review, J. Polymer Sci. B Polymer Phys. 44(2006) 2201-2225. https://doi.org/10.1002/polb.20861

[22] N. Miyake, J. Wainright, R. Savinell, Evaluation of a sol-gel derived nafion/silica hybrid membrane for proton electrolyte membrane fuel cell applications: I. Proton conductivity and water content, J. Electrochem. Soc. 148 (2001) A898-A904. https://doi.org/10.1149/1.1383071

[23] N. Miyake, J. Wainright, R. Savinell, Evaluation of a sol-gel derived Nafion/silica hybrid membrane for polymer electrolyte membrane fuel cell applications: II. Methanol uptake and methanol permeability, J. Electrochem. Soc. 148 (2001) A905-A909. https://doi.org/10.1149/1.1383072

[24] D. Jung, S. Cho, D. Peck, D. Shin, J. Kim, Performance evaluation of a Nafion/silicon oxide hybrid membrane for direct methanol fuel cell, J. Power Sour.106 (2002) 173-177. https://doi.org/10.1016/s0378-7753(01)01053-9

[25] Q. Deng, R. Moore, K.A. Mauritz, Nafion®/(SiO2, ORMOSIL, and dimethylsiloxane) hybrids via in situ sol–gel reactions: characterization of fundamental properties, J. Appl. Polymer Sci. 68 (1998) 747-763. https://doi.org/10.1002/(sici)1097-4628(19980502)68:5<747::aid-app7>3.0.co;2-o

[26] N.H. Jalani, K. Dunn, R. Datta, Synthesis and characterization of Nafion®-MO2 (M= Zr, Si, Ti) nanocomposite membranes for higher temperature PEM fuel cells, Electrochim.Acta. 51 (2005) 553-560. https://doi.org/10.1016/j.electacta.2005.05.016

[27] C.C. Yang, Y.J. Li, T.H. Liou, Preparation of novel poly (vinyl alcohol)/SiO2 nanocomposite membranes by a sol–gel process and their application on alkaline DMFCs, Desalination. 276 (2011) 366-372. https://doi.org/10.1016/j.desal.2011.03.079

[28] Y.M. Sun, W.F. Huang, C.C. Chang, Spray-coated and solution-cast ethylcellulose pseudolatex membranes, J.Membr. Sci.157 (1999) 159-170. https://doi.org/10.1016/s0376-7388(98)00369-x

[29] P. Prapainainar, Z. Du, P. Kongkachuichay, S.M. Holmes, C. Prapainainar, Mordenite/Nafion and analcime/Nafion composite membranes prepared by spray method for improved direct methanol fuel cell performance, Appl. Surf. Sci. 421 (2017) 24-41. https://doi.org/10.1016/j.apsusc.2017.02.004

[30] V. Tricoli, F. Nannetti, Zeolite–Nafion composites as ion conducting membrane materials, Electrochim.Acta. 48 (2003) 2625-2633. https://doi.org/10.1016/s0013-4686(03)00306-2

[31] A. Wolz, S. Zils, M. Michel, C. Roth, Structured multilayered electrodes of proton/electron conducting polymer for polymer electrolyte membrane fuel cells assembled by spray coating, J.Power Sour.195 (2010) 8162-8167. https://doi.org/10.1016/j.jpowsour.2010.06.087

[32] J. Lee, K.Y. Kong, C.R. Jung, E. Cho, S.P. Yoon, J. Han, T.-G. Lee, S.W. Nam, A structured Co–B catalyst for hydrogen extraction from NaBH4 solution, Catal.Today. 120 (2007) 305-310. https://doi.org/10.1016/j.cattod.2006.09.019

[33] S. Mondal, S. Soam, P.P. Kundu, Reduction of methanol crossover and improved electrical efficiency in direct methanol fuel cell by the formation of a thin layer on Nafion 117 membrane: Effect of dip-coating of a blend of sulphonated PVdF-co-HFP and PBI, J.Membr. Sci.474 (2015) 140-147. https://doi.org/10.1016/j.memsci.2014.09.023

[34] S. Kwak, D. Peck, Y. Chun, C. Kim, K. Yoon, New fabrication method of the composite membrane for polymer electrolyte membrane fuel cell, J. New. Mater. Electrochem. Syst. 4 (2001) 25-30.

[35] H. Ito, T. Maeda, A. Nakano, H. Takenaka, Properties of Nafion membranes under PEM water electrolysis conditions, Int. J. Hydrogen Energ. 36 (2011) 10527-10540. https://doi.org/10.1016/j.ijhydene.2011.05.127

[36] P.C. Rieke, N.E. Vanderborgh, Temperature dependence of water content and proton conductivity in polyperfluorosulfonic acid membranes, J.Membr. Sci.32 (1987) 313-328. https://doi.org/10.1016/s0376-7388(00)85014-0

[37] J.T. Hinatsu, M. Mizuhata, H. Takenaka, Water uptake of perfluorosulfonic acid membranes from liquid water and water vapor, J. Electrochem. Soc.141 (1994) 1493-1498. https://doi.org/10.1149/1.2054951

[38] K. Broka, P. Ekdunge, Oxygen and hydrogen permeation properties and water uptake of Nafion® 117 membrane and recast film for PEM fuel cell, J.Appl. Electrochem.27 (1997) 117-123.

[39] T.A. Zawodzinski, C. Derouin, S. Radzinski, R.J. Sherman, V.T. Smith, T.E. Springer, S. Gottesfeld, Water uptake by and transport through Nafion® 117 membranes, J. Electrochem. Soc. 140 (1993) 1041-1047. https://doi.org/10.1149/1.2056194

[40] A. Parthasarathy, S. Srinivasan, A.J. Appleby, C.R. Martin, Temperature dependence of the electrode kinetics of oxygen reduction at the platinum/Nafion®

interface—a microelectrode investigation, J. Electrochem. Soc. 139 (1992) 2530-2537. https://doi.org/10.1149/1.2221258

[41] M. Yoshitake, M. Tamura, N. Yoshida, T. Ishisaki, Studies of perfluorinated ion exchange membranes for polymer electrolyte fuel cells, Denki Kagaku oyobi Kogyo Butsuri Kagaku. 64 (1996) 727-736.

[42] P. von Schroeder, Uber Erstarrungs-und quellugserscheinungen von gelatine, Zeitschrift für physikalische Chemie. 45 (1903) 75-117. https://doi.org/10.1515/zpch-1903-4503

[43] A.Z. Weber, J. Newman, Transport in polymer-electrolyte membranes I. Physical model, J. Electrochem. Soc 150 (2003) A1008-A1015. https://doi.org/10.1149/1.1580822

[44] T.E. Springer, T. Zawodzinski, S. Gottesfeld, Polymer electrolyte fuel cell model, J. Electrochem. Soc 138 (1991) 2334-2342.

[45] R.W. Kopitzke, C.A. Linkous, H.R. Anderson, G.L. Nelson, Conductivity and water uptake of aromatic-based proton exchange membrane electrolytes, J. Electrochem. Soc 147 (2000) 1677-1681. https://doi.org/10.1149/1.1393417

[46] M. Doyle, M.E. Lewittes, M.G. Roelofs, S.A. Perusich, R.E. Lowrey, Relationship between ionic conductivity of perfluorinated ionomeric membranes and nonaqueous solvent properties, J.Membr. Sci.184 (2001) 257-273. https://doi.org/10.1016/s0376-7388(00)00642-6

[47] K. Onda, T. Murakami, T. Hikosaka, M. Kobayashi, K. Ito, Performance analysis of polymer-electrolyte water electrolysis cell at a small-unit test cell and performance prediction of large stacked cell, J. Electrochem. Soc. 149 (2002) A1069-A1078. https://doi.org/10.1149/1.1492287

[48] F. Marangio, M. Santarelli, M. Pagani, M.C. Quaglia, Direct High Pressure Hydrogen Production: a Laboratory Scale PEM Electrolyser Prototype, ECS Transac.17 (2009) 555-567. https://doi.org/10.1149/1.3142786

[49] C. Prapainainar, S. Kanjanapaisit, P. Kongkachuichay, S.M. Holmes, P. Prapainainar, Surface modification of mordenite in Nafion composite membrane for direct ethanol fuel cell and its characterizations: Effect of types of silane coupling agent, J.Environ. Chem. Eng.4 (2016) 2637-2646. https://doi.org/10.1016/j.jece.2016.05.005

[50] P. Prapainainar, A. Theampetch, P. Kongkachuichay, N. Laosiripojana, S. Holmes, C. Prapainainar, Effect of solution casting temperature on properties of nafion composite membrane with surface modified mordenite for direct methanol fuel cell, Surf. Coating. Tech. 271 (2015) 63-73. https://doi.org/10.1016/j.surfcoat.2015.01.021

Nanomaterials for Alcohol Fuel Cells
Materials Research Foundations **49** (2019) 177-192

Materials Research Forum LLC
doi: https://doi.org/10.21741/9781644900192-6

Chapter 6

Carbon Polymer Supports Hybrid for Alcohol Oxidation

M. Harikrishna Kumar[1], S. Mahalakshmi[2], K.V. Mahesh Kumar[3], P.Sathishkumar[4], C. Moganapriya[1], R.Rajasekar[1]

[1]Department of Mechanical Engineering, Kongu Engineering College, Erode, TamilNadu, India

[2]Department of Mechanical Engineering, KonguPolytechnic College, Erode, Tamil Nadu, India

[3]Department of Mechatronics Engineering, Kongu Engineering College, Erode, Tamil Nadu, India

[4]Department of Mining Engineering, Indian Institute of Technology, Kharagpur, West Bengal, India

rajasekar.cr@gmail.com

Abstract

Many works have been reported by using polymer-supported reagents organic synthesis of alcohol oxidation. New functional materials have been developed by coordination of polymers which is used in a wide variety of applications. To carry out oxidization of alcohol many oxidizing agents are used. Coordination of polymers through catalyst is more advantageous due to high stability and a large number of active metals. In recent days many nanomaterials are used as a catalyst which increases the oxidation rate in the base matrix. Synthesis of the base matrix is also possible for enhancing alcohol oxidation.

Keywords

Polymers, Synthesis Technique, Catalyst, Alcohol Oxidation

Contents

1.   Introduction

Many research works have been concentrated on the evolution of new functional materials by coordination of polymers for application such as storage materials, molecular recognition devices, etc. [1-4]. More works have been reported by utilizing different oxidizing agents for oxidation of alcohols in various media. However, less attention is given to catalyst for coordinating polymer. A number of works have been carried out using a catalyst in a metal matrix, and it purely depends on their structures and electronic properties. Researchers have presented numerous works on catalytic systems for the oxidation of alcohols by adding metal composites and nanoparticles [5]. By means of altering the base material, catalyst, nanoparticles, etc. tender a different kind of molecular structures in the polymer matrix [1]. To offer a localized property within a polymer, the d orbitals of some transition materials will interact with organic elements. Coordination of polymers through catalyst is more advantageous due to its stability and number of active metals present in it [1, 6]. Structure of the catalyst should also be characterized to activate the catalyst [1]. In order to convert chemical energy into electricity, nanomaterial plays a significant purpose in ethanol fuel cells as a catalyst in oxidizing organic material with high energy density, energy conversion rate, low

doi: https://doi.org/10.21741/9781644900192-6

operating temperature and eco-friendliness [7]. In the case of fuel cell applications palladium based nanomaterials are used as a catalyst for ethanol oxidation [7-10]. Catalyst would possess high activity and stability properties for the controlled shape, size, and composition of nanomaterials [7, 11]. Two-dimensional nanomaterials possess high electronic properties, chemical reactivity and high specific surface areas in thin sheets [7, 12, 13]. Graphene sheets, metal chalcogenide nanosheets, and metal nanosheets are chemically stable and have a wide application such as biomedical, energy conversion and storage devices [7, 14, 15]. Carbon-based conductive substrates have to selected to disperse aggregated nanoparticles in order to improve the electro-catalytic activity of electrodes [7]. For example, palladium based nanoparticles are dispersed in carbon nanotubes (CNT), to enhance ethanol electro-oxidation [7, 16, 17].

2. Benzyl alcohol oxidation

Huayun Han et al. reported that over the last decade's demand for benzaldehyde has increased since it is used as a raw material in perfume, pharmaceutical, and fertilizer industry [1, 18, 19]. For both lab scale experiments and manufacturing processes, oxidation of alcohol is the primal reaction. The wide investigation has been carried out on benzyl alcohol oxidation by catalytic vapor phase method [1, 20, 21].

2.1 Materials

A number of works have been reported on oxidation of benzyl alcohol by adopting different strategies such as selective oxidation of benzyl alcohol, catalytic vapor-phase oxidation of benzyl alcohol and liquid-phase oxidation of benzyl alcohol. Based on the above strategies Huayun Han et al. has synthesized three coordination polymers, Iron coordination polymer (Polymer-1), $[[Cu(fcz)_2(H_2O)]\cdot SO_4\cdot DMF\cdot 2CH_3OH\cdot 2H_2O]n$ (Polymer-2), and $[[Cu(fcz)_2Cl_2]\cdot 2CH3OH]n$ (Polymer-3) as shown in figure 1. By altering the metal ions, ligands, anions, coordination modes, etc., coordination polymer extend a diverse number of molecular structures and electronic properties. Synthesized polymers are structured in an aqueous medium with H_2O_2 as oxidant and catalyzed in benzyl alcohol oxidation process. Huayun Han et al. has also used catalysts, or solvent selection of benzaldehyde remains the same [1].

2.2 Synthesis root and chemicals used

All the three polymers Iron coordination polymer(Polymer-1), $\{[Cu(fcz)_2(H_2O)]\cdot SO_4\cdot DMF\cdot 2CH_3OH\cdot 2H_2O\}n$ (Polymer-2), and $\{[Cu(fcz)2Cl_2]\cdot 2CH_3OH\}n$ (Polymer-3) were synthesized by adding methanol into an aqueous solution dropwise to give a clear solution. The clear solution is kept at room temperature until crystals are formed. Resultant formed crystals are insoluble in water and organic solvents [1].

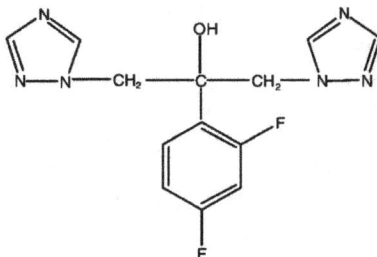

Figure 1. Scheme of Synthesis of three polymers[1].

2.3 Evaluation methodology and results

Huayun Han et al. chosen the benzyl alcohol oxidation as a probe reaction to determine the catalytic properties. The crystal is cut into appropriate sizes. The amount of 85 mg/dL of benzyl alcohol in 10 mL aqueous medium is kept constant for determining the effect of several reaction conditions such as temperature, the quantity of catalyst and oxidant. Figure 2 depicts the consequence of temperature in the oxidation process with the 1st polymer. As the temperature increases, the conversion of benzyl alcohol also increases. Huayun has stated that significant conversion rate is observed for the temperature range 30°C to 50°C. Selectivity of benzaldehyde has declined as the temperature increases. It is concluded that 40°C is the optimum temperature for the conversion of benzyl alcohol.

Fig. 3 shows the result of the quantity of catalyst in polymer 1. 0.005 mmol and 0.01 mmol of polymer 1 was employed to study the reaction system. As 0.005 mmol of catalyst in polymer 1 was employed, it is observed that 48% of benzyl alcohol was converted in four hours. As the quantity of catalyst was raised to 0.01 mmol, 51% of benzyl alcohol was converted in one hour, and 87% of benzyl alcohol was converted in four hours. For polymer 2 and 3, the trend is similar to as polymer 1.

Huayun Han et al. have also studied the consequence of the quantity of oxidant by using hydrogen peroxide in different concentrations such as 4.6, 6.0 and 6.9 mmol. The conversion rate of benzyl alcohol was increased as the concentration of oxidant increases. In addition to those three factors, the solvent was also considered for benzyl alcohol oxidation. The selection and conversion rate are more beneficial in the aqueous medium than in acetonitrile.

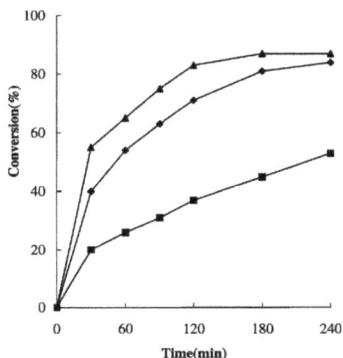

Figure 2. Effect of reaction temperature on benzyl alcohol oxidation on 1st polymer[1].

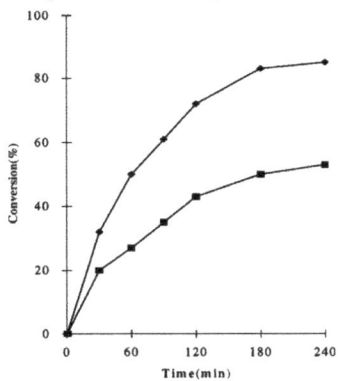

Figure 3. Effect of amount of catalyst on benzyl alcohol oxidation on 1st polymer[1].

3. Aerobic alcohol oxidation

Among both researchers and industrialist Haruta's et al. invention has made a significant impact by using gold nanoclusters as a catalyst for CO oxidation [22]. Based on this invention many works have been reported on CO oxidation and alcohol oxidation using gold nanoclusters as a catalyst[22-24].

ShokyokuKanaoka et al. have reported work using gold clusters as a catalyst in Vinyl Ether Star Polymers. During the examination of the catalytic mechanism disperse mechanism of gold and size dependence plays a significant role in catalytic reactions [25, 26]. Aggregation of Au nanoclusters (NCs) causes a major problem during the reaction and work up procedures. In order to prevent aggregation of nanoclusters, a stable micelle structure is required. A polymer linking mechanism was employed to produce polymers with hydrophilic arms and a hydrophobic core to form a unimolecular micelle. Figure 4 represents the preparation and use of Au nanoclusters as a catalyst in polymer cross-linking. Shokyoku Kanaoka et al. have also studied the formulation of gold nanoclusters in water using star polymers comprising of EOEOVE. Resulting clusters are employed as a catalyst for oxidation without deterioration. Thermosensitivity technique is applied to recover and reuse the catalyst [25].

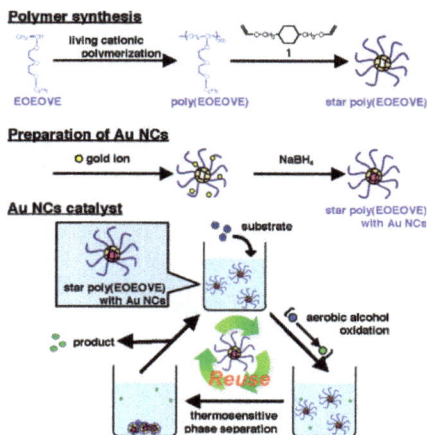

Figure 4. Preparation and use Au nanoclusters as catalyst[25].

Hironori Tsunoyama et al. has also reported work on aerobic alcohol oxidation in water using gold nanoclusters as a catalyst [26]. It is accepted that both experimentally and theoretically adsorbed molecular oxygen from gold nanoclusters plays a vital role in CO oxidation process [22, 23]. Hironori Tsunoyama et al. have used hydrophilic polymer say PVP for oxidation of benzylic alcohols concentrating on size effects of gold nanoclusters. To have more reaction on the nanoclusters surface, the catalysts weakly stabilized through multiple coordination of PVP polymer. Kinetic measurements have been compared between smaller gold coated PVP nanoclusters of size 1.3 nm and larger nanoclusters of size 9.5 nm. Gold nanoclusters with smaller size nanoclusters show higher catalytic activities than the larger nanoclusters. In this work, palladium nanoclusters are also used for aerobic alcohol oxidation, and it is compared with gold nanoclusters catalyst.

Through TEM images for the prepared specimen gold coated PVP-1 (poly(N-vinyl-2-pyrrolidone) with nanoparticle size ranging between 1.3 ± 0.3 nm), gold coatedPVP-2 (poly(N-vinyl-2-pyrrolidone) with nanoparticle size ranging between 9.5 ± 1.0 nm), palladium based PVP-1,and palladium based PVP-2 it is observed that average diameter 1.3 ± 0.3 nm, 9.5 ± 1.0 nm, 1.5 ± 0.3 nm and 2.2 ± 0.4 nm respectively as shown in figure 5. Increase in diameter of nanocluster particles is due to coagulation or agglomeration. Catalytic activities were also determined for all the four prepared specimens in water. As a result aldehydes or carboxylic acids are made in the reaction. Benzoic acid, benzyl benzoate, and hydroxybenzaldehydes were prepared [26]. Following this Hironori Tsunoyama et al. has concluded that oxygen adsorption into the gold nanoclusters is the fundamental element for the catalytic activities.

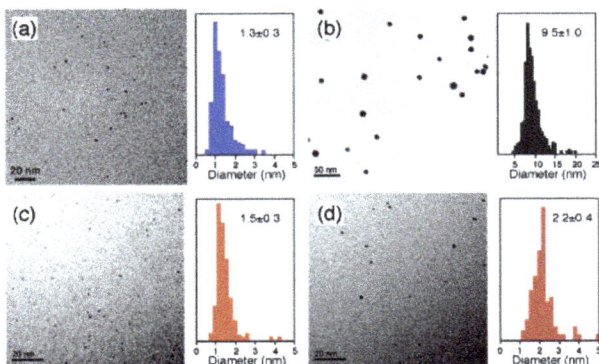

Figure 5. TEM images of Au:PVP-1 (a), Au:PVP-2 (b), Pd:PVP-1 (c) and Pd:PVP-2 (d)[26].

4.   Aerobic alcohol oxidation by using multiple polymers

Chung et al. have also reported on aerobic alcohol oxidation by using multiple polymeric ligands to coordinate catalytically active species. In organic synthesis, many works have been reported by using polymer-supported reagents [27, 28]. Sheldon et al. have concluded the inclusion of bipyridine and TEMPO (2,2,6,6-tetramethyl-piperidyl-1-oxy) chemical group to several polymers showed enhanced properties[29]. With this invention, Chung et al. has concluded that two different polymer ligands can be used simultaneously. A number of works have been reported using broad varieties of TEMPO reagents, such as a polyamine -TEMPO, silica-TEMPO, MCM-41- TEMPO, sol-gel - TEMPO, ROMP polymer- TEMPO, polystyrene-TEMPOs, polymeric TEMPO,and polyethyleneglycol TEMPOs.

4.1   Synthesis root

Sheldon et al. have described steps for aerobic alcohol oxidation. The substrate should dissolve in the acetonitrile-water mixture [29, 30]. Monomethyl ether PEG is water-soluble and can be well precipitated with diethyl ether is selected with molecular weight =~ 5000 Daand it acts as the supporting agent for the ligands. MPEG-TEMPO and MPEG-bipyridine were selected as the ligands and procedure for the synthesis of the same are depicted in figure 6. The multiple elements used in synthesis are MPEG-TEMPO (1) and MPEG-bipyridine (2), MPEG-OMs (3), pyridinium salt (4), methyl-bipyridine (5)carboxylic acid (6), ester (7) and alcohol (8). As per the procedure described by Pozzi et al. [31], Ligand 1 was prepared hydroxy-TEMPO and MPEG-OMs. For the synthesis of ligand 2, first pyridinium salt (4) was prepared by heating acetylpyridine with iodine in pyridine as the procedure described by Edwin et al. [32]. Pyridinium salt (4) was admitted to reacting with methacrolein in formamide in the presence of ammonium acetate to produce methyl-bipyridine (5). As a result of oxidation of methyl-bipyridine (5) with KMnO4, it yields carboxylic acid (6). Esterification of carboxylic acid (6) along with ethanol in the presence of p-Toluenesulfonicacidresulted in the formation of ester (7).Resulted ester was reduced by sodium borohydride which in turn produces alcohol (8). Eventually, deprotonation of alcohol followed by alkylation with MPEG-OMs (3) was performed to form ligand 2.

4.2   Steps for aerobic alcohol oxidation

Reagents used for aerobic oxidation process are potassium tert-butoxide, cupric bromide, MPEG-bipyridine and MPEG-TEMPO (0.26 g, 0.05 mmol) with the weight of 11.2 mg, 11.2 mg, 0.26 g, 0.26 g respectively. The reagents were added consecutively 1 mmol of alcohol in Acetonitrile with water in the ratio of 2:1. Oxygen is passed into the reagent mixture for 10 min, and the mixture temperature is raised to 80 °C for 18 hours. During the process, Methylene chloride the crude mixture was extracted and dried magnesium sulfate filtered and concentrated under low pressure. Polymers MPEG-TEMPO and MPEG-bipyridine were removed by adding diethyl ether. To obtain pure aldehyde product, it is concentrated under low pressure. Recovered polymers may be used in reused in future experiments [29, 30].

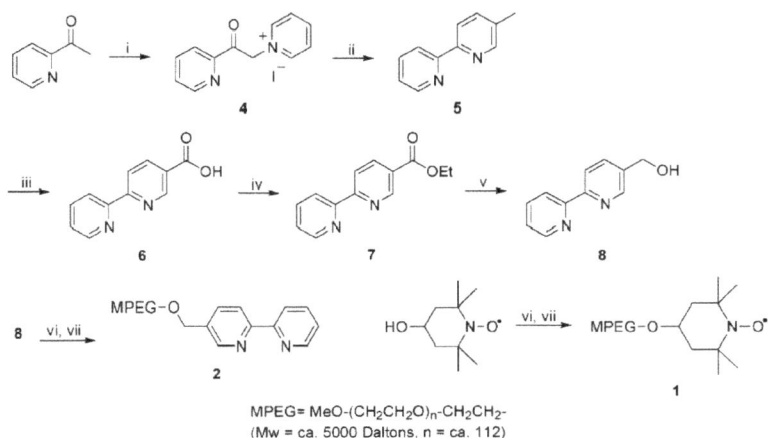

Figure 4.1. Synthesis of MPEG-TEMPO (1) and MPEG-2,2'-bipyridine polymer [30].

5. Multi-polymer system for organocatalytic alcohol oxidation

Yuen Sze But et al. utilized two soluble supported polymer reagents for organocatalytic oxidation of alcohols to produce aldehydes or ketones These polymeric reagents can be reused. Here the two polymers react with each other before reacting with substrate. Among the two polymers, the non-cross linked polymer becomes soluble as the reaction progresses and other polymeric reagent remains insoluble throughout the process. Also, nitroxyl catalyst is used to oxidize alcohols to form either aldehydes or ketones [28]. Yuen Sze But et al. have experimented, whether PSDIB (polystyrenesupportedDiacetoxyiodosobenzene) could be used together with nitroxyl reagent and the results are shown in figure 7.

5.1 Procedure for alcohol oxidation

JJ-TEMPO (JandaJelbased 2,2,6,6-tetramethyl-1-piperidinyloxy) and PSDIB (polystyrenesupportedDiacetoxyiodosobenzene)are added to alcohol in Ethylene dichloride. Until alcohol was no longer available, the mixture was stirred at 70°C. An excess mixture of JJ-TEMPO and PSDIB were removed during the reaction time by the filtering process. Reacted PSDIB was extracted by filtering through silica gel. As a result of removing solvent from the desired product, it becomes pure. The molecular mass of all products are analyzed by LRMS,and molecular mass of all products are found to be spectroscopically identical.

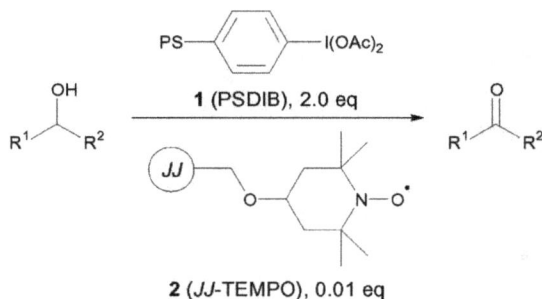

Figure 5.1. Scheme of PSDIB and nitroxyl reagent for alcohol oxidation [28]

6. Oxidation of aryl alcohol using water as a catalyst

The oxidation of alcohols to ketones by utilizing catalyst has a wide range of application in chemical, drug and food industries [33, 34]. Researchers mainly focus on the usage of molecular oxygen as an oxidizing media to accomplish energy-effective and sustainable oxidation processes [5, 35]. However, all the research works which have been carried out by using solvents are based on the polarity of the polymers. Based on these criteria, Jack et al. have started an investigation of oxidation of alcohols such as phenylethanol, benzyl alcohol,and octanol by employing nanoparticle embedded rhodium polymer. The investigation has been carried out using a solvent/co-solvent system which comprises toluene, water, and atmospheric air as the oxidant. Satoh et al. have proved that Rh being metal are more effective in oxidative reactions [36]. Rh is mainly used for reducing nitric oxide in the exhaust system of automobiles. However, still, it is so far neglected for alcohol oxidation of metals like palladium, platinum,and gold. Catalytic tests were carried out by the researcher for phenylethanol oxidation by Rh incarcerated nanoparticles, and Rh incarcerated alumina in the presence of toluene. The catalytic tests were performed at a temperature of 100 °C for four hours with molar metal to substrate ratio as 1: 100. A sample result of the catalytic test carried out is shown in Table 1.

Two solvents such as water and toluene were also used to study the performance of the catalytic test for the oxidation of phenylethanol to acetophenone for both Rh incarcerated nanoparticles, and Rh incarcerated alumina. Results are depicted in Table 2. When compared to the first category water and toluene category results were impressive.

Presence of water enhances the activity of catalysts and turning them to materials increases the conversion efficiency by 50% as compared to alumina. The activity of a catalyst can be changed by changing the reaction medium. However, in this case, the effect is media dependent while using nanoparticles [5].

The investigation was also carried out by varying volume ratios of water by keeping a constant volume of toluene of 1 mL for the reactivity of Rh incarcerated nanoparticles (Rh-PI),andRhsupported on alumina catalysts (Rh/Al$_2$O$_3$). From Table 3, as the volume of water increases the percentage of conversion is also increases.

This scenario is observed for both Rh-PI and Rh/Al$_2$O$_3$. Re-usability tests were also carried out for Rh/Al$_2$O$_3$and Rh-PI in water. Two catalystsRh/Al$_2$O$_3$and Rh-PI possessed better re-usability with high conversion and selectivity. Rhin carcerated nanoparticles were confirmed to be an effective catalytic system for oxidation of aryl alcohols such as phenylethanol and benzyl alcohol [5].

7.   Oxidation of alcohol by using Poly(ethylene glycol)-Supported Nitroxyls

In organic synthesis, the transformation of alcohols into aldehydes or ketones is a promising parameter for oxidation [37]. Oxidation of alcohol in the presence of hypochlorite or bromide under the biphasic condition is carried out for PEG-supported nitroxyl catalysts. High solubility is required for oxidation protocol, and PEG is well suited for this oxidation. The catalyst is prepared by attaching TEMPO residue to the polymer material using ether linkage mechanism. This catalyst comprises PEG with different molecular weights and hydroxyl functionalities per polymer chain. Hereby it is referred to as the linear "linker-less" [37].

7.1   Synthesis procedure for PEG-supported catalysts

Polyethylene glycol dissolves in organic solvents and water. Synthesis of linear linker-less PEG-supported nitroxyl elements is well breakable by two-step synthesis process. Janda et al. have depicted that mesyl ester was formed with the help of hydroxyl functionalities in the PEG chains [38]. Macromolecular sulfonyl esters could be detached by removing excess methanesulfonyl chloride. When sulfonyl esters are treated with an alkoxide of hydroxy-TEMPO, it results in the formation of ether.

7.2   Steps for the oxidation of alcohols as described by Ferreira

A 25 mL nitroxyl catalyst was taken in the round-bottom flask containing the alcohol substrate and is maintained at 0°C. At time t = 0 min, sodium hypochlorite, and sodium bromide were added into the round-bottom flask. The reaction is allowed to take place by means stirring after which 50 mL of dichloromethane was extracted. Extracted medium is dried over magnesium sulfate, and it is filtered, concentrated under low pressure. Gas Chromatography finally analyzes the resulting medium with Flame-Ionization Detection technique [37].

Conclusion

Many oxidizing agents are used for alcohol oxidation. Catalyst plays a significant role in activating the oxidization process. Controlled shape, size, and composition of catalyst would possess high activity and stability properties. Nano-materials like gold clusters, palladium clusters are used as a catalyst which yielded high conversion rate. Carbon polymers are also used as an oxidizing agent. In order to mix two or three polymers synthesis techniques are also used for alcohol oxidizing. In order to measure the oxidation rate first, the temperature is kept constant and finds the optimum temperature. Next amount of the catalyst is kept constant to determine the optimum volume of catalyst and optimum oxidant level can be determined. Catalyst with smaller size and optimum composition yields better results in the conversion of alcohol oxidation.

References

[1] H. Han, S. Zhang, H. Hou, Y. Fan, Y. Zhu, Fe (Cu)-containing coordination polymers: syntheses, crystal structures, and applications as benzyl alcohol oxidation catalysts, Eur. J. Inorg. Chem. 2006 (2006) 1594-1600. https://doi.org/10.1002/ejic.200500808

[2] H. Hou, Y. Song, H. Xu, Y. Wei, Y. Fan, Y. Zhu, L. Li, C. Du, Polymeric complexes with "piperazine– pyridine" building blocks: synthesis, network structures, and third-order nonlinear optical properties, Macromolecules. 36 (2003) 999-1008. https://doi.org/10.1021/ma025787n

[3] X. Meng, G. Li, H. Hou, H. Han, Y. Fan, Y. Zhu, C. Du, A series of novel metal-ferrocenedicarboxylate coordination polymers: crystal structures, magnetic and luminescence properties, J. Organomet. Chem. 679 (2003) 153-161. https://doi.org/10.1016/s0022-328x(03)00516-3

[4] S.A. Bourne, J. Lu, A. Mondal, B. Moulton, M.J. Zaworotko, Self-assembly of nanometer-scale secondary building units into an undulating two-dimensional network with two types of hydrophobic cavity, Angew. Chem. Int. Ed.40 (2001) 2111-2113. https://doi.org/10.1002/1521-3773(20010601)40:11<2111::aid-anie2111>3.3.co;2-6

[5] J.O. Weston, H. Miyamura, T. Yasukawa, D. Sutarma, C.A. Baker, P.K. Singh, M. Bravo-Sanchez, N. Sano, P.J. Cumpson, Y. Ryabenkova, Water as a catalytic switch in the oxidation of aryl alcohols by polymer incarcerated rhodium nanoparticles, Catal. Sci. Technol. 7 (2017) 3985-3998. https://doi.org/10.1039/c7cy01006k

[6] S.J. Hong, J.Y. Ryu, J.Y. Lee, C. Kim, S.-J. Kim, Y. Kim, Synthesis, structure and heterogeneous catalytic activities of Cu-containing polymeric compounds: anion effect and comparison of homogeneous vs. heterogeneous catalytic activity, Dalton Trans. (2004) 2697-2701. https://doi.org/10.1039/b406877g

[7] S. Ghosh, A.L. Teillout, D. Floresyona, P. de Oliveira, A. Hagege, H. Remita, Conducting polymer-supported palladium nanoplates for applications in direct alcohol oxidation, Int. J. Hydrogen Energ. 40 (2015) 4951-4959. https://doi.org/10.1016/j.ijhydene.2015.01.101

[8] F. Favier, E.C. Walter, M.P. Zach, T. Benter, R.M. Penner, Hydrogen sensors and switches from electrodeposited palladium mesowire arrays, Science. 293 (2001) 2227-2231. https://doi.org/10.1126/science.1063189

[9] H.P. Liang, N.S. Lawrence, L.J. Wan, L. Jiang, W.G. Song, T.G. Jones, Controllable synthesis of hollow hierarchical palladium nanostructures with enhanced activity for proton/hydrogen sensing, J. Phys. Chem. C. 112 (2008) 338-344. https://doi.org/10.1021/jp0752320

[10] C. Langhammer, I. Zorić, B. Kasemo, B.M. Clemens, Hydrogen storage in Pd nanodisks characterized with a novel nanoplasmonic sensing scheme, Nano Letters. 7 (2007) 3122-3127. https://doi.org/10.1021/nl071664a

[11] C. Koenigsmann, A.C. Santulli, E. Sutter, S.S. Wong, Ambient surfactantless synthesis, growth mechanism, and size-dependent electrocatalytic behavior of high-quality, single crystalline palladium nanowires, ACS nano. 5 (2011) 7471-7487. https://doi.org/10.1021/nn202434r

[12] X. Huang, C. Tan, Z. Yin, H. Zhang, 25th Anniversary article: Hybrid nanostructures based on two-dimensional nanomaterials, Adv. Mater. 26 (2014) 2185-2204. https://doi.org/10.1002/adma.201304964

[13] Q.H. Wang, K. Kalantar-Zadeh, A. Kis, J.N. Coleman, M.S. Strano, Electronics and optoelectronics of two-dimensional transition metal dichalcogenides, Nat. Nanotechnol. 7 (2012) 699. https://doi.org/10.1038/nnano.2012.193

[14] F. Ksar, G. Sharma, F. Audonnet, P. Beaunier, H. Remita, Palladium urchin-like nanostructures and their H2 sorption properties, Nanotechnology. 22 (2011) 305609. https://doi.org/10.1088/0957-4484/22/30/305609

[15] M. Chhowalla, H.S. Shin, G. Eda, L.-J. Li, K.P. Loh, H. Zhang, The chemistry of two-dimensional layered transition metal dichalcogenide nanosheets, Nat. Chem. 5 (2013) 263. https://doi.org/10.1038/nchem.1589

[16] J. Zhang, Y. Mo, M. Vukmirovic, R. Klie, K. Sasaki, R. Adzic, Platinum monolayer electrocatalysts for O_2 reduction: Pt monolayer on Pd (111) and on carbon-supported Pd nanoparticles, J. Phys. Chem. B. 108 (2004) 10955-10964. https://doi.org/10.1021/jp0379953

[17] N. Mackiewicz, G. Surendran, H. Remita, B. Keita, G. Zhang, L. Nadjo, A. Hagege, E. Doris, C. Mioskowski, Supramolecular self-assembly of amphiphiles on carbon nanotubes: a versatile strategy for the construction of CNT/metal nanohybrids, application to electrocatalysis, J. Am. Chem. Soc. 130 (2008) 8110-8111. https://doi.org/10.1021/ja8026373

[18] D.V. McGrath, R.H. Grubbs, J.W. Ziller, Aqueous ruthenium (II) complexes of functionalized olefins: the x-ray structure of Ru $(H_2O)_2$ (. eta. 1 (O):. eta. 2 (C, C')-OCOCH$_2$CH=CHCH$_3$)$_2$, J. Am. Chem. Soc. 113 (1991) 3611-3613. https://doi.org/10.1021/ja00009a069

[19] D. Andrew Knight, T.L. Schull, Rhodium catalyzed allylic isomerization in water, Synthetic Comm.33 (2003) 827-831. https://doi.org/10.1081/scc-120016328

[20] J. Farkas, S. Békássy, J. Madarász, F. Figueras, Selective oxidation of benzylic alcohols to aldehydes with metal nitrate reagents catalyzed by BEA zeolites or clays, New. J. Chem. 26 (2002) 750-754. https://doi.org/10.1039/b106252m

[21] M. Arai, S. Nishiyama, S. Tsuruya, M. Masai, Effect of alkali-metal promoter on silica-supported copper catalysts in benzyl alcohol oxidation, J. Chem. Soc. Faraday Trans.92 (1996) 2631-2636. https://doi.org/10.1039/ft9969202631

[22] M. Haruta, N. Yamada, T. Kobayashi, S. Iijima, Gold catalysts prepared by coprecipitation for low-temperature oxidation of hydrogen and of carbon monoxide, J. Catal. 115 (1989) 301-309. https://doi.org/10.1002/chin.198920020

[23] M. Valden, X. Lai, D.W. Goodman, Onset of catalytic activity of gold clusters on titania with the appearance of nonmetallic properties, science. 281 (1998) 1647-1650. https://doi.org/10.1126/science.281.5383.1647

[24] C. Milone, R. Ingoglia, G. Neri, A. Pistone, S. Galvagno, Gold catalysts for the liquid phase oxidation of o-hydroxybenzyl alcohol, Appl. Catal. Gen.211 (2001) 251-257. https://doi.org/10.1016/s0926-860x(00)00875-9

[25] S. Kanaoka, N. Yagi, Y. Fukuyama, S. Aoshima, H. Tsunoyama, T. Tsukuda, H. Sakurai, Thermosensitive gold nanoclusters stabilized by well-defined vinyl ether star polymers: reusable and durable catalysts for aerobic alcohol oxidation, J. Am. Chem. Soc. 129 (2007) 12060-12061. https://doi.org/10.1021/ja0735599

[26] H. Tsunoyama, H. Sakurai, Y. Negishi, T. Tsukuda, Size-specific catalytic activity of polymer-stabilized gold nanoclusters for aerobic alcohol oxidation in water, J. Am. Chem. Soc. 127 (2005) 9374-9375. https://doi.org/10.1021/ja052161e

[27] S.V. Ley, I.R. Baxendale, R.N. Bream, P.S. Jackson, A.G. Leach, D.A. Longbottom, M. Nesi, J.S. Scott, R.I. Storer, S.J. Taylor, Multi-step organic synthesis using solid-supported reagents and scavengers: Anew paradigm in chemical library generation, J. Chem. Soc., Perkin Trans. 1.(2000) 3815-4195. https://doi.org/10.1039/b006588i

[28] T.Y.S. But, Y. Tashino, H. Togo, P.H. Toy, A multipolymer system for organocatalytic alcohol oxidation, Org. Biomol. Chem. 3 (2005) 970-971. https://doi.org/10.1039/b500965k

[29] P. Gamez, I.W. Arends, R.A. Sheldon, J. Reedijk, Room temperature aerobic copper–catalysed selective oxidation of primary alcohols to aldehydes, Adv. Synth. Catal.346 (2004) 805-811. https://doi.org/10.1002/adsc.200404063

[30] C.W.Y. Chung, P.H. Toy, Multipolymer reaction system for selective aerobic alcohol oxidation: Simultaneous use of multiple different polymer-supported ligands, J. Combin. Chem. 9 (2007) 115-120. https://doi.org/10.1021/cc060111f

[31] G. Pozzi, M. Cavazzini, S. Quici, M. Benaglia, G. Dell'Anna, Poly (ethylene glycol)-supported TEMPO: an efficient, recoverable metal-free catalyst for the selective oxidation of alcohols, Org. Lett. 6 (2004) 441-443. https://doi.org/10.1002/chin.200424056

[32] E.C. Constable, J. Lewis, The preparation and coordination chemistry of 2, 2': 6', 2 ″-terpyridine macrocycles—1, Polyhedron 1. (1982) 303-306. https://doi.org/10.1016/s0277-5387(00)87169-7

[33] S. Caron, R.W. Dugger, S.G. Ruggeri, J.A. Ragan, D.H.B. Ripin, Large-scale oxidations in the pharmaceutical industry, Chem. rev.106 (2006) 2943-2989. https://doi.org/10.1021/cr040679f

[34] C.J. Dillard, J.B. German, Phytochemicals: nutraceuticals and human health, J. Sci. Food Agr. 80 (2000) 1744-1756.

[35] T. Punniyamurthy, S. Velusamy, J. Iqbal, Recent advances in transition metal catalyzed oxidation of organic substrates with molecular oxygen, Chem. Rev.105 (2005) 2329-2364. https://doi.org/10.1021/cr050523v

[36] T. Satoh, M. Miura, Oxidative coupling of aromatic substrates with alkynes and alkenes under rhodium catalysis, Chem. Eur. J. 16 (2010) 11212-11222. https://doi.org/10.1002/chem.201001363

[37] P. Ferreira, E. Phillips, D. Rippon, S.C. Tsang, W. Hayes, Poly (ethylene glycol)-supported nitroxyls: branched catalysts for the selective oxidation of alcohols, J. Org. Chem.69 (2004) 6851-6859. https://doi.org/10.1021/jo0490494

[38] F. Sieber, P. Wentworth, J.D. Toker, A.D. Wentworth, W.A. Metz, N.N. Reed, K.D. Janda, Development and application of a poly (ethylene glycol)-supported triarylphosphine reagent: Expanding the sphere of liquid-phase organic synthesis, J. Org. Chem.64 (1999) 5188-5192. https://doi.org/10.1021/jo9903712

Nanomaterials for Alcohol Fuel Cells
Materials Research Foundations 49 (2019) 193-230

Materials Research Forum LLC
doi: https://doi.org/10.21741/9781644900192-7

Chapter 7

Polymer Electrolyte Membrane Methanol Fuel Cells: Technology and Applications

P. Thomas[1], N.P. Rumjit[2], C.W. Lai[1,*], M.R.B. Johan[1]

[1]Nanotechnology & Catalysis Research Centre (NANOCAT), Institute for Advanced Studies (IAS), University of Malaya (UM), Level 3, Block A, 50603 Kuala Lumpur, Malaysia

[2]Department of Environmental & Water Resources Engineering, School of Civil Engineering, Vellore Institute of Technology, Vellore, India

* cwlai@um.edu.my

Abstract

This chapter discusses the recent development of polymer electrolyte membrane methanol fuel cells and its applications, including the fuel cell's components, models and comparison studies concerning the direct methanol fuel cell. Indeed, polymer electrolyte membrane methanol fuel cells undergo electrochemical reaction for the conversion of chemical energy into useful electrical energy, suiting its applications role as mobile and stationary energy sources due to its outstanding power density, energy production, and fewer emissions. Thus, polymer electrolyte membrane methanol fuel cells open up new dimensions in the energy storage industry lately.

Keywords

Polymer Electrolyte Membrane Methanol Fuel Cells, Electrochemical Reaction, Chemical Energy, Electrical Energy, Energy Storage

Contents

1.   Introduction

In the past few decades, concerns over environmental pollution gave attraction towards new green technologies such as fuel cells. The fuel cells have numerous advantages compared to conventional power sources such as batteries or internal combustion engines. Some of its main advantages are high efficiency, eliminate pollution and greenhouse gases, less maintenance requirements and mostly suitable towards decentralised power grids [1]. Fuel cells are electrochemical devices that utilize hydrogen as a fuel source, which reacts with oxygen to generate both water and electricity. In recent years numerous research investigations were carried out in the fuel cell technology, resulting in further upgradation of the system to more efficient [2–5]. Fuel cells undergo electrochemical reaction for the conversion of chemical energy into electrical energy, suiting its applications role as mobile and stationary energy sources due to its outstanding power density, energy production, and fewer emissions. Based on operating temperature and electrolyte, the fuel cell is categorized into mainly five types: Solid oxide fuel cell, phosphoric acid fuel cell, alkaline fuel cell, molten carbonate fuel cell, direct methanol fuel cell, and polymer electrolyte fuel cell [6]. Table 1 is enlisting the details of various fuel cell types. Voluminous efforts have been carried out for commercialization and industrial production for polymer electrolyte membrane fuel cells [6,7]. However, the cost of production and durability remains as the main hurdle in commercialization [9]. Recently numerous attempts been carried out to decline the cost of production and to enhance durability, still many improvements needed to be carried out before its industrial-scale production [9,10]. The main hurdles in the path of the commercialization are its life duration, and the life period mainly gets affected due to the occurrence of contamination in membrane electrode assembly, water management system, power, and thermal cycling [12–15].

This chapter focuses on the technology of polymer electrolyte membrane methanol fuel cells and its applications. These include a brief discussion on components, models and comparison study concerning the direct methanol fuel cell.

Table 1. Overview of various fuel types.

Types of Fuel Cell	Advantages	Operating Temperature	Efficiency	Drawbacks	Applications
DMFC	Fast starting; Easy to handle fuel supply; Low operating temperature	50–120 °C	30%	Performance issues; High Catalyst loading	Portable Applications
MCFC	Thermal losses; Flexibility in fuel; Cheap Catalyst	650 °C	37–42%	The issue in Life duration; Molten and corrosive electrolyte; Expensive material	Stationary Applications
AFC	Cheap Catalyst and Electrolyte; Better cathode performance	70 °C	60%	Need to remove water from the anode; Need to replenish electrolyte	Stationary and Transport Applications

PEMFC	Fast startup; High power density; Low operating temperature	Below 100 °C (Low-Temperature PEMFC) 200 °C (High-Temperature PEMFC)	40 – 60% (Low Temperature PEMFC) 40 % (High-Temperature PEMFC)	Expensive electrolyte and catalyst; Poor carbon monoxide tolerance	Stationary, Portable and Transport Applications
PAFC	Cheap electrolyte; Mature technology and reliability	180°C	80%	Expensive Catalyst; Low carbon monoxide tolerance; Need periodic replacement of electrolyte	Stationary Applications
SOFC	Cheap catalyst; Flexibility in fuel; Solid Electrolyte	800 °C – 1000 °C.	80%	Expensive; Issues regarding sealing	Stationary Applications

2. Overview of polymer electrolyte membrane fuel cell

The polymer electrolyte fuel cells open up new dimensions in the energy storage industry. The polymer electrolyte fuel cell was initially reported by Wilard T. Grubb and his teammates in 1957 at General Electric [16]. The polymer electrolyte fuel cell developed by General Electric mainly for the space exploration programmes. In years 1967 to 1969 many projects were carried out regarding the suitability of fuel cell for spaceships, however, in the 1970s, these projects were discontinued due to the high production costs associated with a fuel cell. In 1980s research was restarted by industry giants such as Ergenics Power Systems, Ballard, Treadwell, and Engelhard.

The polymer electrolyte fuel cells operate at a temperature range of below 80 °C. In most of the polymer electrolyte fuel cells, utilizes hydrogen source as fuel at anode end and the air is provided as an oxidant at the cathode end [6]. The basic principle involves the transportation of hydrogen towards polymer electrolyte fuel cells from its storage point, later there occurred diffusion through the diffusion layer and then breakups on the catalyst surface as electrons and protons. Protons get easily shifted towards cathode via polymer electrolyte membrane; however, the membrane acts as a barrier for electron

Nanomaterials for Alcohol Fuel Cells   Materials Research Forum LLC
Materials Research Foundations **49** (2019) 193-230   doi: https://doi.org/10.21741/9781644900192-7

migration. Accumulated electrons carried away by connecting an external load. The chemical equation for polymer electrolyte membrane fuel cell as listed below:

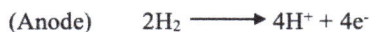

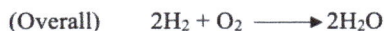

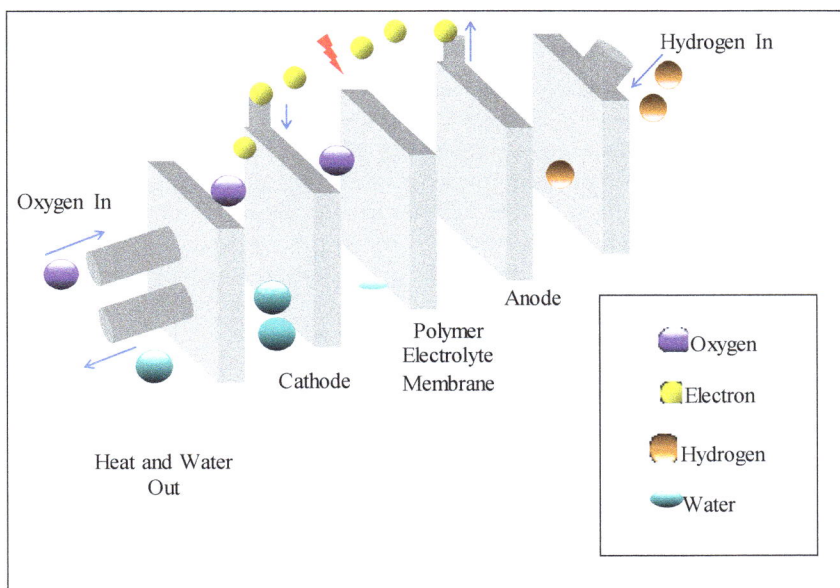

(Anode) $2H_2 \longrightarrow 4H^+ + 4e^-$

(Cathode) $O_2 + 4H^+ + 4e^- \longrightarrow 2H_2O$

(Overall) $2H_2 + O_2 \longrightarrow 2H_2O$

Figure 1. Working Principle of polymer electrolyte membrane fuel cell.

Still, there are limitations for commercialization polymer electrolyte membrane fuel cell especially in terms of hydrogen fuel. There are numerous issues associated with hydrogen, production, storage, and technological barriers. Steam reforming is a traditional technique to produce hydrogen [17]. One of the main drawbacks of this process is the emission of carbon dioxide to the atmosphere causing serious issues towards the environment. This process produces low purity hydrogen fuel, traces of carbon monoxide present in hydrogen fuel results poisoning of anode catalyst. The costs associated with storage and transportation raises a question towards its economic viability. The expected costs associated with this process range up to $2.5/Kg for conventional hydrogen generation from coal gasification. In another hand, the electrolysis process utilized to derive high purity hydrogen [18]. This method followed the redox reaction to produce hydrogen, and the main drawback is the efficiency and cost of production. The cost is approximated as $11/Kg for solar driven electrolysis process. These issues need to be addressed before the commercialization of polymer electrolyte membrane fuel cell.

3.   Polymer electrolyte methanol fuel cell

In methanol fuel cell the electrical energy is produced when methanol catalytically oxidized at the anode side generates both electrons and protons. Protons diffuse via polymer electrolyte membranes to the cathode; the combination of electrons and protons results in the generation of electrical energy. The power generation associated with two main by-products carbon dioxide and water. The schematic diagram of Polymer Electrolyte Membrane fuel cell is illustrated in Fig. 1. The membrane electrode assembly referred to as a cathode catalyst/ Polymer Electrolyte Membrane/ Anode catalyst. Catalyst adhered to carbon fibre cloth and membrane or gas diffusion layer placed on both sides of the methanol fuel cell. The main components associated with the Polymer Electrolyte Methanol fuel cell are explained below.

3.1   Components

The Polymer Electrolyte Membrane Methanol fuel cell has membrane electrolyte sandwiched between flow field plates. It is usually termed as a monopolar plate for single cell configuration also known as a bipolar plate for stack configuration [19]. The components of Polymer Electrolyte Methanol fuel cell demonstrated in Fig. 2. The flow field plate has numerous roles; these include the reactants supply, providing adequate structure support, and works as current collector also facilitate the removal of by-products such as water and heat. The membrane electrolyte assembly further categorized into five main layers cathodic catalyst layer, an anodic catalyst layer, cathodic diffusion layer, and an anodic diffusion layer. The diffusion layer on both cathode and anode side assists in the mobilization of electrons, providing adequate structural support also aids water management and acts as gas diffuser [20–22]. Similarly, the polymer electrolyte membranes also play a significant role acts as a separator for reactant gas, mobilizing agent for protons and acts as electron barrier insulator [23].

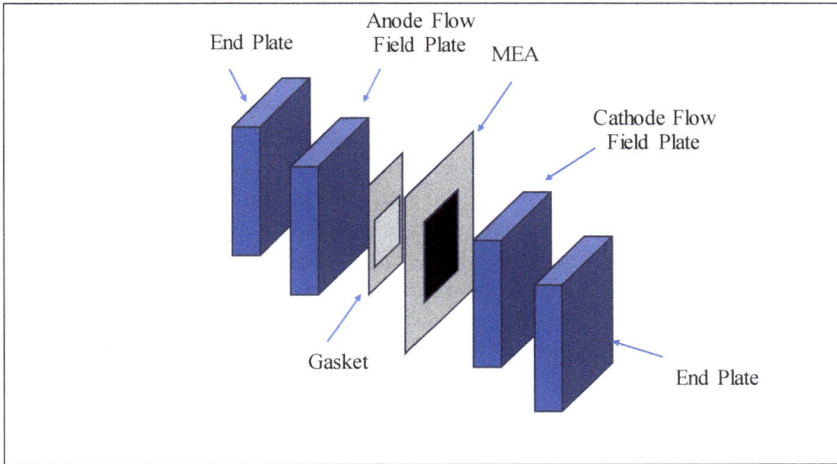

Figure 2.Components of Polymer Electrolyte Methanol fuel cell.

3.1.1 Methanol fuel source

Hydrogen is the most durable fuel for fuel cell application, however, as explained above the production and storage of hydrogen has its challenges. It is better to store and utilize fuel in a liquid form that inherits better kinetic performance and thermodynamic when oxidized in a fuel cell. Methanol is recognized as a promising source and can be utilized directly for direct methanol fuel cell and also can be utilize as a hydrogen carrier. Methanol is a flammable, clear, colourless liquid with a density of 0.79 g cm^{-2}and the moderate boiling point of 67 °C. As methanol is in liquid form, the significant challenges regarding transportation and storage will not be an issue. For low-temperature fuel cell applications, methanol is the appropriate option, especially for portable and stationary applications.

3.1.2 Flow field plate

The flow field plate is one of the principal components in Polymer Electrolyte Methanol fuel cell in terms of weight and cost [25]. Two main material categories of flow field plate are metallic flow field plate and carbon-based flow field plate [25,26]. In transportation and mobile applications flat metallic flow field plates are preferred.

However, for commercial and residential applications polymer carbon compounds are an appropriate choice as it inherits a corrosion resistant property [28].

Graphite which is one of the prime materials was utilized for flow field plate construction due to their superior electrical conductivity and exceptional chemical stability. Even though the graphite material meets several required characteristics of the flow field plate for the Polymer Electrolyte Methanol fuel cell but the size needs to be bulky due to its porous and brittle nature [29]. However, the graphite-based flow field plate is not suitable for lightweight applications. For lightweight transportation and mobile applications, a thin metallic flow field plate would be an ideal choice [30] for superior corrosion resistivity and electrical conductivity various precious elements such as gold and platinum would be the obvious choice. However, the cost of the material would not be economically viable [31]. Nickel, titanium, and aluminum are other alternative option for flow field plates in Polymer Electrolyte Methanol fuel cell. These materials have numerous advantages such as high strength, low cost, and density. Stainless steel is also an effective option, there will be the formation of the passive oxide film which reduces surface conductivity and is the main drawback.

Protective layers are coated with metallic flow field plates to overcome various drawbacks associated with it. Protective layer coatings are categorized into two main types; carbon and metal-based coating. Different stainless-steel materials such as 321, 304, 316Ti and 316L coated with gold, chromium nitride, and titanium nitride been utilized to improve the corrosion resistant property [31,32]. Polymer coating films such as polypyr role and polyaniline are the foremost common choice for coating materials [34]. Composite material opens new options in flow field plates. They demonstrated superior characteristics compared to metallic and graphite bipolar plate in terms of low weight, flexibility and corrosion resistant. One of the main disadvantages of the flow field plate is their poor electrical conductivity compared to metallic and graphite plates. Thermoplastic and thermoset resins been utilized for the manufacturing of composite bipolar plates. Thermoset resins such as epoxies, phenolics, vinyl ester, and polyester are commonly employed. In thermoplastic resins such as polyphenylene sulfide, polypropylene, fluoropolymer, and polyvinylidene fluoride.

Table 2. Exhibits a detailed comparative analysis of thermoset and thermoplastic composites.

Types of Composites	Advantages	Drawbacks	Processing Option
Thermoplastic Composite	Fast Cycling; Low contact resistance; Flow field introduced during molding; Easy to mold and manufacture	Less chemical stable; Performance issues in low temperature; Poor Electrical Conductivity	Post-molding CNC milling of the blank; Injection molding; Compression molding
Thermoset Composite	Fast cycling; High-temperature operation; Low contact resistance; Flow field introduced during molding	Poor Electrical Conductivity	Post-molding CNC milling of the blank; Compression molding

3.1.3 Membrane electrolyte assembly

Membrane Electrolyte Assembly fabricated individually followed by molding at high pressure and temperature. The Membrane Electrolyte Assembly components are mainly categorized into two main components; catalyst coated membrane and catalyst coated substrate [35]. The catalyst coated membrane combined with catalyst coated substrate boost the performance by reducing membrane interfacial resistance. The polymer electrolyte membrane must meet various specifications such as; better electrical insulation, proton conductivity, thermal stability, mechanical support, cost-effectiveness, low swelling stress, hydrolytic stability, and better barrier property. General Electric developed the first prototype polymer electrolyte membrane under the Gemini space program. The detailed structure of the membrane electrolyte assembly was shown in Fig.3. The prototype polymer membrane based on sulfonated polystyrene divinylbenzene copolymer had some significant drawbacks such as high cost and limited life period [36].

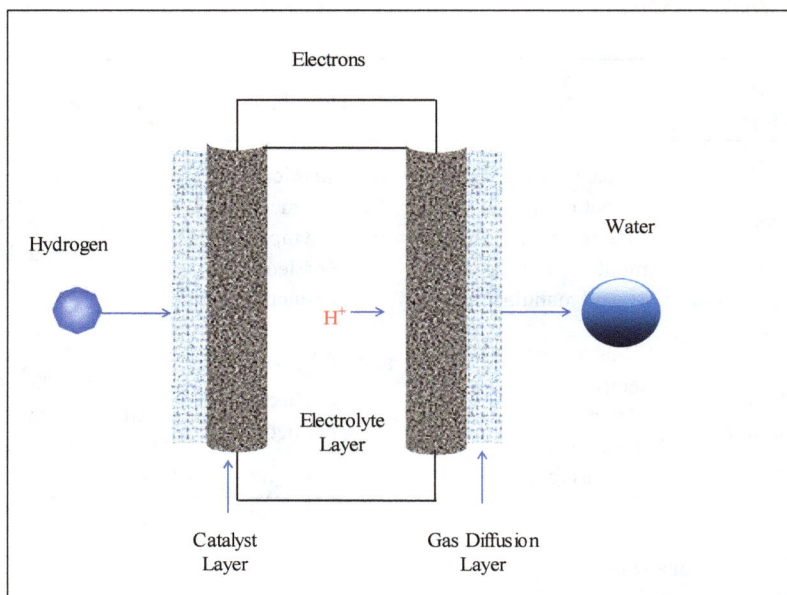

Figure 3. Detailed structure of Membrane Electrolyte Assembly.

Nafion is most commonly utilized membrane electrolyte assembly for Polymer Electrolyte Methanol fuel cell has numerous advantages such as electrochemical stability and excellent proton conductivity [37]. Different types of Nafion have been commonly utilized such as Nafion N110, N112, N117, N115, NR 211 and 212 and detailed technical characteristics of Nafion is listed in Table 3.

Table 3.Detailed characteristics of Nafion.

Nafion Types	Equivalent Weight	Dry Thickness	R ($\Omega \cdot cm^2$)	E_0	Exchange Current Density (mV/dec)
Nafion 112	1100	50	0.11	0.99	56.0
Nafion 115	1100	125	0.26	1.01	62.0
Nafion 117	110	175	0.33	1.02	68.0

The membrane conductivity of Nafion can be measured using steady-state I-V measurement and ac impedance spectroscopy. The conductivity of the membrane has a linear relation concerning membrane thickness [38]. The Nafion NR-211due to its tinier membrane nature demonstrated superior characteristics compared to other Nafion membranes. Even though Nafion membranes have several advantages the performance under high temperature is not satisfactory due to poor ion conductivity also the cost of production is not feasible. To overcome these limitations, newly modified membranes such as polyvinylidene fluoride, polyether ether ketone, polypropylenesulfone, polyphenyloxide, polyimide, and polybenzimidazole were developed [38-41].

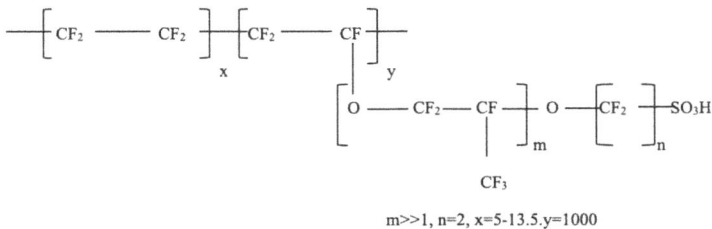

$m>>1, n=2, x=5-13.5.y=1000$

Figure 4. Chemical Structure of Nafion.

In catalyst, platinum is the prime choice of electrocatalyst for Polymer Electrolyte Methanol fuel cell [43]. The cost of platinum always has a negative side in the final manufacturing and production cost for Polymer Electrolyte Methanol fuel cells. Various studies have been carried out to reduce the cost and to discover other alternative options [44–47]. Several new modified catalysts such as platinum-monolayer catalysts, platinum-alloy catalysts, core-shell structured platinum-based catalysts, hollow platinum-based Nanocatalysts and non-noble catalysts [48–51]. The membrane electrolyte assembly consists of two main catalytic layers where both hydrogen oxidation and oxygen reduction reaction takes place. The hydrogen oxidation reaction occurring at the anode surface is intrinsically quicker while the oxygen reduction reaction occurring at the cathode surface is sluggish. High concentration of platinum required at cathode end. The presence of carbon monoxide results in catalyst poisoning at anode end significantly drops the performance of Polymer Electrolyte Methanol fuel cell [52]. The catalyst poisoning can be controlled by supplying a pure fuel source, utilization of carbon monoxide tolerant catalyst and operating at high temperature. Carbon monoxide tolerant catalysts have been developed to overcome complications of Polymer Electrolyte Methanol fuel cells. Platinum-based mixtures such as platinum-cobalt, platinum-ruthenium, platinum-molybdenum, platinum-tungsten, platinum-nickel and platinum-iron [46,52]. The platinum-based catalyst has certain limitations such as gradual degradation, sensitivity contaminants, and capital costs. In bifunctional mechanism these metal-OH species acts as oxygen source required for the oxidation of adsorbed CO to CO_2, liberating active platinum sites on the surface of the catalyst material where the oxidation and adsorption of the gaseous hydrogen occur [54]. During the methanol oxidation the CO gets generated as an intermediate, CO gets adsorbed on the platinum surface results in performance degradation. In the Direct methanol fuel cell Pt-Ru used as catalyst instead of platinum. The CO oxidation gets enhanced by the second metal. Stable $Ru(OH)_{ads}$ are generated by water. When the $Ru(OH)_{ads}$ react with an adjacent $PtCO_{ads,}$ CO is oxidized to CO_2. The methanol can be oxidized by the free Platinum surface.

$$Pt + CH_3OH \longrightarrow PtCO_{ads} + 4H^+ + 4e^-$$

$$Ru + H_2O \longrightarrow Ru(OH)_{ads} + H^+ + e^-$$

$$PtCO_{ad} + Ru(OH)_{ads} \longrightarrow CO_2 + Pt + Ru + H^+ + e^-$$

Table 4. Platinum-based catalyst advantages and drawbacks.

Features	Platinum-Based	Non-Platinum based	ModifiedPlatinum-Based
Advantages	Superior characteristics, Better fuel cell reactions	More availability, Cheaper	Cheaper, Better durability,
Drawbacks	High Cost and Stability Issues	Performance issue for long-term	Performance issues

Table 5. Lists platinum-based carbon and non-carbon support catalysts.

Catalysts	Synthesis Technique	Performance and Durability	Fuel Cell	References
Platinum/Carbon Nanotubes	Reduction and Wet oxidation	Accelerated durability test: Pt/CNT- ESA decreases 26.1%.	PEMFC	[55]
Pt/MWCNT	Ethylene glycol reduction	Loss of Platinum surface (MWCNT) is 37% and Corrosion only 5.7%.	PEMFC	[56]
Pt/NCNT	In situ reduction	Specific activity is 0.141 vs 0.150 mA cm^{-2}, which is four times higher than Pt/C.	PEMFC	[57]
Pt/NG	In situ one-step method	Specific oxidation current 41 mA mg^{-1}.	DMFC	[58]
Pt/Boron-doped graphene	Modified polyol method	3.3 times higher performance compared to Pt/G, the maximum power density of 565 mW cm^{-2}.	PEMFC	[59]
Pt nanowire/SG	Solvothermal	The high specific activity of 0.675 mA cm^{-2}, 3000 cycles.	PEMFC	[60]

PtRu/SWCNT	Single-pot	Power density of 45 mW cm^{-2}; methanol oxidation current is 50–60 mA cm^{-2}.	DMFC	[61]
PtRu/BDD	Sequential electrodeposition	Maximum current densities for methanol oxidation were in the range of 190 mA cm^{-2}.	DMFC	[62]
Pt/G/CB	Ethylene glycol reduction	65% improvement in the performance	PEMFC	[63]
Pt/SiC/C	Ethylene glycol reduction	ESA value of 48 m2/g	PEMFC	[64]
Pt/TiO$_2$/C	Phase transfer synthesis	Enhanced double layer capacitance in the electrode; Improvised ORR current density.	PEMFC	[65]
Pt/Ir-IrO$_2$	Microwave-assisted polyol	Highest ECSA 24.74 m^{2}g^{-1}; Improvised ORR current density	URFC	[66]

The diffusion layer is sandwiched between the flow field plate, and the catalyst layer consists of the isolated macroporous layer. The macroporous layer fabricated either using metal or carbon-based materials. Carbon-based materials are having some advantages such as better stability under acidic environment, flexibility, better electrical conductivity, and high gas permeability. Carbon cloth and paper are the common forms of utilizing carbon-based materials in Polymer Electrolyte Methanol fuel cells. On the other hand, the metal-based macroporous layer is utilized in the form of micro-machined metal substrate, metal form and metal mesh gained attention due to its high stability and superior mechanical strength. Performance analysis was carried out on different materials regarding the suitability of the diffusion layer; stainless steel wire cloth demonstrated better performance compared to carbon paper and cloth [67–70]. The stainless-steel wire cloth demonstrated excellent electronic conductivity and mobility of carbon dioxide and methanol from the catalytic layer [57,58].

Hydrophobic treatment is necessary to be carried out in the macroporous layer to aids mobility of oxygen and to prevent water flooding. Main hydrophobic agents are fluorinated ethylene propylene, polytetrafluoroethylene and polyvinylidenefluoride are commonly employed. The performance of Polymer Electrolyte Methanol fuel cells with hydrophobic agents ranges between 10 to 15% [73].

3.2   Principle and working process

The working principle of Polymer Electrolyte Methanol fuel cells operates by supplying a fuel mixture of methanol and water on the anode side, and the air is supplied over to the cathode side. During the methanol oxidation process, carbon dioxide, electrons, and protons are produced at the anode end. The generated electrons and protons shifted via polymer member electrolyte and an external electric circuit. The carbon dioxide gets eradicated from the catalytic layer while water produced at the cathode end. The overall reaction of Polymer Electrolyte Methanol fuel cells at the anode, cathode end is explained below:

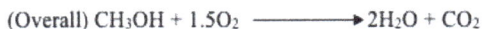

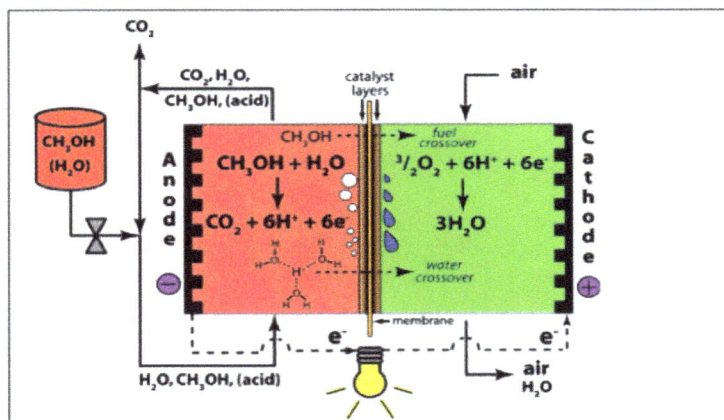

(Anode) $CH_3OH + H_2O \longrightarrow 6H^+ + 6e^- + CO_2$

(Cathode) $1.5O_2 + 6H^+ + 6e^- \longrightarrow 3H_2O$

(Overall) $CH_3OH + 1.5O_2 \longrightarrow 2H_2O + CO_2$

Figure 5. Working Principle of Polymer Electrolyte Methanol fuel cell [74].

Nanomaterials for Alcohol Fuel Cells   Materials Research Forum LLC
Materials Research Foundations **49** (2019) 193-230   doi: https://doi.org/10.21741/9781644900192-7

Methanol crossover is one of the major drawbacks for the conventional direct methanol fuel cell. The methanol crossover results in mixing potential at cathode end and further declines the fuel efficiency. Due to these factors, the overall power output and open circuit voltage get reduced. To overcome these issues regarding methanol crossover thicker membranes have been employed in Polymer Electrolyte Membrane Methanol fuel cells. To enhance performance, optimised methanol feeding is supplied with a low concentration methanol solution (1-2 mol/L) at the anode end. Studies were carried out by supplying vapor phase methanol in a fuel cell, and the vapor phase observed less methanol crossover compared to the direct application of liquid methanol [75–78]. The power density detected 34 mW/cm^2 much better compared to the direct utilization of liquid methanol.

$$j = DH\Delta C/L \ + C_2K\Delta P/\mu L + E_DMi/F$$

The methanol oxidation consists of several intermediates results in sluggishness in reaction. Carbon monoxide is one of the intermediates results in the methanol oxidation process, and the produced carbon monoxide gets absorbed on platinum.

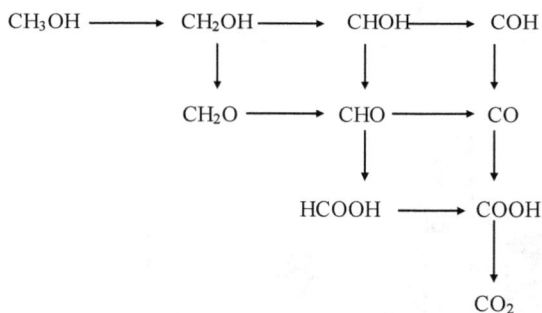

Figure 6. Oxidation of Methanol fuel.

This causes performance issues in methanol based fuel cell, and several investigations were carried out to find an alternative metal for platinum [79–82]. Studies reported the utilization of Pt-Ru alloy in methanol fuel cells and observed them to be more suitable compared to other metals.

The polymer electrolyte presents in methanol fuel cell also manages the cathode flooding, controls the degradation of the electrolyte membrane and also removes carbon dioxide. The performance of Polymer Electrolyte Methanol fuel cell depends on the reactant gas distribution and thermal management of the fuel cell system.

3.2.1　Flow field design

The flow field design is having a potential impact on the performance of Polymer Electrolyte Methanol fuel cells. In case the failure of uniform distribution of reactant gas in membrane electrolyte assembly cause hotspots, which affects the degradation of the material? Many parameters need to be taken care of while selecting the flow field design [83]. Various parameters such as shape, configuration, number of channels, channel depth, and width.

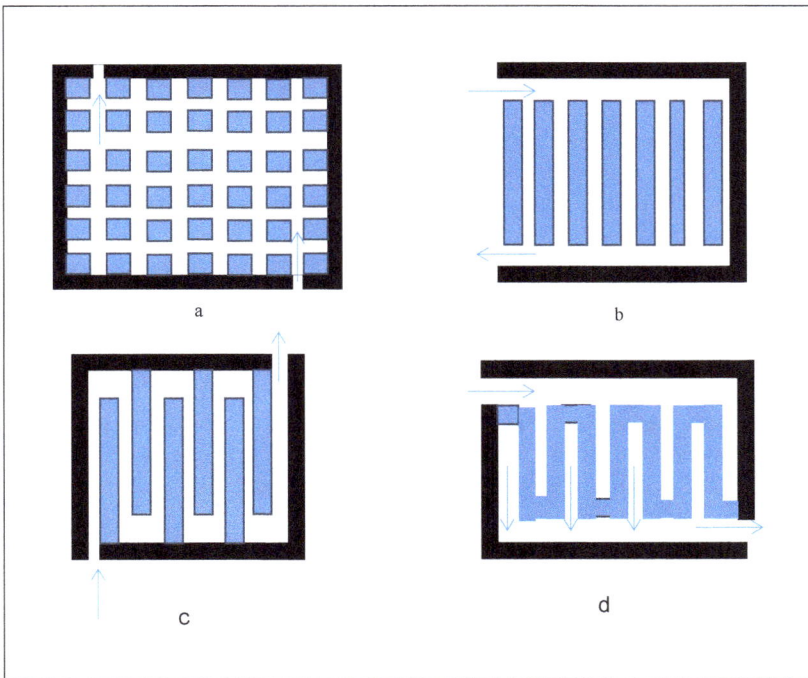

Figure 7. Schematic diagram of typical flow field design: (a) pin-type, (b) series-parallel, (c) serpentine and (d) interdigitated type.

The flow field design is mainly categorized into two types; interdigitated channel and open channel [84–86]. Open channel design consists of the cascade, pin type, serpentine and straight – parallel flow fields. Multiple paths present in pin and parallel type flow designs are effective models for the effective distribution of low feed pressure reactants on membrane electrolyte assembly. As a result, the parasitic power loses associated comparatively lower. The presence of water drops in the channel could cause asymmetric flow resistance, and non-uniform distribution can be induced to flow the reactants along the path of low resistance preferentially. However, for the parallel flow design, the removal of blockage due to lack of sufficient pressure is a big issue. In single serpentine design, the high-pressure presence in a single flow channel helps to remove the water droplets, but pumping loss is induced. New multiple serpentine channels were designed by combining the advantages of both parallel and single channels. Several investigations were carried out to further optimize various parameters and geometries for multiple serpentine channels [87–90].

Another main classification of flow field design is interdigitated flow field which indirectly connects both inlet and outlet channels. The diffusion layer influences the reactant flow via convection principle. The interdigitated flow channel improvises the utilization of the catalyst by efficient removal of water present in the diffusion layer. However, the main drawback is the high-power consumption. The current distribution studies were carried out in single serpentine and standard interdigitated channel. Studies proved that the interdigitated flow channel was observed to be more uniform [91]. New bioinspired interdigitated flow design demonstrated improved pressure distribution, improvised the power density by 30%, decrease in pressure loss as compared to both standard and single interdigitated designs [92].

3.2.2 Thermal management

Heat often generated depends on four major factors. The main factors include ohmic resistance, water condensation, heat generated due to electrochemical reactions and entropic heat of reactions [79,80]. The overall heat generation rate produced in a single cell can be estimated using the following equation;

$$q'' = (E_{tn} - E_{cell}) \times i$$

I, E_{cell} and E_{tn} expressed as thermos neutral or thermal voltage denotes imaginary maximum voltage for fuel cell systems considering various factors such as current density, operating voltage and enthalpy change of reactions. As the current density increases the rate of heat generation also rises correspondingly. When the current density is high the rate of heat generation is also at a higher level. This requires the necessity of adequate cooling systems to control the temperature of fuel cell stacks while operating at high current density. Experimental studies were carried out to identify the thermal characteristics of each component.

Table 6. Thermal and chemical properties of different components.

Properties	Flow field Plate	Diffusion Layer	Cathode Layer	Polymer Electrolyte Membrane
Chemical Composition	Graphite	Consists of a carbon fiber substrate and a microporous layer	A mixture of ionomer materials, carbon Nano powders, and Platinum nanoparticles	PFSA Membrane
Heat Conductivity	30	0.22	0.27	0.2

The heat conduction has a significant influence over the heat transfer mechanism in the polymer electrolyte membrane, while both convection and conduction contributes to the heat transfer mechanism in both diffusion and conduction layer [95]. Effective cooling plays a vital role in the performance of a Polymer Electrolyte Methanol fuel cell. Various cooling techniques and strategies have been developed and applied. Table 6. lists various cooling techniques [96].

Table 7. List of cooling techniques utilized in fuel cells.

Cooling Technique	Description	Advantages	Disadvantages
Edge Cooling or Heat Spreaders	Utilizing heat pipes for heat spreading	Good Thermal conductivity, Simple design, Less parasitic power	Integration of heat pipes in flow field plates Small low weight and thickness
	Utilizing better thermal conductive material	No need for separate internal coolant, Simple Design	Limited availability of materials
Liquid Cooling	Channels Integrated in BPPs	Better cooling capability, Simple Design, Flexible control	Requires more parasitic power, Coolant degradation
Airflow Cooling	Separate channels required for cooling	Less parasitic power, Simple Design	Performance issues
Phase Changing Cooling	Cooling through boiling	Simple design, No need for separate coolant pump	Flow Instability
	Evaporation Cooling	Simple Design, Internal humidification, and cooling	Thermal Instability, Issue in dynamic control

4. Applications

The Polymer Electrolyte Membrane Methanol fuel cell has shown a wide range of applications in various areas such as mobile phones, power generators and in public transports due to its potential in generating high power from the mW-100 kW range [97]. Various applications of Polymer Electrolyte Membrane Methanol fuel cells for various sectors are explained below.

4.1 Transportation industry

In the past, few decades lot of research been carried out in the area of sustainable transportation to minimize the over dependency of fossil fuels and to reduce its harmful emissions towards the environment [98–100]. To meet these requirements, the transportation industry started research on replacement of traditional IC engines over to electric and hybrid engines. The fuel cell based automobile research stated in the 1970s and made a lot of advances in recent years [101,102]. The main challenges for utilization of the fuel cell in automobiles are freeze start, cost, and durability.

4.1.1 Light weight automobiles

Hence identifying suitable PEMFC technology is always a high demand in the energy sector. Top automobile brands such as Honda, Hyundai, Nissan, Toyota, and Ford are using PEMFC technology due to the high potential impact on the transportation sector by reducing system cost and enhancing system durability.

4.1.2 Heavy automobiles

In the heavyweight category, most of the successful models are released in Buses. Daimlers made a revolution in Europe continent with the launch of 33 fuel cell based buses [103]. Seeing its success in the European continent, many other nations from Asia and African continents also launched fuel cell based vehicles for transportation. For electric buses, the fuel cell stacks were designed to supply the power of 250 kW requirement of hydrogen storage [104]. Recently, developing countries like India and China as the policy of reducing emissions started introducing electric vehicles for public transportation [105].

4.1.3 Other automobiles

The application of polymer electrolyte membrane based fuel cells is not limited to buses and light automobiles. The polymer electrolyte membrane based fuel cells also employed for weight lifting vehicles such as forklifts, electric powered bicycles, yachts, unmanned ariel vehicles, Locomotives, etc. [102]. Recently Alstom has successfully tested a fuel

Materials Research Forum LLC
doi: https://doi.org/10.21741/9781644900192-7

cell based locomotive in Germany it was reported to reach the speed of over 80 km/h [106–108]. There were also reported successful testing of polymer electrolyte membrane based fuel cells in yachts and submarine engines [93,94]. Table. 8 lists various industries associated with the transportation industry.

Table 8. Lists of various industries associated with the transportation industry.

Manufacturer	Product Details	Location	References
Volvo	APU	Sweden	[110]
Hyundai	SUV automobiles: Hyundai Nexo; Hyundai Tucson FCEV; Fuel Cell Bus	South Korea	[111]
Honda	Automobiles: Honda FCX Clarity; Honda Clarity Fuel cell	Japan	[112]
Toyota	Automobiles: Toyota Mirai; Honda Clarity Fuel cell; Toyota FCHV bus	Japan	[113]
Daimler	Bus: Mercedes-Benz Citaro fuel cell bus;Mercedes-Benz Sprinter fuel cell van	Germany	[114]
Tata Motors	Bus: Tata Star Bus Fuel cell	India	[115]
Skoda	Bus: SkodaTriHybus	Germany	[116]
Howaldtswerke-Deutsche Werft	Vessels and Submarines: Type 212 and 214 Submarines for German and Italian Navy	Germany	[117]
Yamaha	Motorcycle: Yamaha FC	Japan	[118]

4.2   Portable applications

The longer duration for charging and low energy capabilities are the main issues for portable energy storage devices. These issues are resolved by the utilization of portable polymer electrolyte membrane based fuel cells. Globally the production of fuel cell been inclined, in which polymer electrolyte membrane based methanol fuel cell production been increased [119]. The portable electronic devices power ranges from 5-50W appropriate for several applications from micro to the mega range [120]. Companies such as Sony, Toshiba use polymer electrolyte membrane methanol fuel cells for recharging laptops, cell phones. Samsung DSI company use polymer electrolyte methanol fuel cells as a battery for military applications with shows 800% durability and generate 54% more power [121].

Table 9. The list of major manufactures associated with portable polymer electrolyte membrane based fuel cells.

Manufacturer	Product Details	Location	References
Sony	Methanol based fuel cell unit for Mobile and Laptop devices	Japan	[122]
Toshiba	10 W Methanol fuel cell charging unit for telecommunication devices	Japan	[123]
Samsung DSI	Direct Methanol Fuel cell stacks for Military Application.	Korea	[124]
Horizon	Polymer Electrolyte Fuel cell based power source for gadgets and toys	China	[125]
CMR fuel cells	25 W hybrid methanol-based fuel cell for Laptop	UK	[126]
Jadoo Power System	100 W portable electric power supply for aeromedical evacuation applications	USA	[121]
Neah power systems	Both Residential and Industrial application fuel cell systems	USA	[127]

4.3   Stationary applications

The PEMFC power system is to distribute power for household purposes at ranges less than 10kW and 25-50kW for decentralized uses. Both fuel cell and ICEs can utilize for decentralized power generation. The fuel cell has several advantages compared to ICEs such as low noise, high power conversion efficiency, easy scale up and zero emissions [128]. PEMFC can act as a portable power resource in delivering uninterrupted power supply for electronic devices such as mobile phones, laptops, and radio communication in military applications. Polymer electrolyte methanol fuel cells are used as a potential power source for the diverse applications in transportation, power backup, and portable devices because of an easy repository of liquid methanol and can be used as a gasoline substitute [129]. Polymer electrolyte methanol fuel cells can support small portable devices for a lifetime of over 2000h. The back-up power required for various sectors such as banks, telecom companies, and hospitals requires an uninterrupted power supply for smooth operations [129–131].

Table 10. The list of major industries associated with a stationary fuel cell provider.

Manufacturer	Product Details	Location	References
Ballard Power Systems	Manufactures heavy-duty fuel cell modules, PEM fuel cell stacks, and Backup Power systems	Canada	[132]
Altergy	PEM fuel cell stacks for backup power systems	USA	[133]
Bloom Energy	Fuel cell stacks for backup power systems.	USA	[134]
NuveraFuel Cells	Supplying Fuel cells for off-grid and back up supply.	USA	[135]
SFC Energy	Off-grid power supply with EFOY Pro - stationary and mobile	Germany	[136]
Toshiba	Residential Fuel Cell system to support cogeneration	Japan	[137]
Acumentrics	Both Residential and Industrial application fuel cell systems; UPS	USA	[138]
Viaspace	Methanol based fuel cell cartridges manufacturing unit	USA	[139]
Electrochem	Off-grid power supply fuel cell for - stationary and mobile	USA	[140]
Matsushita Electric Industrial Company	Both Residential application fuel cell production	Japan	[141]

Conclusion

In summary, polymer electrolyte membrane methanol fuel cell has shown a wide range of applications in various areas such as mobile phones, power generators and in public transports due to its potential in generating high power from the mW-100 kW range. Continuous efforts have been carried out for commercialization aspect and industrial production for polymer electrolyte membrane fuel cells all over the world. Nevertheless, the high production cost, as well as the durability issue, remains the main challenge in the commercialization aspect. Moreover, the occurrence of contamination in membrane electrode assembly, water management system, power, and thermal cycling appeared as the main hurdles in the path of the commercialization, which affects their life span significantly.

Acknowledgments

The authors would like to thank the University of Malaya for funding this research work under the University Malaya Research Grant (RP045B-17AET) and University Malaya Research Fund Assistance (BK096-2016).

References

[1] Fuel Cell Benefits - Fuel Cell Today, (n.d.). http://www.fuelcelltoday.com/about-fuel-cells/benefits (accessed February 15, 2019)

[2] C.M. Branco, S. Sharma, M.M. de Camargo Forte, R. Steinberger-Wilckens, New approaches towards novel composite and multilayer membranes for intermediate temperature-polymer electrolyte fuel cells and direct methanol fuel cells, J. Power Sources. 316 (2016) 139–159. https://doi.org/10.1016/j.jpowsour.2016.03.052

[3] M. Zakeri, E. Abouzari-Lotf, M.M. Nasef, A. Ahmad, M. Miyake, T.M. Ting, P. Sithambaranathan, Fabrication and characterization of supported dual acidic ionic liquids for polymer electrolyte membrane fuel cell applications, Arab. J. Chem. (2018). https://doi.org/10.1016/j.arabjc.2018.05.010

[4] P. Prapainainar, S. Maliwan, K. Sarakham, Z. Du, C. Prapainainar, S.M. Holmes, P. Kongkachuichay, Homogeneous polymer/filler composite membrane by spraying method for enhanced direct methanol fuel cell performance, Int. J. Hydrogen Energy. 43 (2018) 14675–14690. https://doi.org/10.1016/j.ijhydene.2018.05.173

[5] S.R. Yang, S.K. Kim, D.H. Jung, T. Kim, H.S. Kim, D.H. Peck, Effects of ethanol in methanol fuel on the performance of membrane electrode assemblies for direct

methanol fuel cells, J. Ind. Eng. Chem. 66 (2018) 100–106.
https://doi.org/10.1016/j.jiec.2018.05.018

[6] K. Kwon, D. Kim, Polymer electrolyte membrane and methanol fuel cell, in: nanostructured polym. membr., John Wiley & Sons, Inc., Hoboken, NJ, USA, 2016: pp. 209–249. https://doi.org/10.1002/9781118831823.ch5

[7] H. Tsuchiya, Mass production cost of PEM fuel cell by learning curve, Int. J. Hydrogen Energy. 29 (2004) 985–990. https://doi.org/10.1016/j.ijhydene.2003.10.011

[8] Polymer Electrolyte Membrane Fuel Cells Market | Growth, Trends, and Forecast (2018 - 2023), (n.d.)

[9] DOE, B.M. Institute, Manufacturing Cost Analysis of Polymer Electrolyte Membrane (PEM) Fuel Cell Systems for Material Handling Applications, DOE Rep. (2017) 1–137

[10] M. Sgroi, F. Zedde, O. Barbera, A. Stassi, D. Sebastián, F. Lufrano, V. Baglio, A. Aricò, J. Bonde, M. Schuster, Cost Analysis of Direct Methanol Fuel Cell Stacks for Mass Production, Energies. 9 (2016) 1008. https://doi.org/10.3390/en9121008

[11] H. Junoh, J. Jaafar, M.N.A. Mohd Norddin, A.F. Ismail, M.H.D. Othman, M.A. Rahman, N. Yusof, W.N. Wan Salleh, H. Ilbeygi, A Review on the Fabrication of Electrospun Polymer Electrolyte Membrane for Direct Methanol Fuel Cell, J. Nanomater. 2015 (2015) 1–16. https://doi.org/10.1155/2015/690965

[12] W. He, G. Lin, T. Van Nguyen, Diagnostic tool to detect electrode flooding in proton-exchange-membrane fuel cells, AIChE J. 49 (2003) 3221–3228. https://doi.org/10.1002/aic.690491221

[13] W. He, J.S. Yi, T. Van Nguyen, Two-phase flow model of the cathode of PEM fuel cells using interdigitated flow fields, AIChE J. 46 (2000) 2053–2064. https://doi.org/10.1002/aic.690461016

[14] H. Li, C. Song, J. Zhang, J. Zhang, Catalyst contamination in PEM fuel cells, in: PEM Fuel Cell Electrocatal. Catal. Layers Fundam. Appl., Springer London, London, 2008, pp. 331–354. https://doi.org/10.1007/978-1-84800-936-3_6

[15] R.M. Darling, J.P. Meyers, Mathematical Model of Platinum Movement in PEM Fuel Cells, J. Electrochem. Soc. 152 (2005) A242–A247. https://doi.org/10.1149/1.1836156

[16] K.A. Page, B.W. Rowe, An Overview of Polymer Electrolyte Membranes for Fuel Cell Applications, in: ACS Symp. Ser., 2012: pp. 147–164. https://doi.org/10.1021/bk-2012-1096.ch009

[17] I. Dinçer, C. Zamfirescu, Sustainable energy systems and applications, in: Springer, Springer, 2011: p. 816

[18] J.H. Wee, Applications of proton exchange membrane fuel cell systems, Renew. Sustain. Energy Rev. 11 (2007) 1720–1738. https://doi.org/10.1016/j.rser.2006.01.005

[19] S. Satyapal, Hydrogen and Fuel Cells Overview, U.S. Dep. Energy Fuel Cell Technol. Off. (2017) 39

[20] American Institute of Chemical Engineers., 2004 AIChE Spring National Meeting : conference proceedings, New Orleans.04., in: AIChE, New York :, 2004

[21] R. Sood, S. Cavaliere, D.J. Jones, J. Rozière, Electrospun nanofibre composite polymer electrolyte fuel cell and electrolysis membranes, Nano Energy. 26 (2016) 729–745. https://doi.org/10.1016/j.nanoen.2016.06.027

[22] T. Maiyalagan, V.S. Saji, Electrocatalysts for low temperature fuel cells : fundamentals and recent trends, 2017

[23] I. Nitta, Inhomogeneous Compression of pemfc gas diffusion layers, Helsinki University of TechnologyFaculty of Information and Natural Sciences, 2008

[24] L. Cindrella, A.M. Kannan, J.F. Lin, K. Saminathan, Y. Ho, C.W. Lin, J. Wertz, Gas diffusion layer for proton exchange membrane fuel cells—A review, J. Power Sources. 194 (2009) 146–160. https://doi.org/10.1016/j.jpowsour.2009.04.005

[25] G. Sivasubramanian, K. Hariharasubramanian, P. Deivanayagam, J. Ramaswamy, High-performance SPEEK/SWCNT/fly ash polymer electrolyte nanocomposite membranes for fuel cell applications, Polym. J. 49 (2017) 703–709. https://doi.org/10.1038/pj.2017.38

[26] V.M.O. Martínez, M.J.S. García, F.J.H. Fernández, A. Pérez de los Ríos, Organic–inorganic membranes impregnated with ionic liquid, in: Org. Compos. Polym. Electrolyte Membr., Springer International Publishing, Cham, (2017)1–23. https://doi.org/10.1007/978-3-319-52739-0_1

[27] Colleen Spiegel, Low-temperature fuel cell membrane electrode assembly processing techniques, (2017)

[28] M. Aliofkhazraei, N. Ali, W.I. (William I.. Milne, C.S. Ozkan, S. Mitura, J.L. Gervasoni, Graphene science handbook. Applications and industrialization, 2016

[29] A. Kraytsberg, Y. Ein-Eli, Review of advanced materials for proton exchange membrane fuel cells, Energy & Fuels. 28 (2014) 7303–7330. https://doi.org/10.1021/ef501977k

[30] C. Wang, S. Wang, L. Peng, J. Zhang, Z. Shao, J. Huang, C. Sun, M. Ouyang, X. He, Recent progress on the key materials and components for proton exchange membrane fuel cells in vehicle applications, Energies. 9 (2016) 603. https://doi.org/10.3390/en9080603

[31] Rafi-Ud-Din, M. Arshad, A. Saleem, M. Shahzad, T. Subhani, S. Hussain, Fabrication and characterization of bipolar plates of vinyl ester resin/graphite-based composite for polymer electrolyte membrane fuel cells, J. Thermoplast. Compos. Mater. 29 (2016) 1315–1331. https://doi.org/10.1177/0892705714563124

[32] B. Jiang, C. Li, The Synthesis and the Catalytic properties of graphene-based composite materials, in: carbon-related mater. recognit. Nobel Lect. by Prof. Akira Suzuki ICCE, Springer International Publishing, Cham, 2017: pp. 3–26. https://doi.org/10.1007/978-3-319-61651-3_1

[33] A. Masand, M. Borah, A.K. Pathak, S.R. Dhakate, Effect of filler content on the properties of expanded- graphite-based composite bipolar plates for application in polymer electrolyte membrane fuel cells, Mater. Res. Express. 4 (2017) 095604. https://doi.org/10.1088/2053-1591/aa85a5

[34] P. Lettenmeier, R. Wang, R. Abouatallah, B. Saruhan, O. Freitag, P. Gazdzicki, T. Morawietz, R. Hiesgen, A.S. Gago, K.A. Friedrich, Low-cost and durable bipolar plates for proton exchange membrane electrolyzers, Sci. Rep. 7 (2017) 44035. https://doi.org/10.1038/srep44035

[35] H. Kahraman, I. Cevik, F. Dündar, F. Ficici, The Corrosion resistance behaviors of metallic bipolar plates for pemfc coated with physical vapor deposition (PVD): an experimental study, Arab. J. Sci. Eng. 41 (2016) 1961–1968. https://doi.org/10.1007/s13369-016-2058-x

[36] J. Jang, C. Choi, J. Kim, Y.-D. Park, N. Kang, Y.S. Choi, D.G. Nam, Surface characterization of chromium nitrided low carbon steel as bipolar plate for polymer electrolyte membrane fuel cell, Sci. Adv. Mater. 10 (2018) 206–209. https://doi.org/10.1166/sam.2018.2952

[37] Y. Wang, S. Zhang, Z. Lu, P. Wang, X. Ji, W. Li, Preparation and performance of electrically conductive Nb-doped TiO_2/polyaniline bilayer coating for 316L stainless steel bipolar plates of proton-exchange membrane fuel cells, RSC Adv. 8 (2018) 19426–19431. https://doi.org/10.1039/C8RA02161A

[38] S. Jang, Y.G. Yoon, Y.S. Lee, Y.W. Choi, One-step fabrication and characterization of reinforced microcomposite membranes for polymer electrolyte membrane fuel cells, J. Memb. Sci. 563 (2018) 896–902. https://doi.org/10.1016/J.MEMSCI.2018.06.060

[39] K. Pourzare, Y. Mansourpanah, S. Farhadi, Advanced nanocomposite membranes for fuel cell applications: a comprehensive review, Biofuel Res. J. 3 (2016) 496–513. https://doi.org/10.18331/BRJ2016.3.4.4

[40] Sigma-Aldrich, Nafion® perfluorinated membrane, Online Cat. (2016) 1

[41] M. Bodner, B. Cermenek, M. Rami, V. Hacker, The Effect of platinum electrocatalyst on membrane degradation in polymer electrolyte fuel cells, Membranes (Basel). 5 (2015) 888–902. https://doi.org/10.3390/membranes5040888

[42] X. Li, Y. Song, Z. Liu, P. Feng, S. Liu, Y. Yu, Z. Jiang, B. Liu, Triple-layer sulfonated poly(ether ether ketone)/sulfonated polyimide membranes for fuel cell applications, High Perform. Polym. 26 (2014) 106–113. https://doi.org/10.1177/0954008313499803

[43] H. Yang, H. Wu, X. Shen, Y. Cao, Z. Li, Z. Jiang, Enhanced proton conductivity of proton exchange membrane at low humidity based on poly(methacrylic acid)-loaded imidazole microcapsules, RSC Adv. 5 (2015) 9079–9088. https://doi.org/10.1039/C4RA13616K

[44] S. Banerjee, Handbook of Specialty Fluorinated Polymers, Elsevier, 2015. https://doi.org/10.1016/C2014-0-01271-3

[45] H. Pu, Polymers for PEM Fuel Cells, John Wiley & Sons, Inc., Hoboken, New Jersey, New Jersey, 2014. https://doi.org/10.1002/9781118869345

[46] K.I. Ozoemena, Nanostructured platinum-free electrocatalysts in alkaline direct alcohol fuel cells: catalyst design, principles and applications, RSC Adv. 6 (2016) 89523–89550. https://doi.org/10.1039/C6RA15057H

[47] T. Asset, Hollow nanoparticles for low cost, high oxygen reduction reaction activity and durability for proton exchange membrane fuel cell application, Université de Liège, Liège, Belgique, 2017

[48] X. Deng, S. Yin, X. Wu, M. Sun, Z. Xie, Q. Huang, Synthesis of PtAu/TiO2 nanowires with carbon skin as highly active and highly stable electrocatalyst for oxygen reduction reaction, Electrochim. Acta. 283 (2018) 987–996. https://doi.org/10.1016/j.electacta.2018.06.139

[49] L. Cao, G. Zhang, W. Lu, X. Qin, Z. Shao, B. Yi, Preparation of hollow PtCu nanoparticles as high-performance electrocatalysts for oxygen reduction reaction in the absence of a surfactant, RSC Adv. 6 (2016) 39993–40001. https://doi.org/10.1039/C6RA04619C

[50] P. Chandran, A. Ghosh, S. Ramaprabhu, High-performance platinum-free oxygen reduction reaction and hydrogen oxidation reaction catalyst in polymer electrolyte membrane fuel cell, Sci. Rep. 8 (2018) 3591. https://doi.org/10.1038/s41598-018-22001-9

[51] S.M. Alia, C. Ngo, S. Shulda, M.-A. Ha, A.A. Dameron, J.N. Weker, K.C. Neyerlin, S.S. Kocha, S. Pylypenko, B.S. Pivovar, Exceptional oxygen reduction reaction activity and durability of platinum–nickel nanowires through synthesis and post-treatment optimization, ACS Omega. 2 (2017) 1408–1418. https://doi.org/10.1021/acsomega.7b00054

[52] H. Wang, R. Liu, Y. Li, X. Lü, Q. Wang, S. Zhao, K. Yuan, Z. Cui, X. Li, S. Xin, R. Zhang, M. Lei, Z. Lin, Durable and efficient hollow porous oxide spinel microspheres for oxygen reduction, Joule. 2 (2018) 337–348. https://doi.org/10.1016/j.joule.2017.11.016

[53] N. Lindahl, E. Zamburlini, L. Feng, H. Grönbeck, M. Escudero-Escribano, I.E.L. Stephens, I. Chorkendorff, C. Langhammer, B. Wickman, High specific and mass activity for the oxygen reduction reaction for thin film catalysts of sputtered Pt $_3$Y, Adv. Mater. Interfaces. 4 (2017) 1700311. https://doi.org/10.1002/admi.201700311

[54] S. Lee, S. Mukerjee, E. Ticianelli, J. McBreen, Electrocatalysis of CO tolerance in hydrogen oxidation reaction in PEM fuel cells, Electrochim. Acta. 44 (1999) 3283–3293. https://doi.org/10.1016/S0013-4686(99)00052-3

[55] Y. Shao, G. Yin, Y. Gao, P. Shi, Durability Study of Pt/C and Pt/CNTs catalysts under simulated pem fuel cell conditions, J. Electrochem. Soc. 153 (2006) A1093. https://doi.org/10.1149/1.2191147

[56] X. Wang, W. Li, Z. Chen, M. Waje, Y. Yan, Durability investigation of carbon nanotube as catalyst support for proton exchange membrane fuel cell, J. Power Sources. 158 (2006) 154–159. https://doi.org/10.1016/J.JPOWSOUR.2005.09.039

[57] L. Guo, W.J. Jiang, Y. Zhang, J.S. Hu, Z.D. Wei, L.J. Wan, Embedding Pt nanocrystals in N-doped porous carbon/carbon nanotubes toward highly stable electrocatalysts for the oxygen reduction reaction, ACS Catal. 5 (2015) 2903–2909. https://doi.org/10.1021/acscatal.5b00117

[58] L.S. Zhang, X.Q. Liang, W.G. Song, Z.Y. Wu, Identification of the nitrogen species on N-doped graphene layers and Pt/NG composite catalyst for direct methanol fuel cell, Phys. Chem. Chem. Phys. 12 (2010) 12055. https://doi.org/10.1039/c0cp00789g

[59] A. Pullamsetty, M. Subbiah, R. Sundara, Platinum on boron doped graphene as cathode electrocatalyst for proton exchange membrane fuel cells, Int. J. Hydrogen Energy. 40 (2015) 10251–10261. https://doi.org/10.1016/J.IJHYDENE.2015.06.020

[60] M.A. Hoque, F.M. Hassan, D. Higgins, J.-Y. Choi, M. Pritzker, S. Knights, S. Ye, Z. Chen, Multigrain Platinum nanowires consisting of oriented nanoparticles anchored on sulfur-doped graphene as a highly active and durable oxygen reduction electrocatalyst, Adv. Mater. 27 (2015) 1229–1234. https://doi.org/10.1002/adma.201404426

[61] G. Girishkumar, T.D. Hall, K. Vinodgopal, Prashant V. Kamat, Single wall carbon nanotube supports for portable direct methanol fuel cells, (2005). https://doi.org/10.1021/JP054764I

[62] I. González-González, C. Lorenzo-Medrano, C.R. Cabrera, Sequential electrodeposition of platinum-ruthenium at boron-doped diamond electrodes for methanol oxidation, Adv. Phys. Chem. 2011 (2011) 1–10. https://doi.org/10.1155/2011/679246

[63] S. Park, Y. Shao, H. Wan, P.C. Rieke, V. V. Viswanathan, S.A. Towne, L. V. Saraf, J. Liu, Y. Lin, Y. Wang, Design of graphene sheets-supported Pt catalyst layer in PEM fuel cells, Electrochem. Commun. 13 (2011) 258–261. https://doi.org/10.1016/J.ELECOM.2010.12.028

[64] H. Lv, S. Mu, N. Cheng, M. Pan, Nano-silicon carbide supported catalysts for PEM fuel cells with high electrochemical stability and improved performance by addition of carbon, Appl. Catal. B Environ. 100 (2010) 190–196. https://doi.org/10.1016/J.APCATB.2010.07.030

[65] S. von Kraemer, K. Wikander, G. Lindbergh, A. Lundblad, A.E.C. Palmqvist, Evaluation of TiO2 as catalyst support in Pt-TiO2/C composite cathodes for the proton exchange membrane fuel cell, J. Power Sources. 180 (2008) 185–190. https://doi.org/10.1016/J.JPOWSOUR.2008.02.023

[66] F.D. Kong, S. Zhang, G.P. Yin, N. Zhang, Z.B. Wang, C.Y. Du, Preparation of Pt/Irx(IrO2)10 − x bifunctional oxygen catalyst for unitized regenerative fuel cell, J. Power Sources. 210 (2012) 321–326. https://doi.org/10.1016/J.JPOWSOUR.2012.02.021

[67] D.Y. Chung, H. Kim, Y.-H. Chung, M.J. Lee, S.J. Yoo, A.D. Bokare, W. Choi, Y.E. Sung, Inhibition of CO poisoning on Pt catalyst coupled with the reduction of toxic hexavalent chromium in a dual-functional fuel cell., Sci. Rep. 4 (2014) 7450. https://doi.org/10.1038/srep07450

[68] Y. Nabae, S. Nagata, T. Hayakawa, H. Niwa, Y. Harada, M. Oshima, A. Isoda, A. Matsunaga, K. Tanaka, T. Aoki, Pt-free carbon-based fuel cell catalyst prepared from spherical polyimide for enhanced oxygen diffusion, Sci. Rep. 6 (2016) 23276. https://doi.org/10.1038/srep23276

[69] N.V. Long, C.M. Thi, Y. Yong, M. Nogami, M. Ohtaki, Platinum and palladium nano-structured catalysts for polymer electrolyte fuel cells and direct methanol fuel cells, J. Nanosci. Nanotechnol. 13 (2013) 4799–824

[70] S. Kaushal, P. Negi, A.K. Sahu, S.R. Dhakate, Upshot of natural graphite inclusion on the performance of porous conducting carbon fiber paper in a polymer electrolyte membrane fuel cell, Mater. Res. Express. 4 (2017) 095603. https://doi.org/10.1088/2053-1591/aa8517

[71] R.C.T. Slade, J.P. Kizewski, S.D. Poynton, R. Zeng, J.R. Varcoe, Alkaline membrane fuel cells, in: Fuel Cells, Springer New York, NY, 2013: pp. 9–29. https://doi.org/10.1007/978-1-4614-5785-5_2

[72] V.K. Mathur, J. Crawford, Fundamentals of gas diffusion layers in pem fuel cells, in: Recent Trends Fuel Cell Sci. Technol., Springer New York, NY, 2007: pp. 116–128. https://doi.org/10.1007/978-0-387-68815-2_4

[73] K.M. Tenny, V.S. Lakhanpal, R.P. Dowd, V. Yarlagadda, T. Van Nguyen, Impact of multi-walled carbon nanotube fabrication on carbon cloth electrodes for hydrogen-vanadium reversible fuel cells, J. Electrochem. Soc. 164 (2017) A2534–A2538. https://doi.org/10.1149/2.1151712jes

[74] A.A. Rashad, E.A. Rashad, A.A. Ali, E. Akram, M.M. Al-rubaye, E. Yousif, N. Hairunisa, Hydrogen in fuel cells : An overview of promotions and demotions, Interdiscip. J. Chem. 2 (2017) 1–6. https://doi.org/10.15761/IJC.1000119

[75] D.M. Zhang, L. Guo, L.T. Duan, Z.Y. Wang, Preparation of multi-layer film on stainless steel as bipolar plate for polymer electrolyte membrane fuel cell, Adv. Mater. Res. 113–116 (2010) 2255–2261. https://doi.org/10.4028/www.scientific.net/AMR.113-116.2255

[76] Y. Suzuki, M. Watanabe, T. Toda, T. Fujii, Development of electrically conductive DLC coated stainless steel separators for polymer electrolyte membrane fuel cell, J. Phys. Conf. Ser. 441 (2013) 012027. https://doi.org/10.1088/1742-6596/441/1/012027

[77] X. Chen, Z. Zhang, J. Shen, Z. Hu, Micro direct methanol fuel cell: functional components, supplies management, packaging technology and application, Int. J. Energy Res. 41 (2017) 613–627. https://doi.org/10.1002/er.3634

[78] X. Li, A. Faghri, Review and advances of direct methanol fuel cells (DMFCs) part I: Design, fabrication, and testing with high concentration methanol solutions, J. Power Sources. 226 (2013) 223–240. https://doi.org/10.1016/J.JPOWSOUR.2012.10.061

[79] M.A. Abdelkareem, N. Morohashi, N. Nakagawa, Factors affecting methanol transport in a passive DMFC employing a porous carbon plate, J. Power Sources. 172 (2007) 659–665. https://doi.org/10.1016/J.JPOWSOUR.2007.05.015

[80] S. Eccarius, F. Krause, K. Beard, C. Agert, Passively operated vapor-fed direct methanol fuel cells for portable applications, J. Power Sources. 182 (2008) 565–579. https://doi.org/10.1016/J.JPOWSOUR.2008.03.091

[81] V.S. Bagotzky, Y.B. Vassiliev, O.A. Khazova, Generalized scheme of chemisorption, electrooxidation and electroreduction of simple organic compounds on platinum group metals, J. Electroanal. Chem. Interfacial Electrochem. 81 (1977) 229–238. https://doi.org/10.1016/S0022-0728(77)80019-3

[82] G. Yang, Y. Sun, P. Lv, F. Zhen, X. Cao, X. Chen, Z. Wang, Z. Yuan, X. Kong, Preparation of Pt–Ru/C as an Oxygen-reduction electrocatalyst in microbial fuel cells for wastewater treatment, Catalysts. 6 (2016) 150. https://doi.org/10.3390/catal6100150

[83] K. Hengge, T. Gänsler, E. Pizzutilo, C. Heinzl, M. Beetz, K.J.J. Mayrhofer, C. Scheu, Accelerated fuel cell tests of anodic Pt/Ru catalyst via identical location TEM: New aspects of degradation behavior, Int. J. Hydrogen Energy. 42 (2017) 25359–25371. https://doi.org/10.1016/J.IJHYDENE.2017.08.108

[84] C. Li, R. Dai, R. Qi, X. Wu, J. Ma, Electrodeposition of Pt–Ru alloy electrocatalysts for direct methanol fuel cell, Int. J. Electrochem. Sci. 12 (2017) 2485–2494. https://doi.org/10.20964/2017.03.13

[85] D. Wu, K. Kusada, H. Kitagawa, Recent progress in the structure control of Pd-Ru bimetallic nanomaterials., Sci. Technol. Adv. Mater. 17 (2016) 583–596. https://doi.org/10.1080/14686996.2016.1221727

[86] M. Mehrpooya, M. Kheir Rouz, A. Nikfarjam, Optimum Design of the Flow-field channels and fabrication of a micro-pem fuel cell, Ind. Eng. Chem. Res. 54 (2015) 3640–3647. https://doi.org/10.1021/ie5049675

[87] L. Peng, P. Yi, X. Lai, Design and manufacturing of stainless steel bipolar plates for proton exchange membrane fuel cells, Int. J. Hydrogen Energy. 39 (2014) 21127–21153. https://doi.org/10.1016/j.ijhydene.2014.08.113

[88] K.I. Sainan, N. Arsyad, M.E. Salleh, F. Mohamad, Development of 1:1 and 2:1 channel ratio bipolar plates (BPPs) for PEMFC, Appl. Mech. Mater. 799-800(2015)799–800 105–109. https://doi.org/10.4028/www.scientific.net/AMM.799-800.105

[89] R. Zeis, Materials and characterization techniques for high-temperature polymer electrolyte membrane fuel cells, Beilstein J. Nanotechnol. 6 (2015) 68–83. https://doi.org/10.3762/bjnano.6.8

[90] J. Larminie, A. Dicks, Fuel Cell Systems Explained, John Wiley & Sons, Ltd, West Sussex, England, England, 2003. https://doi.org/10.1002/9781118878330

[91] M. Wen, K. He, P. Li, L. Yang, L. Deng, F. Jiang, Y. Yao, Optimization design of bipolar plate flow field in PEM stack, in: IOP Conf. Ser. Mater. Sci. Eng. (2017)012148. https://doi.org/10.1088/1757-899X/274/1/012148

[92] M. Kim, C. Kim, Y. Sohn, Application of metal foam as a flow field for PEM fuel cell stack, Fuel Cells. 18 (2018) 123–128. https://doi.org/10.1002/fuce.201700180

[93] E. Afshari, M. Ziaei-Rad, N. Jahantigh, Analytical and numerical study on cooling flow field designs performance of PEM fuel cell with variable heat flux, Mod. Phys. Lett. B. 30 (2016) 1650155–71. https://doi.org/10.1142/S0217984916501554

[94] T. Chen, S. Liu, S. Gong, C. Wu, Development of bipolar plates with different flow channel configurations based on plant vein for fuel cell, Int. J. Energy Res. 37 (2013) 1680–1688. https://doi.org/10.1002/er.3033

[95] G. Zhang, L. Guo, B. Ma, H. Liu, Comparison of current distributions in proton exchange membrane fuel cells with interdigitated and serpentine flow fields, J. Power Sources. 188 (2009) 213–219. https://doi.org/10.1016/J.JPOWSOUR.2008.10.074

[96] S.G. Kandlikar, Z. Lu, Thermal management issues in a PEMFC stack – A brief review of current status, Appl. Therm. Eng. 29 (2009) 1276–1280. https://doi.org/10.1016/J.APPLTHERMALENG.2008.05.009

[97] D. Watzenig, B. Brandstätter, Comprehensive energy management : safe adaptation, predictive control and thermal management, Springer, 2018

[98] A.S. Aricò, V. Baglio, F. Lufrano, A. Stassi, I. Gatto, V. Antonucci, L. Merlo, Modifications of sulfonic acid-based membranes, in: High Temp. Polym. Electrolyte Membr. Fuel Cells, Springer International Publishing, Cham, 2016, pp. 5–36. https://doi.org/10.1007/978-3-319-17082-4_2

[99] J.Y. Hwang, K.Y. Shin, S.H. Lee, K. Kang, H. Kang, J.H. Lee, D.H. Peck, D.H. Jung, J.H. Jang, Periodic fuel supply to a micro-DMFC using a piezoelectric linear

actuator, J. Micromechanics Microengineering. 20 (2010) 085023-30.
https://doi.org/10.1088/0960-1317/20/8/085023

[100] B. Andreaus, A.J. McEvoy, G.G. Scherer, Analysis of performance losses in
polymer electrolyte fuel cells at high current densities by impedance spectroscopy,
Electrochim. Acta. 47 (2002) 2223–2229. https://doi.org/10.1016/S0013-
4686(02)00059-2

[101] D. Wang, H.L. Xin, Y. Yu, H. Wang, E. Rus, D.A. Muller, H.D. Abruña, Pt-
Decorated PdCo@Pd/C core–shell nanoparticles with enhanced stability and
electrocatalytic activity for the oxygen reduction reaction, J. Am. Chem. Soc. 132
(2010) 17664–17666. https://doi.org/10.1021/ja107874u

[102] D. Nag, S.K. Paul, S. Saha, A.K. Goswami, Sustainability assessment for the
transportation environment of Darjeeling, India, J. Environ. Manage. 213 (2018) 489–
502. https://doi.org/10.1016/J.JENVMAN.2018.01.042

[103] D.L. Greene, Sustainable transportation, in: Int. Encycl. Soc. Behav. Sci. Second
Ed., Elsevier, 2015,pp. 845–849. https://doi.org/10.1016/B978-0-08-097086-8.91073-
0

[104] R. Gerike, C. Koszowski, Sustainable urban transportation, in: Encycl. Sustain.
Technol., Elsevier, 2017,pp. 379–391. https://doi.org/10.1016/B978-0-12-409548-
9.10176-9

[105] T. Lipman, D. Sperling, Market concepts, competing technologies and cost
challenges for automotive and stationary applications, in: Handb. Fuel Cells, John
Wiley & Sons, Ltd, Chichester, UK, UK, 2010,pp. 3345.
https://doi.org/10.1002/9780470974001.f313110

[106] Alstom fuel cell trains enter service in Germany, Fuel Cells Bull. 2018 (2018) 1.
https://doi.org/10.1016/S1464-2859(18)30310-9

[107] H.S. Das, C.W. Tan, A.H.M. Yatim, Fuel cell hybrid electric vehicles: A review on
power conditioning units and topologies, Renew. Sustain. Energy Rev. 76 (2017) 268–
291. https://doi.org/10.1016/j.rser.2017.03.056

[108] W.C. II, G. Cooke, Smart green cities: Toward a Carbon Neutral World, 2016

[109] K. Haraldsson, A. Folkesson, P. Alvfors, Fuel cell buses in the Stockholm CUTE
project—First experiences from a climate perspective, J. Power Sources. 145 (2005)
620–631. https://doi.org/10.1016/j.jpowsour.2004.12.081

[110] Statoil, Volvo link up to push truck fuel cell APUs, Fuel Cells Bull. 2005 (2005) 1.
https://doi.org/10.1016/S1464-2859(05)70663-5

[111] SGL, Hyundai expand cooperation in fuel cell components for NEXO, Fuel Cells Bull. 2018 (2018) 15. https://doi.org/10.1016/S1464-2859(18)30182-2

[112] Honda begins Clarity Fuel Cell deliveries in Europe, California, Fuel Cells Bull. 2017 (2017) 2. https://doi.org/10.1016/S1464-2859(17)30002-0

[113] Toyota tech in fuel cell buses for Caetanobus, Japan rail partnership, Fuel Cells Bull. 2018 (2018) 2. https://doi.org/10.1016/S1464-2859(18)30351-1

[114] Daimler launches next-generation hybrid fuel cell bus, Fuel Cells Bull. 2009 (2009) 2–3. https://doi.org/10.1016/S1464-2859(09)70239-1

[115] J.W. Richmond, A Tata motors perspective for sustainable transportation and the development of the Tata Vista EV, Innov. Fuel Econ. Sustain. Road Transp. (2011) 47–60. https://doi.org/10.1533/9780857095879.1.47

[116] P.Motor, Skoda electric launch first triple-hybrid fuel cell passenger bus, Fuel Cells Bull. 2009 (2009) 1. https://doi.org/10.1016/S1464-2859(09)70101-4

[117] M. Pein, Fuel cells ideal for demanding maritime applications, Fuel Cells Bull. 2012 (2012) 14–15. https://doi.org/10.1016/S1464-2859(12)70184-0

[118] Yamaha demos fuel cell scooters, links with Yuasa, Fuel Cells Bull. 2003 (2003) 2. https://doi.org/10.1016/S1464-2859(03)00904-0

[119] S. Zhang, Y. Zhang, J. Chen, C. Yin, X. Liu, Design, fabrication and performance evaluation of an integrated reformed methanol fuel cell for portable use, J. Power Sources. 389 (2018) 37–49. https://doi.org/10.1016/J.JPOWSOUR.2018.04.009

[120] Y.J. Sohn, G.G. Park, T.H. Yang, Y.G. Yoon, W.Y. Lee, S.D. Yim, C.S. Kim, Operating characteristics of an air-cooling PEMFC for portable applications, J. Power Sources. 145 (2005) 604–609. https://doi.org/10.1016/J.JPOWSOUR.2005.02.062

[121] Samsung SDI develops military portable DMFC, Fuel Cells Bull. 2009 (2009) 6–7. https://doi.org/10.1016/S1464-2859(09)70182-8

[122] NEC, Toshiba, and Sony developing ever-smaller fuel cells to replace batteries, Focus Catal. 2003 (2003) 2. https://doi.org/10.1016/S1351-4180(03)00003-5

[123] N.R.C. and N.A. of Engineering, The Hydrogen Economy, National Academies Press, Washington, D.C. 2004. https://doi.org/10.17226/10922

[124] A. Higier, L. Hsu, J. Oiler, A. Phipps, D. Hooper, M. Kerber, Polymer electrolyte fuel cell (PEMFC) based power system for long-term operation of leave-in-place sensors in Navy and Marine Corps applications, Int. J. Hydrogen Energy. 42 (2017) 4706–4709. https://doi.org/10.1016/J.IJHYDENE.2016.09.066

[125] Horizon Fuel Cell in global climate change education initiative, Fuel Cells Bull. 2012 (2012) 9. https://doi.org/10.1016/S1464-2859(12)70107-4

[126] Toshiba Launches Direct Methanol Fuel Cell in Japan as External Power Source for Mobile Electronic Devices, Green Car Congr. (2009)

[127] Neah Power offers fuel cell manufacturing license to customers, Fuel Cells Bull. 2011 (2011) 9. https://doi.org/10.1016/S1464-2859(11)70349-2

[128] Global Direct Methanol Fuel Cells Market Report 2016-2020 - Analysis, Technologies & Forecasts - Vendors: DuPont Fuel Cell, Hitachi, Panasonic - Research and Markets, BusinessWire. (2016)

[129] O.Z. Sharaf, M.F. Orhan, An overview of fuel cell technology: Fundamentals and applications, Renew. Sustain. Energy Rev. 32 (2014) 810–853. https://doi.org/10.1016/J.RSER.2014.01.012

[130] T. Wilberforce, A. Alaswad, A. Palumbo, M. Dassisti, Advances in stationary and portable fuel cell applications, Int. J. Hydrogen Energy. 41 (2016) 16509–16522. https://doi.org/10.1016/J.IJHYDENE.2016.02.057

[131] B. Sundén, J. Fu, B. Sundén, J. Fu, Fuel Cells, Heat Transf. Aerosp. Appl. (2017) 145–153. https://doi.org/10.1016/B978-0-12-809760-1.00008-9

[132] Ballard supplies stacks to Infintium to power materials handling at Mercedes-Benz in US, Fuel Cells Bull. 2019 (2019) 3. https://doi.org/10.1016/S1464-2859(19)30005-7

[133] 101 Telco Solutions offers fuel cell backup power from Altergy, Fuel Cells Bull. 2018 (2018) 7. https://doi.org/10.1016/S1464-2859(18)30365-1

[134] Z. Ma, J. Eichman, J. Kurtz, Fuel Cell Backup Power Unit Configuration and Electricity Market Participation: A Feasibility Study, Golden, CO (United States), 2017. https://doi.org/10.2172/1347197

[135] Nuvera fuel cell powers electrochemical industry, Fuel Cells Bull. 2006 (2006) 6. https://doi.org/10.1016/S1464-2859(06)71304-9

[136] SFC wins Bundeswehr order for vehicle-based and stationary power, Fuel Cells Bull. 2019 (2019) 12–13. https://doi.org/10.1016/S1464-2859(19)30029-X

[137] Toshiba hydrogen fuel cell system out to sea, builds hydrogen centre, Fuel Cells Bull. 2016 (2016) 4–5. https://doi.org/10.1016/S1464-2859(16)30342-X

[138] Acumentrics delivers 250+ SOFC units to remote power users, Fuel Cells Bull. 2015 (2015) 5. https://doi.org/10.1016/S1464-2859(15)30147-4

[139] Viaspace fuel cell cartridges for Samsung, Fuel Cells Bull. 2009 (2009) 6. https://doi.org/10.1016/S1464-2859(09)70117-8

[140] ElectroChem integrates fuel cell technology in India, South Asia, Fuel Cells Bull. 2007 (2007) 7. https://doi.org/10.1016/S1464-2859(07)70318-8

[141] M. Kodama, Innovation through boundary management—a case study in reforms at Matsushita electric, Technovation. 27 (2007) 15–29. https://doi.org/10.1016/J.TECHNOVATION.2005.09.006

Materials Research Forum LLC
doi: https://doi.org/10.21741/9781644900192-8

Chapter 8

Nanomaterials for Oxygen Reduction Reaction

S.Mahalakshmi[1], S.Bhuvanesh[2], R.Aravindan[2], K.S.Abisek[2], M. Harikrishnakumar[2], R. Rajasekar[2*],

[1]Department of Mechatronics Engineering, Kongu Polytechnic College - Erode, Tamil Nadu – 638 052, India

[2]Department of Mechanical Engineering, Kongu Engineering College - Erode, Tamil Nadu – 638 052, India

Abstract

Fuel cells are the best alternatives to meet the growing energy demand with no pollution. Nanomaterials play a significant role in applications as electrocatalysts for fuel cells. Among them, nanomaterials of platinum (Pt) have been mainly used as electrocatalytic material in fuel cells for a long time due to its excellent electrochemical properties towards oxygen reduction reaction (ORR). Bearing the limitations like scarcity and higher cost, Pt nanomaterials have to be replaced by novel materials for ORR. Carbon-based nanomaterials are in research owing to their electrocatalytic activity at par with that of Pt nanomaterials, which are applied as bifunctional electrocatalyst, transition metal oxides and also as supports for various dopants. This article focuses on recent research in nanomaterials for ORR in fuel cells.

Keywords

Fuel Cells, Oxygen Reduction Reaction, Electrocatalysts, Nanomaterials, Platinum, Carbon

List of abbreviations
Pt - Platinum
ORR - oxygen reduction reaction
H_2O_2 -hydrogen peroxide
CNT - carbon nanotube
GO - graphene oxide
RG-O - reduced graphene oxide
EOR - ethanol oxidation reaction
HER - hydrogen evolution reaction
OER - oxygen evolution reaction
MWCNT - multiwall carbon nanotube
XPS - X-ray photoelectron spectroscopy

RHE - reversible hydrogen electrode
HRTEM - high-resolution transmission electron microscopy
WAXS - X-ray diffraction spectroscopy at wide angles
DFA - Debye function analysis
MOR - methanol oxidation reaction
TEM - transmission electron microscopy
NC - nitrogen-doped carbon
Pt/NC - Pt nanoparticles decorated NC
PEM - proton exchange membrane
PtFe/C - carbon supported chemically ordered PtFe nanoparticles
Pt_3Re - Platinum rhenium
DFT - density functional theory
UPD - underpotential deposition
PVP - poly(vinylpyrrolidone)
FM - Frankevan der Merwe
SK - StranskieKrastanov
XRD - X-ray diffraction
HAADF - STEM high angle annular dark field scanning transmission electron microscopy
EDS - energy dispersive spectroscopy
RDE - rotating disk electrode
Pt/MWNT - multiwalled carbon nanotube-supported Pt
EG - ethylene glycol
PIL - ionic liquid polymer
PtNTs - platinum nanotubes
PtPdNts - platinum palladium alloy nanotubes
AgNWs - nanowires of silver
PBA - prussian blue analogs
ND-GLC - 3D porous graphene-like carbon nanomaterial
BG - graphene doped with boron
NPMCs - non-precious metal catalysts
VA-NCT - vertically aligned nitrogen-doped carbon nanotube
N-G-CNT - nitrogen-doped graphene-carbon nanotube
N-Fe-CNT/CNP - nitrogen-doped carbon nanotube/nanoparticle
EDA-NCNT - ethylenediamine nitrogen-doped carbon nanotubes
Py-CNT - pyridine nitrogen doped carbon nanotubes
POMC - P-doped ordered mesoporous carbon
AG amino-functionalized graphene
PXRD - powder X-ray diffraction
DTA-DTG - differential thermal gravimetric analysis
Mo_2C-Pd - Pd nanoparticles grown on Mo_2C

Contents

1.   Introduction

Traditional sources of energy emit harmful greenhouse gases and are a threat to the environment. Provided, they are subjected to depletion. Hence there is a need for pollution free and sustainable energy source. Fuel cells can be considered as a great boon amidst emerging energy crisis for the production of clean energy. Fuel cells comprise of an electropositive anode, electronegative cathode, and an electrolyte. Here fuel oxidation occurs at the anode, reduction occurs at the cathode and ions carry current between the electrodes through the electrolyte [1]. While all other energy production systems leave behind pollutants in some form or another to the environment, fuel cells are promising candidates in the energy market with no pollution. Hydrogen is the fuel to be supplied, fuel cells can be operated at all places of an abundant source of water for hydrogen production. Moreover, the silent operation of fuel cells and their ease to be made in compact sizes makes it feasible to be used in almost all places of energy necessity. The introduction of nanomaterials as catalysts in fuel cells have significantly reduced the material required for the same. Even though Pt being efficient and electrocatalytically active and playing a significant part in the fabrication of catalyst material for fuel cells, it poses limitations regarding cost and abundance due to its scarcity. Nanomaterials of Pt hence serve as an alternative to reduce the material requirement, thereby considerably decreasing the cost. Also, Pt on various supports provides a larger area of the material to be exposed to electrocatalytic activity. Provided nanomaterials of carbon have been found to replace Pt electrodes regarding efficiency.

The oxygen reduction reaction is a mechanism that occurs in fuel cells during which hydrogen molecules combines with oxygen molecules to form H_2O. The mechanism may take place through two different pathways namely, 4-electron pathway and 2-electron pathway. During the 4-electron pathway, oxygen is directly reduced to water. Whereas in 2-electron pathway oxygen is reduced to hydrogen peroxide (H_2O_2)

This review highlights the performance of Pt nanomaterials when applied as catalytic materials towards ORR for fuel cells. Various dopants like nitrogen N, Au, Mo, Co, V, Fe, Cr, W, Mn, etc. incorporated on Pt surface were found to increase the Pt catalyst's electrocatalytic activity [2–6]. Similarly, bimetallic nanoparticles of Pt formed with different metals proved to be efficient as well. The alloying materials produced synergistic effects and favorable interatomic distance as a result of alloying [7]. Monodispersed nanoparticles have been reported to exhibit better activity towards ORR due to their uniform size [8]. Core-shell like structures obtained by various processes such as galvanic displacement and seed-mediated growth have been reported [9–11]. In such cases, the electrocatalytic activity of the resulting electrocatalyst greatly depended on the shell thickness and also the interaction between the shell and the core materials. The shape of the Pt nanoparticles, also have great influence in determining the catalyst's electrocatalytic activity. For that matter, Pt-based nanoparticles with different shapes like nanocubes, polyhedral, octahedral, cuboctahedral have been examined after synthesis for their performance towards ORR [12–14]. The performances of such nanoparticles were found to mostly shape dependent. Pt nanoparticles deposited on various carbon-based supports exhibited remarkable activity towards ORR. The supports provides a large area of the surface for the uniform distribution of the Pt nanoparticles and aids in the growth of the same [15]. Alloying Pt with various metals caused the resulting materials to display unique properties towards ORR, due to the alloying effect [16].

Carbon-based nanomaterials such as graphene, carbon nanotube (CNT), graphene oxide (GO), reduced graphene oxide (RG-O) and functionalised graphene which serves as an alternative to Pt nanomaterials have also been reviewed whose research has been carried out at par with that of Pt. The unique characteristics of carbon-based materials such as vast surface area, high interaction and their remarkable electronic properties make them the best alternative to noble Pt nanoparticles. Dopants such as Co, N, B, S, H, etc. introduced to these carbon-based nanomaterials have found to increase the activity towards ORR to a considerable amount [17–21]. Here, the presence of dopant is considered vital in improving the catalytic activity of the carbon-based nanomaterials. Non-precious metal-free carbon-based electrocatalysts have been synthesized through various facile and novel methods [22–24]. Electrocatalysts can be involved in reactions such as Ethanol Oxidation Reaction (EOR), Hydrogen Evolution Reaction (HER) and Oxygen Evolution Reaction (OER) other than ORR. Bifunctional electrocatalysts are those catalysts that additionally involve in any one of the above-said reactions together with ORR. Carbon-based electrocatalysts were found to exhibit excellent bifunctional electrocatalytic activities [25–26].

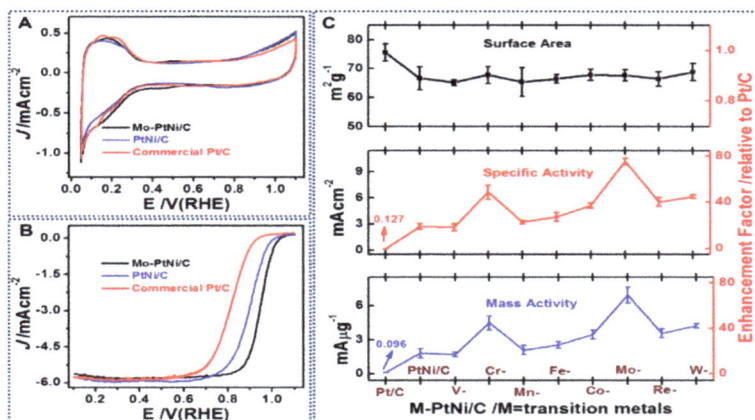

Figure 1. *Electrocatalytic properties of high-performance transition-metal doped octahedral PtNi/C catalysts and commercial Pt/C catalyst*

(A) Cyclic voltammograms of octahedral Mo-PtNi/C, octahedral PtNi/C and commercial Pt/C catalysts recorded at room temperature in N2-purged 0.1 M HClO4 solution with a sweep rate of 100 mV/s. (B) ORR polarization curves of octahedral Mo-PtNi/C, octahedral PtNi/C and commercial Pt/C catalysts recorded at room temperature in an O2-saturated 0.1 M HClO4 aqueous solution with a sweep rate of 10 mV/s and a rotation rate of 1600 rpm. (C) The electrochemically active surface area (ECSA, up panel), specific activity (middle panel) and mass activity (bottom panel) at 0.9 V versus RHE for these transition-metal doped PtNi/C catalysts, which are given as kinetic current

*densities normalized to the ECSA and the loading amount of Pt, respectively. In (A) and
(B), current densities were normalized in reference to the geometric area of the RDE
(0.196 cm^2).*

*From Huang, X., et al., High-performance transition metal–doped Pt₃Ni octahedra for
oxygen reduction reaction. Science, 2015. **348**(6240): p. 1230-1234. Reprinted with
permission from AAAS.*

2.   Platinum-based nanomaterials as fuel cell electrocatalysts

2.1   Bimetallic platinum nanoparticles

Synthesis of Pt: Au bimetallic nanoparticles on multiwall carbon nanotube (MWCNT) by
inverse microemulsion method were reported [27]. Studies on the electrocatalysts for
ORR were done by rotating disk electrode (RDE) system. Higher Pt loading increased
ORR and produced H_2O through four electrons pathway, compared to higher Au loading.
Increase in Au leads to H_2O_2 formation following a two electron transfer mechanism.
Hence the study revealed that the ORR depends mainly on Pt: Au ratio. Bimetallic AuPt
alloyed nanoparticles were synthesized employing peptide Z1 as template/ligand through
a chemical route [7]. A series of samples of AuPt bimetallic nanoparticles were prepared
among which $Au_{33}Pt_{67}$ was found to demonstrate the best electrocatalytic activities for
ORR and also for HER. Such a performance for the bimetallic nanoparticles was obtained
as a result of the Au alloying Pt induced synergistic effects. Electron transfer
phenomenon studied through X-ray photoelectron spectroscopy (XPS) measurement
revealed that the metal nanoparticles underwent a modified electronic structure due to the
addition of Au atoms. The improved electrocatalytic activity was due to the diluted
binding interaction of Au nanoparticles with the oxygenated intermediates against the
strong interaction of Pt nanoparticles at the atomic level. Superior ORR activity was
observed with +1.02 V positive onset potential vs. Reversible Hydrogen Electrode
(RHE). Synthesis of Pd-Pt bimetallic nanodendrites was reported by Xia et al. [28]. The
Pt branches were formed as a dense array on the Pd core by reducing K_2PtCl_4 in an
aqueous solution with L-ascorbic acid in the presence of uniform Pd nanocrystal seeds.
Formation of Pt branches was visible in High-resolution transmission electron
microscopy (HRTEM) images (Fig. 2). The electrocatalytic performance of the Pd-Pt
nanodendrites were found to be 2.5 times more than the catalyst of Pt/C and also 5 times
more than the Pt-black catalyst of first-generation without support. During the synthesis,
Pt to Pd ratio was varied for optimizing the dimension and also the composition of the
nanodendrites which resulted in enhanced ORR activity.

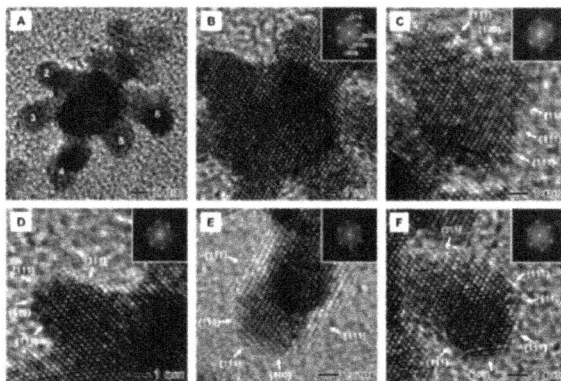

Figure 2. *HRTEM image of a single Pd-Pt nanodendrite. (B) HRTEM image recorded from the center of the Pd-Pt nanodendrite shown in (A). The image clearly shows the continuous lattice fringes from the Pd core to the Pt branches, demonstrating the epitaxial relation between Pd and Pt. (C to F) HRTEM images recorded from Pt branches 1, 2, 4, and 6 marked in (A), respectively. The images reveal that most of the exposed facets on the Pt branches were {111}planes. Some {110} and high-index {311} facets can also be identified in addition to a small fraction of {100} facets. The identical FT patterns shown in the insets indicate that the Pt branches have the same lattice orientation as the Pd core regardless of their different growth directions.*

*From Lim, B., et al., Pd-Pt bimetallic nanodendrites with high activity for oxygen reduction. science, 2009. **324**(5932): p. 1302-1305. Reprinted with permission from AAAS.*

The carbonyl complex route was observed as the efficient method to synthesize nanoparticles of Vulcan XC-72 alloy of Pt-Ni supported on carbon containing a controlled atomic composition and a disordered structure with a narrow particle size distribution [29]. The nanoparticle electrocatalyst was prepared with different Pt/Ni atomic composition. The highly disordered and unique face-centered cubic structure and the decrease in the lattice parameter in the alloys as a result of an increase in the Ni content were revealed through the agreement between X-ray diffraction spectroscopy at wide angles (WAXS) and Debye function analysis (DFA). An alloy of Pt-Ni nanoparticles was found to be dispersed well on the carbon support surface, and with an increase in Ni content, there was a slight decrease in the mean particle size. The alloy of Pt-Ni alloy electrocatalyst was compared with that of Pt-based catalysts concerning ORR activity. The bimetallic catalyst with different Pt/Ni atomic ratios was tested for electrocatalytic activity among which maximum activity was recorded for the alloy with ca. 30~40 % of Ni content which is attributed to the favorable inter-atomic Pt-Pt distance caused as a result of alloying. Enhancement in the electrocatalytic activity of the above alloy was also due to the structure of the alloy that was disordered and also high dispersion of the alloying elements. Mass activity and specific activity was enhanced in the range of ca .2 to 4 at 0.90 V and ca.1.5 to 2 at 0.85 V in comparison to Pt/C catalyst. Previously reported enhancement factor for the electrocatalyst of Pt-Ni nanoparticle was ca. 1.5.

2.2   Doped platinum nanomaterials

Doping of electrocatalysts is done to alter their properties that are favourable for electrocatalytic activity. Doping is the process of adding impurities to semiconductor materials to enhance their electronic properties. During doping the electronic properties can be tuned through temperature control. Also, the synergetic effect of the dopants results in the enhanced electrocatalytic activity of the electrocatalyst. Formation of transitional states, charge transport, tolerance to cross over effect is some of the characteristics incorporated as a result of doping that enables better electrocatalytic activity.

Quaternary catalyst supported with N-doped graphene (PtRuFeCo/NG) was synthesized through polyol method assisted by microwave and tested as a bifunctional catalyst, which proved to be efficient for both methanol oxidation reaction (MOR) and ORR [2].The processing cost of the fuel cell using the above electrocatalyst was seen to be significantly reduced owing to the use of low-cost pre-cursors such as Fe, and Co. Morphology of the electrocatalyst observed through Transmission Electron Microscopy (TEM) revealed that the PtRuFeCo nanoparticles were distributed uniformly without any agglomeration on the graphene substrate. The excellent performance which was observed for the electrocatalyst was as a result of the synergetic effect of Fe and Co alloyed with Pt and Ru and also the catalytic effect of nitrogen doping. The electrocatalyst facilitated a four electron transfer for ORR with a high current to the order of 6.3 mA cm^2. In another method reported, namely polyol reaction technique, the prepared nitrogen-doped carbon (NC) was further decorated with Pt nanoparticles [3]. The resulting Pt nanoparticles decorated NC (Pt/NC) showed efficient ORR characteristics for proton exchange membrane (PEM) fuel cells, where Pt-decorated carbon served as an anode and Pt/NC served as the cathode. Under fuel cell operating conditions, studies on single cell measurement and stability revealed that Pt/NC is an excellent alternative for Pt-carbon

Nanomaterials for Alcohol Fuel Cells   Materials Research Forum LLC
Materials Research Foundations 49 (2019) 231-270   doi: https://doi.org/10.21741/9781644900192-8

composites. The electrocatalyst's performance was due to the uniform Pt nanoparticles distribution and high electrical conductivity of NC.

By adopting microwave-assisted polyol reduction process followed by heat treatment, Zhu et al. were able to synthesize carbon supported chemically ordered PtFe nanoparticles (PtFe/C) with a particle size of ~5 nm average [4]. The as-prepared PtFe/C catalyst was surface doped with Au and Cr to obtain Au-PtFe/C and Cr-PtFe/C catalysts respectively. Compared to the PtFe/C catalyst as well as Johnson Matthey (JM) Pt/C catalyst they showed improved catalytic activity towards ORR. Various transition metal dopants like V, Re, Cr, W, Mn, Mo, Fe, and Co were incorporated over dispersive PtNi/C octahedra surface, and the effect of each dopant was observed for ORR[5]. The results showed the best performance for Mo-PtNi/C, revealing the highest value of specific activity which equaled 10.3 mA/cm^2 and 6.98 A/mg$_{Pt}$ of mass activity (Fig. 1). The study suggests that optimal performance can be achieved by fine-tuning the surface layer chemical and electronic properties. Nitrogen-doped carbon nanotubes (CNT) as supports for Pt NP electrocatalysts were produced from nitrogen-rich precursors by Chen et al.[6]. Nitrogen-rich precursor ethylenediamine produced ethylenediamine-CNTs with high content nitrogen (4.74 N at %) compared to pyridine-based pyridine-CNTs (2.35 N at %). When compared to the nitrogen-free Pt/CNTs, the ORR activity for Pt/ethylenediamine-CNTs and Pt/ pyridine-CNTs was found to increase attributed to the increased nitrogen content. Hence it is explained that nitrogen-rich precursors are highly crucial for the preparation of N-CNT to incorporate a high degree of nitrogen. A 24% increase in current density and a 16.9% increase in peak power density at a 0.6 V cell potential was noted for Pt/ ethylenediamine-CNTs compared to that of Pt/CNTs.Nanoparticles of platinum were found to be dispersed well with uniform size on N-CNT due to increased nitrogen content. The N-CNTs showed enhanced ORR activity compared to that of support materials of platinum owing to their distinct electronic and structural properties.

2.3   Monodispersed platinum-based nanoparticles

Platinum rhenium (Pt$_3$Re) alloy nanoparticles were reported to have been synthesized through an organic solution approach [30]. The synthesized monodisperse and homogeneous Pt$_3$Re nanoparticles showed an improvement factor of 4 in ORR compared to a similar particle sized commercial Pt catalysts through electrochemical studies. Density Functional Theory (DFT) showed that the catalytic enhancement is due to the ligand effect of the subsurface Re that electronically modifies the surface properties of Pt. Structure-dependent FePt nanoparticle catalysts were reported to be a better catalyst for ORR. Here the atoms of Pt and Fe are randomly arranged within the chemically disordered fcc monodispersed FePt nanoparticles[31]. The face-centered cubic (fcc) structure was converted to a face-centered tetragonal (fct) structure that was chemically ordered through high-temperature annealing wherein Fe and Pt were stacked intermetallic alloy. The electrocatalyst was coated with a layer of MgO before annealing to prevent the serious nanoparticle aggregation/sintering. The stability of the fct electrocatalyst was studied in 0.5 M H_2SO_4 solution which was found to be better than that of fcc electrocatalyst.

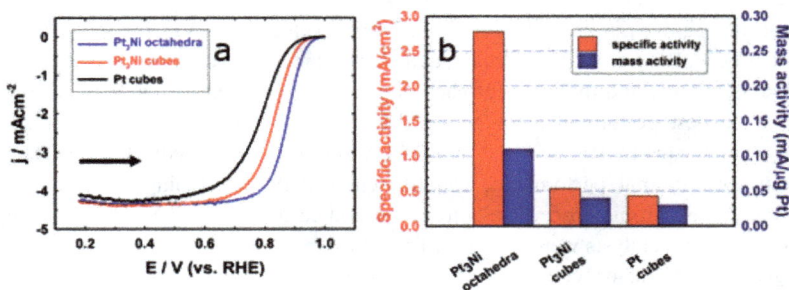

Figure 3. (a) Polarization curves for ORR on Pt_3Ni nanoctahedra, Pt_3Ni nanocubes, and Pt nanocubes supported on a rotating GC disk electrode in O_2 saturated 0.1 M $HClO_4$ solution at 295 K; scan rate, 20 mV s^{-1}; rotation rate, 900 rpm. Catalyst loading in terms of Pt mass: Pt_3Ni octahedra, 3.0 µg; Pt_3Ni cube, 2.0 µg; Pt cube, 1.1 µg. Current density was normalized to the glassy carbon geometric surface area (0.196 cm^2). The arrow indicates the potential scan direction. (b) Comparison of the ORR activities on the three types of catalysts. Specific activity and mass activity were all measured at 0.9 V vs RHE at 295 K.

Reprinted (adapted) with permission from Zhang, J., et al., Synthesis and oxygen reduction activity of shape-controlled Pt_3Ni nanopolyhedra. Nano letters, 2010. **10** (2): p. 638-644. Copyright 2010 American Chemical Society.

Monodisperse Pt nanocubes were also reported to be synthesized by reducing $Pt(acac)_2$ during which oleylamine and oleic acid were present with traces of $Fe(CO)_5$, which aided in the separation of monodisperse Pt nanocubes [32]. Improved catalysis was observed toward ORR with the specific activity twice in comparison with that of the commercial catalyst of Pt. Again through an organic solvothermal approach, size-controlled monodisperse nanoparticles of Pt_3Co were synthesized in which size of the particle varied between 3 and 9 nm [8]. The size-dependent ORR activity of the Pt_3Co nanoparticles was revealed through electrochemical study. The electrocatalytic activity was found to

Nanomaterials for Alcohol Fuel Cells   Materials Research Forum LLC
Materials Research Foundations **49** (2019) 231-270   doi: https://doi.org/10.21741/9781644900192-8

decrease with particle size. The favorable size was observed nearer to 4.5 nm for maximal mass activity, with a balance of activity and specific surface area. In another work reported by Zou et al., the shape dependent characteristics of Pt_3Ni electrocatalysts were demonstrated [33]. The electrocatalytic activity for {111}-facet-terminated nanoctahedra and {100}-bounded nanocubes were studied and compared. Results show that the former exhibited better activity compared to the latter (Fig. 3) At 0.9V, the current density values and mass activity showed increased values for Pt_3Ni nanoctahedra compared to that of Pt and Pt_3Ni nanocubes. Here, the nanoctahedra and nanocubes of Pt_3Ni with terminated facets at {111} and {100} respectively were prepared by a new wet-chemical method. It was observed that decrease in size increased the ORR activity. Since the ORR activity was found to be shape dependent, the same can be improved by controlling the shape of the nanocatalyst. The shape-controlled strategy can be applied for the synthesis of other nanopolyhedral alloys of nonprecious metal-Pt.

2.4   Platinum nanoparticles in a core-shell structure

A core-shell structure consists of an inner core encapsulated in a shell structure. The synthesis process of the core-shell structure involves two steps [88]. First step is the core synthesis and the second being coating of the shell material over the core. The core and the shell may be of different materials, comprising of a core material that is less expensive and a shell material that is more expensive. The core-shell structure helps in minimising the usage of noble materials and also the increased activity of the shell material. For this purpose and to overcome the shortcomings of disordered Pt_3Co alloy nanoparticles, Pt-Co nanocatalysts of a new class were described [34]. The nanocatalyst was composed of Pt shell of 2-3 atomic-layer-thickness containing intermetallic cores of ordered Pt_3Co. In comparison with the disordered Pt_3Co alloy nanoparticles and Pt/C, the above nanoparticles showed over 300% increase of specific activity and a 200% increase of mass activity. The elemental mapping at atomic-scale showed that the ordered core-shell structure was not disturbed and it was observed that even after 5,000 potential cycles there was only a very little loss of activity proving that the electrocatalyst was stable. The Pt-rich shell caused the high activity, and the intermetallic arrangement of Pt_3Co core provided stability for the electrocatalyst. Electrocatalysts with very low content of noble metal were reported. Core-shell nanoparticles of non-noble metal and noble metal on carbon support were synthesized through a chemical route [9]. Cu monolayer that was obtained by underpotential deposition (UPD) was subjected to galvanic displacement by Pt for deposition of a monolayer of the metal on the core-shell nanoparticles supported by carbon. This method greatly reduced the use of noble metal for electrocatalysts with no compromise in performance concerning ORR. A monolayer of Pt on particles of AuNi exhibited enhanced ORR kinetics than monolayer of Pt deposited on Au(111) which is an indication of weak bonding between Pt and AuNi. Also, Pt on AuNi expands less compared to Pt on Au(111). Carbon-supported Cu nanoparticles were prefabricated and subjected to galvanic displacement reaction with Pt^{4+}[10]. The resulting Pt-Cu core electrocatalyst with a shell of Pt (Pt@Cu) demonstrated higher activity compared to Pt catalyst at 110% of Pt@Cu.When the nominal initial Cu content was increased, a linear increase was observed in the specific activities concerning the surface area of Pt@Cu samples attributing to the modification of the outer Pt shell by the inner Pt-Cu core electronically. The Pt@Cu samples evaluated for electrocatalytic activity towards ORR, through linear sweep voltammetric method exhibited a value of 5.0 ± 0.25 mA/cm^2 limiting current. The role of strain effects in enhancing the ORR activity was observed by comparing samples of different wt. ratio but similar lattice pararmeters. The results show varied performances for the samples even though their lattice parameters were similar.Long et al. have reported the modified polyol method using $AgNO_3$ for synthesizing core-shell nanoparticles of Pt-Pd alloy under controlled size and morphology using poly(vinylpyrrolidone) (PVP) as the stabilizer[35]. The core-shell morphology for various prepared samples was visible through TEM and HRTEM images (Fig. 4). The thin Pd nanoshells on Pt nanocores and the thin nanoshell of Pt on the nanocores of Pd were grown by epitaxial growth mode and non-epitaxial growth mode respectively. It was observed that the nanoparticles of Pt-Pd core-shell were grown by both Frankevan der Merwe (FM), StranskieKrastanov (SK) modes. The catalytic characterization of the prepared nanoparticles of the core-shell structure revealed the correlation among morphology, size, and structure. Core/shell-structured Pd/FePt nanoparticles with Pd core of 5 nm and 1 to 3 nm thick FePt shell were synthesized by a unique approach [36]. The ORR activity of the electrocatalyst was found to be FePt shell thickness dependent, and in $HClO_4$ solution of 0.1 M, thin FePt shell of about 1 nm thick is durable and active towards ORR. The Cu monolayer underwent a galvanic displacement through underpotential deposition, which created a monolayer of Pt that was deposited on the nanoparticles of Co-Pd core-shell [37]. The Pt monolayer electrocatalyst showed mass-specific activity that was about three times higher than that of commercial electrocatalysts of Pt/C. This was due to the interaction of Pt atoms with the substrate which results in changes in the property of the d-band. The core/shell structure of FePtM/FePt nanowires (NWs) with a shell thickness of FePt being controlled between 0.3 to 1.3 nm was reported by Sun et al. to have been synthesized through seed-mediated growth [11]. The electrocatalytic activity for ORR of such structures was found to depend mainly on the shell thickness and core composition. Also, the reported core/shell NWs were more durable and active than their alloy counterparts. The FePtPd/FePt NWs of 0.8 nm shell showed 3.47 mA.cm^{-2} specific activity and 1.68 A/mg Pt mass activity at 0.5 V (vs. Ag/AgCl) which was superior to the other NWs and also the commercial Pt. Also, it was observed that the FePtM/FePt NWs were stable for ORR even after a considerable amount of potential sweeps conducted between 0.4 and 0.8 V (vs. Ag/AgCl).

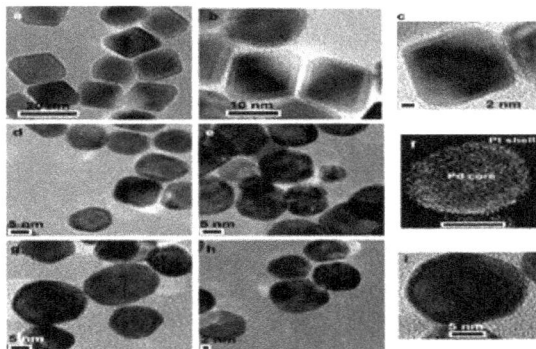

Figure 4. *(a) - (c) TEM and HRTEM images of Pt-Pd core-shell nanoparticles. (d) - (i) TEM and HRTEM images of Pd-Pt core-shell nanoparticles.*

*Reprinted from International Journal of Hydrogen Energy, **36**(14), Long, N.V., et al., Synthesis and characterization of Pt–Pd alloy and core-shell bimetallic nanoparticles for direct methanol fuel cells (DMFCs): Enhanced electrocatalytic properties of well-shaped core-shell morphologies and nanostructures, p. 8478-8491, (2011), with permission from Elsevier*

2.5 Shape-dependent ORR of platinum nanoparticles

The possible different shapes obtained for Pt nanoparticle electrocatalysts are octahedral(8 faces), truncated octahedral(14 faces), polyhedron(more than 6 faces), truncated cube(14 faces), cube(6 faces) and dumbbell shape. The shape of the nanoparticles greatly influence the ORR activity of the electrocatalyst which is explained below.

The shape dependent ORR activity of Pt nanoparticles have been demonstrated by Sun et al. RDE measured current density for 7 nm Pt nanocubes was found to be four times as that of 3 nm polyhedral nanoparticles of Pt which indicates the effect of the shape of the NP towards ORR [12]. Fe epitaxial growth on nanoparticles of Pt and further oxidation of Fe produced dumbbell-like Pt-Fe_3O_4 NPs [38]. Compared to Pt nanoparticle catalyst, more active sites were observed due to the formation of higher electron population owing to the interaction between Pt and Fe_3O_4. This proved that the electrocatalytic activity depends not only on the size and shape of the Pt nanoparticles but also on the interaction of Pt with Fe_3O_4 nanoparticles. Size and shape dependence of the catalytic activity of the nanoparticles of Pt between 1-5 nm in $HClO_4$ solutions were demonstrated by DFT calculations [13]14]. Cuboctahedral models were subjected to DFT calculations without imposing geometric constraints and hence full insight of the shape and size-controlled reactivity was obtained. Maximal mass activity was found at 2.2 nm. The mass activity was found to increase from 1.3 to 2.2 nm and decrease when the size was increased further. For particles, less than 2.2 nm low specific activity was noted which was due to the presence of edge sites which causes oxygen binding energies that were very strong at these sites. Octahedral PtNi NPs were reported to have been synthesized by a facile surfactant-free solvothermal method which exhibited 10x ORR activity gains [39]. The increase in ORR activity is attributed to the surface composition and octahedral shape of the final catalyst. The octahedral shape greatly influences the area of coverage of the hydroxyl atoms and also the Pt surface sites that are catalytically active which in turn affects the ORR activity. During the synthesizing process the surface Pt:Ni composition is tuned by the reaction time, whereas the shape of the PtNi nanoparticles is controlled by the type of precursor ligands. Surface area specific gain was observed to be ~3.14 mA/cm^2Pt and Pt mass based activity gain was ~1.45 A/mg Pt. Alloy of Pt nanoparticles was synthesized by solvothermal synthesis in which N, N-dimethylformamide (DMF) was used both as a reductant and solvent[40]. Well-faceted truncated octahedral and octahedral nanocrystals of PtNi (Fig. 5), and cuboctahedral and cubic nanocrystals of Pt_3Ni were generated. Crystalline morphology and composition were characterized by high angle annular dark field scanning transmission electron microscopy (HAADF-STEM), X-ray diffraction (XRD) and energy dispersive spectroscopy (EDS). ORR activities were recorded by a rotating disk electrode (RDE) technique. The nanoparticle of Pt_3Ni alloy exhibited larger than 1000 $\mu A/cm^2_{Pt}$ specific activities, PtNi exhibited activities nearer to 3000 $\mu A/cm^2_{Pt}$ which was fifteen times as that of Pt/carbon catalysts. Using $W(CO)_6$ as a source of CO and surfactants oleic acid and oleylamine, uniform nanoparticles of Pt-Ni octahedra of 9-nm was synthesized which in the presence of Ni aided the formation of {111} facets [14]. Introduction of the solvent benzyl ether reduced the presence of surfactants on the resultant Pt-Ni surface. Acetic acid treatment further removed the surfactants resulting in a 51 time higher specific activity compared to that of Pt/catalyst. Compared to the highest mass activity of 1.45 A mg_{Pt}^{-1} reported earlier, a very high mass activity at 0.9 V of the order of 3.3 A mg_{Pt}^{-1} was recorded. The enhanced ORR activity was observed as a result of the electronic structure of Pt being modified which caused weak interactions between hydroxyl adsorbates and Pt.

Figure 5. *Bright-field TEMs of Pt₃Ni nanoparticles obtained from a 24h solvothermal DMF reaction.*

*Reprinted (adapted) with permission from Carpenter, M.K., et al., Solvothermal synthesis of platinum alloy nanoparticles for oxygen reduction electrocatalysis. Journal of the American Chemical Society, **134** (20): p. 8535-8542. Copyright (2012) American Chemical Society.*

2.6   Platinum nanoparticles on carbon-based supports

Carbon materials such as graphene, modified graphene, reduced graphene oxide, CNT and multiwalled CNT are used as supports for Pt nanoparticles. The carbon-based supports were seen to improve the electrocatalytic activity of the electrocatalysts that are explained below.

By solution-phase self-assembly method, the synthesized FePt nanoparticles were assembled on graphene [41]. Compared to commercial Pt nanoparticles which exhibited 0.271 and 0.07 mA/cm^2 specific activities at 0.512 V and 0.557 V respectively, the above synthesized G/FePt nanoparticles annealed for 1 h at 100 °C produced 1.6 mA/cm^2 and 0.616 mA/cm^2 specific activities at the same potentials. The results prove graphene to be excellent support for enhancing the activity of the nanoparticle electrocatalysts. Triphenylphosphine modified nanoparticles of Pt were dispersed on the surface of CNTs by a novel method [42]. The organic triphenylphosphine molecule serves as a crosslinker and helps in the strong adhesion of Pt on the CNT surface and protects the Pt nanoparticles from aggregation. Even though aggregation was found at the time of thermal treatment, the Pt nanoparticles size never exceeded 4 nm. Pt nanoparticles supported nitrobenzene-modified graphene (Pt-NB/G) was fabricated, and their electrocatalytic activities were recorded and compared with that of the commercial catalyst of Pt/C[43]. The NB- modified graphene nanosheets served as a good Pt catalyst support exhibiting high stability and excellent electrocatalytic properties. The specific activity at 0.9 V for ORR vs. RHE was 0.184 mA cm^{-2} which was nearer to that of 20 wt.% the catalyst of Pt/C which was about 0.214 mA cm^{-2}. High stability of the catalyst of Pt-NB/G in 0.1 M KOH was observed in a long-term durability test.

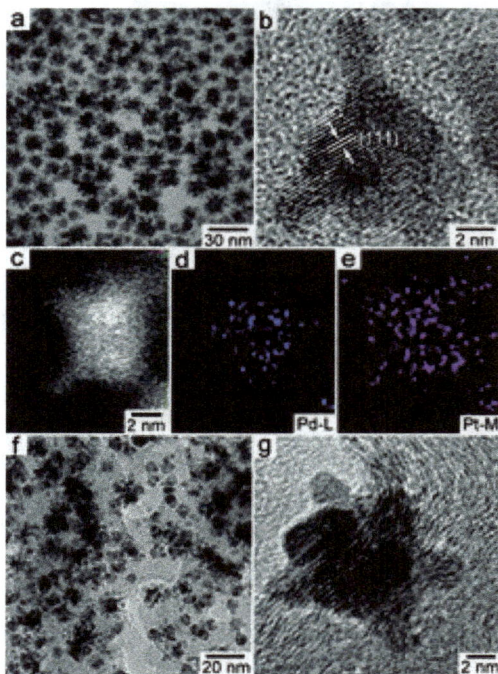

Figure 6. *Representative (a) TEM, (b) HR-TEM, (c) HAADF-STEM images and (d, e) elemental maps for Pd and Pt metals of Pt-on-Pd bimetallic nanoparticles; (f) TEM and (g) HR-TEM images of carbon-supported Pt-on-Pd bimetallic catalysts after the thermal treatments.*

*Reprinted (adapted) with permission from Peng, Z. and H. Yang, Synthesis and oxygen reduction electrocatalytic property of Pt-on-Pd bimetallic heteronanostructures. Journal of the American Chemical Society. **131** (22): p. 7542-7543. Copyright (2009) American Chemical Society.*

Pt-Pd bimetallic nanoparticles were reported to have been formed on a carbon support (Fig. 6) [44]. The size, composition and particle-on-particle morphology were found to be stable for palladium nanoparticles supported Pt catalysts. When the individual Pt catalyst was found to degrade, the larger particle size of Pt-on-Pd nanostructure and the interfacial structures present between supports of Pt and Pd prevented the Pt dissolution in ORR. Similar fcc phase and also 3.89 Å unit length for Pd and 3.92 Å unit length for Pt was a major factor for the synergistic effect of the electrocatalyst. Pre-treated CNTs were used for the preparation of Platinum/carbon nanotubes (Pt/CNT) electrocatalysts [45]. Pre-treatment aided in the formation of reactive sites for the Pt metal particles to adhere. Pre-treatment with HNO_3 resulted in 19.6% of Pt loading while that with ethylene glycol and Pt salt showed an increased Pt loading of 32%. The former exhibited higher fuel cell performance while the performance of the latter was low due to agglomeration and larger Pt particle size. By a simple modified polyol method, uniform diameter nanoparticles of Pt were synthesized on reduced graphene oxide (RG-O) supports [46]. When the Pt/RG-O composite comprising 70 wt % P were compared with JM Pt/C comprising 75 wt %Pt, higher electrochemical surface area and significant catalytic activity was observed towards ORR. Also, the maximum power density was found to be 11 % greater than that of JM Pt/C commercial catalyst which was in the order of 128 mW cm^{-2}. Samples of multiwalled carbon nanotube-supported Pt (Pt/MWNT) nanocomposites were reported to be prepared by two different processes of Pt salt reduction in aqueous solution and Pt ion salt reduction in ethylene glycol (EG) solution [47]. By the EG method, a third sample of Pt/XC-72 nanocomposite was prepared for comparison. Among the three samples, it was observed that the Pt/MWNT nanocomposites prepared by EG method exhibited higher performance characteristics towards ORR. It was reported that the type of support and the reduction method carried out influences the performance of the catalysts. Hence the results proves that the performances of the prepared catalysts are directly attributed to the MWNT support and the EG method of reduction.

Ionic liquid polymer (PIL) was formed on the CNT surface that introduces a considerable amount of functional groups on the surface exhibiting uniform distribution in order to grow metal nanoparticles [15]. The PIL film prevents aggregation of the CNTs by forming a uniform distribution of ionic species with positive charge thereby serving as a medium to stabilize and anchor metal nanoparticles. Characteristics like improved dispersion, better ESA and smaller particle size were observed for nanoparticles of Pt and PtRu on CNTs-PIL supports compared to those without PIL modification. The carbon-supported PtM/C catalysts where M = Fe, Cr, or Co reported by Qian et al. were synthesized by the new reverse micelle method [48]. The nanoparticles of Pt/M obtained were of uniform spherical shape. The performance of the PtM/C catalysts was tested by

RDE. The trend shows the catalyst of PtCo/C-type to have higher electrocatalytic activity towards ORR compared to the other synthesized catalysts. Limitations in the conventional preparation methods of colloidal nanoparticles on carbon support with the size of the particle between 2-5 nm have been overcome by a practical synthesis method [49]. Here, a spherical macromolecular template has been employed with no disorder in molecular weight or structure for the synthesis of ultrafine sub nanometre Pt clusters. The phenylazomethine dendrimer template through stepwise complexion controls the number of metal complexes in the assembly allowing them to accumulate in discrete nano-cages. Further, Pt clusters with a defined number of atoms were formed by reduction of Pt(IV) chloride to Pt(0). Influence of chlorine on ORR of the catalyst of Pt on carbon supports under the presence of the variety of anions was studied [50]. The ORR activity was found to decrease in the order $ClO_4^-> HSO_4^- > Cl^-$ with an increase in the adsorption bond strength of the anions. Ishikawa et al. reported ethylene glycol reduction method for the preparation of alloy of Pt and Pt_3M nanoparticles on graphene support in 2-D [51]. The alloying elements enhanced the ORR activity by the inhibition (hydr) oxy species formation on Pt surface. Also even at 500 cycles, the electrodes were found to be stable.

2.7   Platinum nanoparticles alloy

Alloys are combination of more than one metal that are employed to make use of the benefits of the alloying metals or elements. Pt is alloyed with transition metals in order to decrease the loading of Pt thereby reducing the cost. Moreover these alloys exhibit much better and unique physicochemical properties compared to that of Pt and hence better activity towards ORR. Four to five-fold ORR activity improvements were observed as compared to the previous results for the volumetrically formed Pt-Cu-C0 electrocatalyst alloy nanoparticles [52]. Variation of three alloy precursors of $Pt_{20}Cu_{60}Co_{20}$, $Pt_{20}Cu_{20}Co_{60,}$ and $Pt_{20}Cu_{40}Co_{40}$were considered. From these Pt-poor alloy precursors, active catalysts were formed in situ by voltammetric de-alloying. All three combinations were seen to exhibit high electrocatalytic activity. They showed ORR activity up to 0.5 A mg_{Pt}^{-1} in fuel cells.The ternary alloy catalyst was seen to exhibit better ORR activity compared to that of Pt nanoparticles.Nanoparticles of $PtCo_3$ and $PtCu_3$ electrocatalysts that were deployed were studied for alloy formation and subsequent electrochemical activity [53]. Annealing temperatures in the range of 650 °C, 800 °C, and 900 °C for a period of 7h were required to induce the alloy formation in the alloy of Pt-Co and also for the growth of the particle. In comparison with pure Pt/C catalyst, alloys of Pt-Cu and Pt-Co exhibit improved surface area and mass-specific ORR activities after voltammetric activation. Improvement in ORR activity is due to the geometric strain effects produced on the dealloyed nanoparticles of Pt-Co and $PtCu_3$. Pt and Au though immiscible, can be employed for the preparation of PtAu-alloyed nanosized particles through proper methodology thus preventing severe heat treatment and subsequently avoiding particle agglomeration [16]. Whether it is microemulsion or impregnation, the proper methodology has to be employed. Colloidal technique affording a preferential encapsulation of Pt was reported that generated separate metal phases. The property of the alloy was found to be different than that of segregated phases. Here alloying caused a decrease in oxophilicity of the PtAu samples which caused them to display unique properties in ORR.

2.8 Other Platinum based nanoparticles

A simple one-pot method using a conventional household microwave oven was reported for the synthesize of yttria-stabilized Pt nanoparticles (Pt_xYO_y) [54]. The specific activity of the Yttrium electrocatalyst was seen to be doubled at 3:1 Pt:Y ratio compared to that of commercial Pt-Vulcan. From DFT calculations, oxidation and surface migration of Y which formed a stable superficial yttrium oxide species was detected. Here the surface of yttria promotes more ORR sites that are highly active and acts as a stabilizing agent.A new generation supportless electrocatalyst of PtNTs (Platinum Nanotubes) and PtPdNts (platinum palladium alloy nanotubes) were fabricated by galvanic replacement reaction of nanowires of silver(AgNWs) [55]. Adopting the polyol method, AgNWs were developed which was further heated with $Pt(CH_3COO)_2$at reflux in aqueous solution. Finally, the product was collected through centrifugation. Owing to their unique dimensional combination, the electrocatalyst provides the advantages of the catalysts of platinum-black and Pt/C. The electrocatalyst possesses high durability, better surface area, improved activity, and higher utilization. Simple and practical modified polyol process was employed by Xin et al. without using a stabilizing agent for synthesizing narrow size distributed and highly dispersed Pt-based electrocatalysts [56]. The so formed electrocatalyst of Pt/C with the size of the particle $\sim$ 2.9 nm showed better electrocatalytic activity than commercially available catalyst (Table 1). By a simple method, Pt-C nanohybrid was developed in which, mesoporous carbon served as an entrap for the microporous carbon enveloped Pt nanoparticles [57]. The above hybrid was formed through in situ deposition of Pt nanoparticles followed by polymerization using glucose as the reductant and carbon source. Through polymerization, carbonaceous products were formed in the pores of SBA-15. The large area of the surface of the support together with the highly distributed Pt nanoparticles resulted in high catalytic activity and stability towards methanol-tolerant oxygen electroreduction.

Table 1. Particle size obtained from the different characterizing methods

Samples	TEM	CO-Chemisorption
40 wt% Pt/C-EG	2.9	3.3
40 wt% Pt/C-JM	3.8	4.0

[a] The unit is nm.

Chemical communications by Royal Society of Chemistry (Great Britain) Reproduced with permission of ROYAL SOCIETY OF CHEMISTRY

3.   Carbon based nanomaterials as fuel cell electrocatalysts

3.1   Bifunctional electrocatalysts

As explained earlier bifunctional electrocatalysts are those that involve in more than one reaction towards electrocatalytic cativity. Pd nanoclusters loading on carbon nanotubes served as bifunctional electrocatalysts for both ORR and EOR [26]. Atomically precise $Pd_5(C_{12}H_{25}S)_{13}$ nanoclusters were synthesized through N, N-dimethylformamide (DMF)-mediation and ligand-exchange reaction. Surface-cleaned Pd_5 clusters (Pd5NCs/MWCNTs) were obtained by initially loading the as-prepared Pd_5 nanoclusters on multiwalled carbon nanotubes (MWCNTs) followed by pyrolysis. Pyrolysis was carried out to remove the thiolate ligands. At 0.8 V, the Pd_5 NCs exhibited 5.70 and 4.53 times higher mass and specific activities respectively in comparison to Pd/C catalyst. It was also observed that the Pd_5 NCs showed low values of onset and peak potentials, and high values of current density for the EOR in comparison with the commercially available Pd/C catalyst. Enhanced electrocatalytic activity was due to high ratio of area-to-volume, high density and relatively smaller size of the active atoms that are exposed. Synthesize of bifunctional nonprecious electrocatalyst for ORR and HER were reported by directly pyrolyzing a mixture of Prussian blue analogs (PBA), graphene oxide and graphitic carbon nitride in the presence of silica colloids [58]. Specific surface area as high as 703.26 cm^2 g^{-1} was observed which attributed to the abundant mesoscale pores/holes that were present in the resulting Co@Co-N/RG-O electrocatalyst. During the synthesis, SiO_2 nano templates were introduced which resulted in the increase in surface area. The mesoporous textures were as a result of post-synthesis removal of hard silica templates. The electrocatalysts showed remarkable ORR activity with higher operation stability, stronger immunity to fuel crossover effect, higher limiting current with a more half-wave potential of +0.848 V. The efficiency of HER was proved with ten mA cm^{-2} value for current density that was reached through an overpotential of only 180 mV.

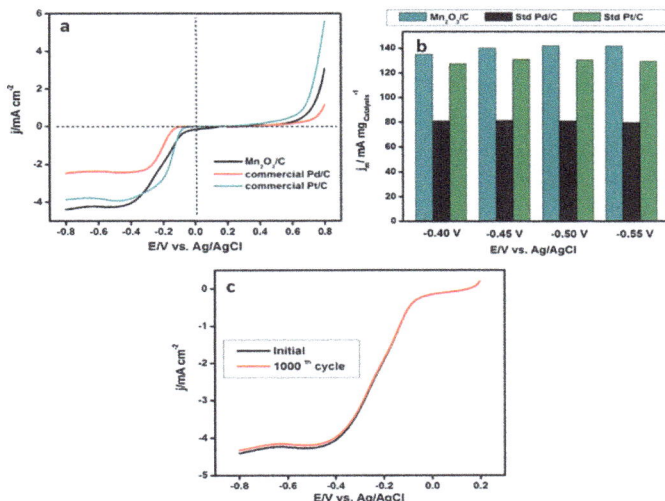

Figure 7. *(a)ORR and OER activities of Mn₂O₃/C (b) Mass specific activities of Mn₂O₃/C at different potentials compared to Pt/C and Pd/C and (c) LSV curve of Mn₂O₃/C after 1000 cycles to check the stability.*

*Reprinted from, Molecular Catalysis, **451,** Hazarika, K.K., et al., Cubic Mn₂O₃ nanoparticles on carbon as bifunctional electrocatalyst for oxygen reduction and oxygen evolution reactions, p. 153-160, Copyright (2018), with permission from Elsevier*

Transition metal oxides can function as a substitute for the catalyst of Pt owing to their highly active nature [59]. To overcome the poor electronic conductivity of transition metal oxides, carbon materials are either mixed or used as supports for such electrocatalysts. Here bifunctional catalyst of Mn oxides was synthesized by growing Mn_2O_3 nanoparticles on carbon which proved to be more active for ORR and also OER. In comparison to the commercially available Pt/C and Pd/C electrocatalysts, Mn_2O_3/C exhibits higher ORR following a 4-electron pathway, due to the better synergistic coupling effect of Mn_2O_3 and carbon (Fig. 7). Even after 1000 continuous redox cycles, the catalyst remained almost stable which shows the outstanding stability of the electrocatalyst. Dai et al. reported bifunctional electrocatalytic material of N, P-doped mesoporous carbon foam having a higher surface area of ~1,663 $m^2\,g^{-1}$ [25]. The material was fabricated through one-step pyrolysis of an aerogel of polyaniline that was synthesized when phytic acid was present. Characteristics like 55 $mW\,cm^{-2}$, of peak power density, 735 mAh gzn^{-1} of specific capacity and an open-circuit potential of 1.48 V was observed in Zn-air batteries. Also, the batteries exhibited a stable operation for 240 h after recharging mechanically. It was observed from the density functional theory calculations that the bifunctional electrocatalytic activity of the material was due to the co-doping of N and P and also the effects of graphene edge combined with a mesoporous network of the carbon foam.

3.2　Doped carbon nanomaterials

Similar to Pt nanoparticles, carbon nanoparticles are also doped in order to enhance their electrocatalytic properties. The various dopants used and their effect on the electrocatalytic activity are discussed, Using inexpensive precursors such as tris-(hydroxymethyl)-aminomethane, citric acid and ferric chloride, mesoporous carbon materials (Fe-N-C) were synthesized by pyrolysis method under a temperature ranging between 750 and 950 °C [17]. Best ORR catalytic activity was noted for Fe-N-C-850 catalyst among the prepared Fe-N-C-750, Fe-N-C-850, and-N-C-950 catalysts. The co-presence of Fe_3C, Fe-N, and pyridinic N active sites were identified through N1s atomic type and X-ray photoelectron spectroscopy (XPS), that resulted in long-time stability, relatively positive onset potential of 0.92 V, better poisoning resistance to methanol, higher current density and a near direct four-electron reaction pathway, resulting in best ORR activity in alkaline solution for the Fe-N-C-850 catalyst. *In-situ* alkaline activation of cellulose with ammonia injection created 3D porous graphene-like carbon nanomaterial (ND-GLC) [60]. In comparison to the commercial catalyst of Pt/C, the ND-GLC demonstrated outstanding ORR performance resulting from the synergistic promoting effect of the defects and pyridinic-N dopants which was revealed through RDE measurements and Zn-air battery applications. Hybrid Cobalt-doped ZnO nanoparticles (Co-doped ZnO NPs@RG-O) synthesized through simple solution-refluxing strategy showed excellent stability, nearly four-electron catalytic pathway and methanol-tolerance capability for ORR [21]. Here Co-doping concentration of 0.38 was found to be optimum.

Carbon-based electrocatalyst of dipolar-doped nitrogen-boron multi-walled carbon nanotubes (N, B-CNTs) was reported that demonstrated to be a low-cost and high-efficiency electrocatalyst for ORR [18]. It was observed that on the surface of doped boron and nitrogen, O_3H^- and O_3H_2 transitional states of oxygen were formed promoting a pathway of four-electron towards ORR. Maximum activity and stability were observed for N,B-CNT in alkaline solutions among the other pristine and doped CNTs towards ORR. The electrocatalyst exhibited characteristics such as a slight negative shift in the onset potential and a doubled peak current density in comparison with the commercially available catalyst of Pt/C. During the oxygen reduction reaction, nitrogen and boron after being doped were observed to form active centers of n-type and p-type in the N, B-CNT respectively which resulted in the weakening of O-O stretching and also the extension of O-O bond length. Dai et al. reported that Co_3O_4 nanoparticles grown on reduced graphene oxide proved to be efficient in acting as a bi-functional electrocatalyst [61]. Co_3O_4 and graphene individually exhibited little electrocatalytic activity. The higher electrocatalytic activity of the hybrid was attributed to the synergetic chemical coupling effects of Co_3O_4 and graphene.

Further, nitrogen doping on graphene to form Co_3O_4/N-doped graphene showed superior stability in alkaline solutions compared to Pt, even though the catalytic activity was similar to Pt. Also graphene doped with nitrogen (N-graphene) that was synthesized through chemical vapor deposition of methane during which ammonia was present was reported for the first time to act as a high metal-free electroactive electrode [62]. The material also showed long-term stability during operation and high tolerance to cross over effect towards ORR compared to Pt. In alkaline solutions, the hybrid exhibited superior stability compared to Pt which may be attributed to the synergistic coupling of the nanomaterials. Graphene doped with boron (BG) was observed to be a promising substitute for metal electrocatalysts in fuel cells [20]. Tests on methanol and CO tolerances showed that the electrocatalytic catalytic effect of BG was not affected by the addition of 1 M methanol or CO to the O_2 saturated electrolyte, on the contrary to the poisoning effect of adsorbed CO at the Pt electrode.

Mesoporous graphene doped with N and S was prepared using a simple one-step doping process employing solid and low-cost precursors [19]. Commercially available colloidal silica was used for introducing the porosity in the material. The material exhibited electrocatalytic characteristics nearer to that of the commercial catalyst of Pt/C. The dual S and N dopants cause the charge and spin densities to redistribute and further cause the carbon atom active sites to increase. The electrocatalyst also exhibited better stability for a more extended period and remarkable tolerance to fuel in comparison with Pt/C in an alkaline environment. Using density DFT, the electrocatalytic mechanism of N-graphene was studied in an acidic environment [63]. The observations revealed that the N-graphene exhibited ORR through a direct four-electron pathway which was similar to that of experimental observations. The active sites of catalytic activity that were identified on single N-graphene showed higher values of positive atomic charge density or higher values of positive spin density which was not observed for pure graphene. Herein after the introduction of H, sequential reactions were observed including chemical O-C bonding between graphene and oxygen, O-O bond breaking and finally water formation. A nitrogen-containing graphene sheet with the addition of hydrogen is shown in Fig. 8.

The doped nitrogen atom is responsible for the introduction of electrons with no pair on the surface of graphene and also the change in the distribution of atomic charge. With an attempt to develop non-precious metal catalysts (NPMCs), which replaces Pt-based catalysts, carbon nanotubes doped with nitrogen and their graphene composites have been reported [64]. The large surface area porous structures of the vertically aligned nitrogen-doped carbon nanotube (VA-NCT) and nitrogen-doped graphene-carbon nanotube (N-G-CNT) electrocatalysts showed enhanced electrolyte/reactant diffusion. Comparable gravimetric power densities and better long-term operational stabilities were demonstrated particularly in alkaline electrolytes.

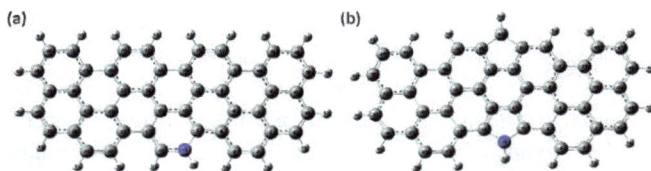

Figure 8. *Nitrogen-containing graphene sheets of (a) $C_{45}NH_{20}$ and (b) $C_{45}NH_{18}$. The larger gray circles are carbon atoms, the larger blue circle is a nitrogen atom, and the smaller light white circles are hydrogen atoms; all the atoms are numbered in the circles.*

Reprinted (adapted) with permission from Zhang, L. and Z. Xia, Mechanisms of oxygen reduction reaction on nitrogen-doped graphene for fuel cells. The Journal of Physical Chemistry C. 115(22): p. 11170-11176. Copyright (2011) American Chemical Society.

Although carbon-based nanostructured materials have shown excellent ORR activity in alkaline media, they have not been implemented in energy storage/conversion devices like metal-air batteries, alkaline fuel cells and certain electrolysers [65]. Using a simple single-step method, composites of nitrogen-doped carbon nanotube/nanoparticle (N-Fe-CNT/CNP) were synthesized from cyanamide which acted as a precursor of carbon nanotube and nitrogen and iron acetate as a precursor of iron. The resulting composite showed excellent ORR activity in alkaline media. They can even surpass the commercial Pt-based catalysts when used at higher loading. Low-cost precursors graphene oxide (GO) and urea were used in the preparation of N-graphene[66]. GO, and urea was well mixed, and pyrolysis was carried out in an inert atmosphere at 800 °C. The process resulted in the reduction of graphene oxide and subsequent doping into the graphitic lattice by the N atoms to produce N-graphene. Characterisation of NG was done through SEM and TEM. With an increased percentage of N-graphene, the content of N was found to be about 7.86%. Electrocatalytic activity of the catalyst favoured a four-electron pathway toward ORR. Superior to Pt/C catalysts these N-graphene electrocatalysts were found to exhibit good stability and anti-crossover property. Ethylenediamine nitrogen-doped carbon nanotubes (EDA-NCNT) showed excellent characteristics towards ORR, due to the effect of nitrogen precursors on nitrogen content [67]. The results were compared to those obtained for pyridine nitrogen-doped carbon nanotubes (Py-CNT). ORR performance concerning limiting current density, H_2O selectivity, potentials of onset and half-wave and the number of electrons that were transferred were found to be inferior for Py-CNT. This was due to the presence of higher nitrogen content of EDA-NCNT, indicating the role of the content of nitrogen in nitrogen-based precursors. The purpose of doping microstructure on metal-free C-based electrocatalysts for ORR were studied by preparing B and N co-doped CNTs with dominant B and N bonding by CVD growth and again the one dominated by the B and N that remained separated by post-treatment method [68]. Here the separated one was found to show excellent ORR activity compared to the bonded one (Fig. 9). This is because the bonded case failed to break the inertness of CNTs.

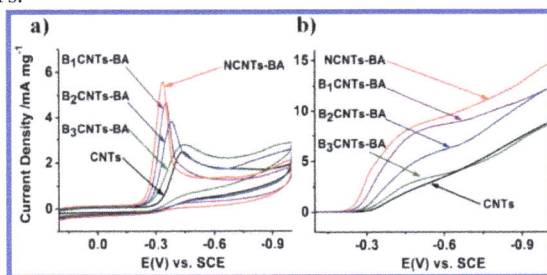

Figure 9. *Electrocatalytic capabilities of the BCNTs-BA catalysts for ORR in O_2-saturated 1 mol L^{-1} NaOH electrolyte. (a) CV at a scan rate of 50 mV s^{-1}. (b) RDE at a scan rate of 10 mV s^{-1} and rotation speed of 2500 rpm. CV and RDE curves for pristine CNTs and NCNTs-BA are also presented for comparison. On the vertical scale, 1 mA mg^{-1} corresponds to 0.1 mA cm^{-2}.*

*Reprinted (adapted) with permission from Zhao, Y., et al., Can boron and nitrogen co-doping improve oxygen reduction reaction activity of carbon nanotubes? Journal of the American Chemical Society. **135**(4): p. 1201-1204. Copyright(2013) American Chemical Society.*

Adopting a simple, cost-effective and readily reproducible metal-free nano casting approach, P-doped ordered mesoporous carbon (POMC) cathodes were fabricated [69].

Due to good electron-donating properties of P, defects in the carbon framework were induced, increasing electron delocalization. Also, the influence of channel length of the POMC on ORR was investigated which showed increased electrocatalytic performance for decreased channel length. This is because shorter channel lengths contributed to lowered resistance and increased surface area. Sulfur-doped graphene clusters were reported to be prepared by Xia et al. with/ without Stone-Wales defects and the effect of the defects on sulfur doping were studied [70]. It was observed that the defects change the crystal lattice and also the local charge distribution thereby facilitating sulfur doping. Also the sulfur doping at the edges of graphene serve as a substitute for carbon atoms either in the form of sulfur or in the form of sulfur oxide and exhibit electrocatalytic activity towards ORR in comparison with Pt. While the sulfur atoms followed a transfer pathway of two-electron, the carbon atoms took a transfer pathway of four-electron. Hybrid of manganese-cobalt spinel $MnCo_2O_4$/graphene proved to be an electrocatalyst of high efficiency towards ORR [71]. Reduced graphene oxide sheets doped with nitrogen were subjected to direct nanoparticle nucleation and growth followed by spinel Co_3O_4 nanoparticle cation substitution for the preparation of the electrocatalyst. Compared to the physical mixture of nanoparticles, the hybrid that was covalently bonded provided better durability and activity.

3.3   Other carbon-based electrocatalysts

Carbon-based materials such as carbides, CNT, CB, CO, rGO, graphene, functionalised graphene, graphene oxide etc. have been employed for the synthesis of electrocatalysts due to their enhanced electronic properties. A series of transition metal-nitrogen modified tungsten carbides (Fe/N/WC/C), were synthesized by a chemical route by varying the weight percentage of WC:C and the pyrolysis temperature [72]. Among the various proportions, WC(30)-Nx/C(700) catalyst was found to exhibit the maximum electrocatalytic activity, where 30 represents the weight percentage of WC:C and 700 represent the pyrolysis temperature. Compared to standard Pt catalyst, the obtained materials showed only a difference of about 0.060 V half-wave potential for ORR in an alkaline medium. The FeN_x moiety in acidic medium and also the N/C moiety in alkaline pHs were considered to be the reasons for the observed significant improvement in ORR, after incorporation of FeN_x and N/C species on WC/C nanomaterials. Another alternative for Pt/C electrocatalyst was reported to have been synthesized by a facile route by dispersing $WMoS_x$ nanoparticles on graphene which showed better durability and tolerance towards methanol than Pt/C [22]. The bimetallic material acted as an effective electrocatalyst in the alkaline electrolyte for ORR. The material was observed to exhibit a positive value of -0.05 V for onset potential and also a value of -0.13 V for half wave potential with a large limit density of current. The high ORR activity was found to be exhibited due to the large area of the surface of graphene and also the unique nanostructure of the $WMoS_x$.

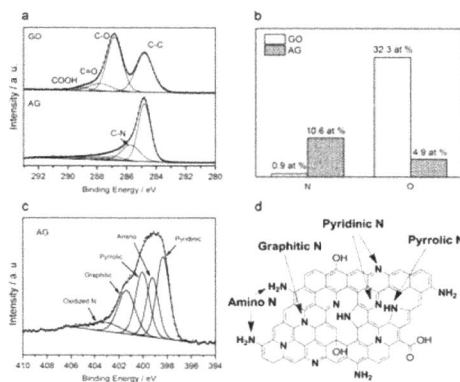

Figure 10. (a) The C 1s XPS spectra of GO and AG. (b) Chart showing the percentages of nitrogen and oxygen in the GO and AG measured by XPS. (c) The N 1s XPS spectrum of AG. (d) Schematic illustration of AG. The various 'N' atoms represent the pyridinic N, pyrrolic N, graphitic N, and amino group in graphene structure.

Reprinted from Nano Energy, 2(1), Zhang, C., et al., Synthesis of amino-functionalized graphene as metal-free catalyst and exploration of the roles of various nitrogen states in oxygen reduction reaction, p. 88-97, Copyright (2013), with permission from Elsevier

NPMCs are an excellent alternative to their metal counterparts due to the detrimental effects of strong reduction and acidic environments at PEM fuel cell cathodes on metal catalysts [64]. Carbon-based metal-free electrocatalysts of N-G-CNT + KB and VA-NCNT were synthesized through chemical route and pyrolysis respectively. Both showed excellent stability as cathodes for PEM fuel cells which are considered an intrinsic character of metal-free carbon-based catalysts. Hydrothermal self-assembly followed by freeze-drying and thermal treatment process was carried out for the fabrication of 3D monolithic Fe_3O_4/N-GAs hybrids [23]. The non-precious metal cathode catalyst owing to the 3D macroporous structure and also a large specific area of the graphene aerogel support showed excellent electrocatalytic activities. A more positive -0.19 V of onset potential and a high cathodic -2.56 mA cm^2 of current density was reported compared to Fe_3O_4/N-GSs (-0.26 V, -1.46 mA cm^{-2}) and Fe_3O_4/N-CB (-0.24 V, -1.99 mA cm^2). Post-treatment of PdCo/C catalysts under flowing H_2 at high temperature was reported to yield

PdCo@Pd core-shell nanoparticles on carbon surface [73]. The resulting nano electrocatalyst was subjected to a spontaneous displacement reaction for decorating with a very small amount of Pt. The process significantly reduces the amount of Pt loading. Higher MeOH tolerance was observed for the Pt-decorated PdCo@Pd/C and PdCo@Pd/C catalysts compared to that of Pt/C. Attempts to prepare non-Pt metal clusters for electrocatalysts suffer problems due to the aggregation tendency of such clusters, the easy occurrence of dissolution and also the presence of large amounts of capping agents on the surface [74]. The shortcomings were tried to be overcome by presenting a novel method for growing metal clusters that were ultrafine on RG-O sheets. For this reason, Au cluster/RG-O hybrids were synthesized that showed convincing electrocatalytic characters such as superior methanol tolerance, high onset potential, and excellent stability. It was observed that the reaction time decides the size of the clusters and that the size of the clusters gradually increased by prolonging the reaction time.

Using graphite oxide as the precursor, in the presence of ammonia solution, amino-functionalized (AG) graphene have been reported by a simple and effective solvothermal method[75]. The nitrogen content in the resulting metal-free electrocatalyst was about 10.6% (atom%) with enhanced electrocatalytic activities towards ORR. The various roles exhibited by different nitrogen states were observed through electrochemical measurements and XPS (Fig. 10). It was reported that the electron transfer number and onset potential were determined by the amino- and graphitic-type of nitrogen components. The mechanism of catalytic activity for ORR for various states of nitrogen can be understood by this economic synthesis of AG. The nanocomposite of Poly(2-fluoroaniline) was subjected to controlled pyrolysis to synthesize high specific surface area mesoporous carbons within which were homogeneously embedded one-pot hydrothermal synthesized FeO(OH) nanorods [76]. Nanorods, when decomposed thermally and further subjected to evaporation, formed mesopores larger in number in the resulting carbon skeletons during the process of carbonization of FeO(OH), besides forming sacrificial templates. Enhancement of electrocatalytic activity of Ru upon modification with increasing amounts of Se has been investigated [24]. Ru nanoparticles on carbon support were subjected to reductive annealing after treatment with SeO_2 for the formation of samples of Se-modified Ru/C with the ratio of Se:Ru ranging between 0 and 1. At Se:Ru = 1, core-shell particles were visualized with high-resolution electron microscopy, which showed the Ru core that was hexagonally packed and the Ru selenide shell having a lamellar morphology. Se inhibits surface oxidation thereby enhancing the electrocatalytic activity towards ORR in the Se modified Ru nanoparticles. Using ferrocene and sodium azide as the precursor materials, one-step synthesis was employed at a temperature of 350°C in N_2 [77]. During the process, the carbon nanotubes formed pea-pod like compartments in which were encapsulated the Fe nanoparticles. Thorough washing in acid solution was done to remove the Fe particles outside of the carbon shells. A well defined pea-pod-like structure was revealed through SEM and TEM images. The number of layers in the graphitic wall was between 1 to 8 with a layer distance of 3.40 ±0.05 Å. This system avoids the direct contact of metal particles with a harsh environment without impeding the activation of O_2. Provided the system exhibits high activity towards ORR and stability even in the presence of SO_2 poison in PEMFC.

Ultrafine nanoparticles of Pd were monodispersed on the surface of graphene oxide (GO) involving redox reaction between precursors of $PdCl_4^{2-}$ and GO [78]. The surfactant-free formation process made the catalyst clean paving the way for higher values of electrocatalytic activity. The alloying effect of Pd with Mo in the as-synthesized carbon-supported $Pd_{100-x}Mo_x$ (o ≤ x ≤ 40) has been studied [79]. The activity was found to increase along with the Mo content, with maximum activity at ten atom % Mo. During alloy formation at 900 °C heat treatment, the size of the particle of the Pd-Mo alloys was found to increase largely in comparison with that of catalysts of Pt. However the $Pd_{90}Mo_{10}$ catalyst was found to exhibit similar tolerance to methanol poisoning as that of catalyst of Pt at a temperature of 80 °C in direct methanol fuel cells. Halogenated graphene edges ClGnP, BrGnP, IGnP have been reported which showed high surface areas of 471, 579 and 662 m^2/g respectively [80]. The samples were prepared by the process of ball-milling graphite in the presence of Cl_2, Br_2, and I_2 respectively. The as-prepared halogenated graphene edges were found to be solution processable with remarkable electrocatalytic activities towards ORR like long-term cycle stability, notable selectivity and considerable tolerance to methanol crossover/CO poisoning effects. Hybrid nanoelectro catalyst of bamboo-like carbon nanotube/Fe_3C nanoparticle was prepared by a one-pot synthesis method. The so formed one dimensional nanoparticle showed better electrocatalytic activity compared to that of the commercial catalyst of Pt/C like better methanol tolerance and higher values of stability in alkaline as well as acidic solutions [81]. The non-precious metal cathodic catalyst of carbon-supported $CoSe_2$ nanoparticles was prepared with conventional heating through the *in situ* method that was surfactant-free [82]. Investigations on the structural and electrochemical properties were done through powder X-ray diffraction (PXRD), RDE and differential thermal gravimetric analysis (DTA-DTG) techniques. The $CoSe_2$ nanoparticles were observed to have two different crystal structure of orthorhombic at 250 and 300 °C and of cubic at 400 and 430 °C, after heat treatment under nitrogen. The latter structure in a 0.5 M H_2SO_4 exhibited a better ORR than the former revealing the structure dependent electrocatalytic activity of the prepared 20 wt.-% $CoSe_2$/C nanoparticles. The non-pt catalyst of magnetite in which M(II) was substituted - $M_xFe_{3-x}O_4$ ($M_xFe_{1-x}O\cdot Fe_2O_3$) where

Fe, Mn, Cu, and Co takes the place of M was synthesized on a commercial carbon support. The catalyst showed the M(II)-dependent characteristics of the catalyst towards ORR when studied for electrocatalytic activity [83]. $Mn_xFe_{3-x}O_4$ was found to be highly active followed by $Co_xFe_{3-x}O_4$, $Cu_xFe_{3-x}O_4$, and Fe_3O_4. Further control of x tuned the ORR activity of $Mn_xFe_{3-x}O_4$ making $MnFe_2O_4$ NPs as high-efficiency non-Pt catalyst of a new class in alkaline media towards ORR. Polymeric carbon precursor and a metal salt, when subjected to simple heat treatment followed by $KMnO_4$ oxidation yielded hollow graphitic nanoparticles [84]. The size of the nanoparticles ranged between 30 nm and 40 nm with a shell thickness ~ 5-8 nm. High electrical conductivity and smaller size of the particles of the hollow graphite, resulted in better power and current densities compared to that of commercial catalysts. This solid-phase synthesis is an economical method for the large scale production of hollow graphitic nanoparticles.

4.   Other nanomaterials for electrocatalysts

In spite of nanomaterials of Pt and carbon being dominantly used as electrocatalytic materials, there is also the application of other nanomaterials such as Cu and Pd as electrocatalysts. In-situ-formed composite electrocatalyst of $Cu_2S@RWY$- $x\%$, (x representing weight percent of Cu_2) was reported to be built by a simple ion exchange process followed by subsequent etching[85]. Performance of the electrocatalyst was recorded which showed best catalytic activities at $Cu_2S@RWY$- 48.6%, with quasi-four-electron transfer pathway, 0.90 V onset potential (vs. RHE) and small Tafel slope (69 mV/dec). Here in higher ORR performance of the apparently-inactive Cu_2S NPs has been explored for the first time which was due to the optimized balance between catalysts' loading and their accessibility and also the host-induced stabilization. Bifunctional electrocatalysts of Pd nanoparticles grown on Mo_2C (Mo_2C-Pd) was reported to be prepared for ORR and HER [86]. The Mo_2C-Pd samples were prepared for varied Pd loadings among which Mo_2C-Pd-9% was found to exhibit the best performance. It was observed that with an increase in Pd loading the electrocatalytic activity increased initially and then decreased. The morphology of the electrocatalyst was studied through X-ray diffraction XRD, TEM, and XPS. Bifunctional catalyst material of thin nanostructured films of manganese oxide for both water oxidation and oxygen reduction was prepared by Jaramillo and Gorlin [87]. Comparison between bifunctional Mn oxide and precious metal oxygen electrodes have been presented in Table 2. The result opens up new avenues for earth-abundant, scalable, nonprecious metal catalysts.

Table 2 .Oxygen Electrode Activities

Catalyst Material	ORR: E(V) at $I = -3$ mA·cm^{-2}	OER: E(V) at $I = 10$ mA·cm^{-2}	Oxygen Electrode Δ(OER-ORR): E (V)
20 wt % Ir/C	0.69	1.61	0.92
20 wt % Ru/C	0.61	1.62	1.01
20 wt % Pt/C	0.86	2.02 (1.88)^a	1.16 (1.02)^a
Mn oxide	0.73	1.77	1.04

^{a} Extrapolated value using the same Tafel slope as that at 1.74 V, where negligible corrosion occurs and the Pt is active for the OER.

Reprinted (adapted) with permission from Gorlin, Y. and T.F. Jaramillo, A bifunctional nonprecious metal catalyst for oxygen reduction and water oxidation. Journal of the American Chemical Society. **132**(39): p. 13612-13614. Copyright (2010) American Chemical Society.

Conclusion

The review explores various nanomaterials that are used as electrocatalysts in fuel cells for ORR. Various novel and simple processes employed for the synthesis of nanomaterials are also highlighted. Particular focus is laid on Pt and carbon-based nanomaterials used as electrocatalysts. Effect of doping on the above two materials have been extensively discussed. It has been analyzed from the study that the electrocatalytic activity of the catalyst not only depends on the individual properties of the catalytic material but also on various other factors like - type of dopant, various supports used, the structure of the arrangement of the nanoparticles - the core-shell structure for eg., the property of the alloying element, the process employed for the synthesis of nanomaterial and also the size and shape of the nanoparticles. Carbon-based materials not only serve as an alternative for the noble Pt nanoparticles but also act as supports on which various other nanoparticles are grown for maximum exposure and interaction to exhibit excellent activity towards ORR. In the future, further study on various novel nanomaterials that can be used as electrocatalysts in fuel cells is expected. Nanomaterials, due to their intrinsic size-dependent properties can replace the traditional materials with comparatively increased efficiency.

References

[1] S. Du, B. Millington, B. G. Pollet, The effect of Nafion ionomer loading coated on gas diffusion electrodes with in-situ grown Pt nanowires and their durability in proton exchange membrane fuel cells,Int. J. Hydrogen Energ.36 (2011) 4386–4393. https://doi.org/10.1016/J.IJHYDENE.2011.01.014.

[2] M. Rethinasabapathy, S.M. Kang, Y. Haldorai, N. Jonna, M. Jankiraman, G.W. Lee, S.C. Jang, B. Natesan, C. Roh, Y. S. Huh, Quaternary PtRuFeCo nanoparticles supported N-doped graphene as an efficient bifunctional electrocatalyst for low-temperature fuel cells,J. Ind. Eng. Chem.69 (2019) 285–294. https://doi.org/10.1016/j.jiec.2018.09.043.

[3] D. Puthusseri, S. Ramaprabhu, Oxygen reduction reaction activity of platinum nanoparticles decorated nitrogen doped carbon in proton exchange membrane fuel cell under real operating conditions, Int. J. Hydrogen Energ.41 (2016) 13163–13170. https://doi.org/10.1016/j.ijhydene.2016.05.146.

[4] Y. Cai, P. Gao, F. Wang, H. Zhu, Surface tuning of carbon supported chemically ordered nanoparticles for promoting their catalysis toward the oxygen reduction reaction,Electrochim. Acta.246 (2017) 671–679. https://doi.org/10.1016/j.electacta.2017.05.068.

[5] X. Huang, Z. Zhao, L. Cao, Y. Chen, E. Zhu, Z. Lin, M. Li, A. Yan, A. Zettl, Y. M. Wang, X. Duan, T. Mueller, Y. Huang, High-performance transition metal-doped Pt3Ni octahedra for oxygen reduction reaction,Science.348 (2015) 1230–1234. https://doi.org/10.1126/science.aaa8765.

[6] D. C. Higgins, D. Meza, Z. Chen, Nitrogen-doped carbon nanotubes as platinum catalyst supports for oxygen reduction reaction in proton exchange membrane fuel cells,J. Phys. Chem. C, 114 (2010) 21982–21988. https://doi.org/10.1021/jp106814j.

[7] W. Wu, Z. Tang, K. Wang, Z. Liu, L. Li, S. Chen, Peptide templated AuPt alloyed nanoparticles as highly efficient bi-functional electrocatalysts for both oxygen reduction reaction and hydrogen evolution reaction,Electrochim. Acta.260 (2018) 168–176. https://doi.org/10.1016/J.ELECTACTA.2017.11.057.

[8] C. Wang, D. van der Vliet, K.C. Chang, H. You, D. Strmcnik, J. A. Schlueter, N. M. Markovic, V. R. Stamenkovic, Monodisperse Pt 3 Co nanoparticles as a catalyst for the oxygen reduction reaction: size-dependent activity,J. Phys. Chem. C. 113 (2009) 19365–19368. https://doi.org/10.1021/jp908203p.

[9] J. Zhang, F. H. B. Lima, M. H. Shao, K. Sasaki, J. X. Wang, J. Hanson, R. R. Adzic, Platinum monolayer on nonnoble metal−noblemetal core−shell nanoparticle electrocatalysts for O_2 reduction,J. Phys. Chem. B. 109 (2005) 22701–22704. https://doi.org/10.1021/jp055634c.

[10] A. Sarkar, A. Manthiram, Synthesis of Pt@Cu core−shell nanoparticles by galvanic displacement of Cu by Pt^{4+}ions and their application as electrocatalysts for oxygen reduction reaction in fuel cells,J. Phys. Chem. C.114 (2010) 4725–4732. https://doi.org/10.1021/jp908933r.

[11] S. Guo, S. Zhang, D. Su, S. Sun, Seed-mediated synthesis of core/shell FePtM/FePt (M = Pd, Au) nanowires and their electrocatalysis for oxygen reduction reaction,J. Am. Chem. Soc.135 (2013) 13879–13884. https://doi.org/10.1021/ja406091p.

[12] C. Wang, H. Daimon, T. Onodera, T. Koda, S. Sun, A general approach to the size- and shape-controlled synthesis of platinum nanoparticles and their catalytic reduction of oxygen,Angew. Chem. Int. Ed.47 (2008) 3588–3591. https://doi.org/10.1002/anie.200800073.

[13] M. Shao, A. Peles, K. Shoemaker, Electrocatalysis on platinum nanoparticles: particle size effect on oxygen reduction reaction activity,Nano Lett.11 (2011) 3714–3719. https://doi.org/10.1021/nl2017459.

[14] S.I. Choi, S. Xie, M. Shao, J. H. Odell, N. Lu, H.C. Peng, L. Protsailo, S. Guerrero, J. Park, X. Xia, J. Wang, M. J. Kim, Y. Xia, Synthesis,andcharacterization of 9 nm Pt–Ni octahedra with a record high activity of 3.3 A/mg Pt for the oxygen reduction reaction,Nano Lett.13 (2013) 3420–3425. https://doi.org/10.1021/nl401881z.

[15] B. Wu, D. Hu, Y. Kuang, B. Liu, X. Zhang, J. Chen, Functionalization of carbon nanotubes by an ionic-liquid polymer: dispersion of Pt and PtRu nanoparticles on carbon nanotubes and their electrocatalytic oxidation of methanol,Angew. Chem. Int. Ed.48 (2009) 4751–4754. https://doi.org/10.1002/anie.200900899.

[16] P. Hernández-Fernández, S. Rojas, P. Ocón, J. L. Gómez de la Fuente, J. San Fabián, J. Sanza, M. A. Peña, F. J. García-García, P. Terreros, J. L. G. Fierro, Influence of the preparation route of bimetallic Pt−Au nanoparticle electrocatalysts for the oxygen reduction reaction,J. Phys. Chem. C.111 (2007) 2913–2923. https://doi.org/10.1021/jp066812k.

[17] C. He, T. Zhang, F. Sun, C. Li, Y. Lin, Fe/N co-doped mesoporous carbon nanomaterial as an efficient electrocatalyst for oxygen reduction reaction,Electrochim. Acta.231 (2017) 549–556. https://doi.org/10.1016/j.electacta.2017.01.104.

[18] T. Huang, S. Mao, M. Qiu, O. Mao, C. Yuan, J. Chen, Nitrogen-boron dipolar-doped nanocarbon as a high-efficiency electrocatalyst for oxygen reduction reaction,Electrochim. Acta.222 (2016) 481–487. https://doi.org/10.1016/j.electacta.2016.10.201.

[19] J. Liang, Y. Jiao, M. Jaroniec, S. Z. Qiao, Sulfur,and nitrogen dual-doped mesoporous graphene electrocatalyst for oxygen reduction with synergistically enhanced performance,Angew. Chem. Int. Ed.51 (2012) 11496–11500. https://doi.org/10.1002/anie.201206720.

[20] Z.H. Sheng, H.L. Gao, W.J. Bao, F.B. Wang, X.H. Xia, Synthesis of boron doped graphene for oxygen reduction reaction in fuel cells,J. Mater. Chem.22 (2012) 390–395. https://doi.org/10.1039/C1JM14694G.

[21] Y. Sun, Z. Shen, S. Xin, L. Ma, C. Xiao, S. Ding, F. Li, G. Gao, Ultrafine Co-doped ZnO nanoparticles on reduced graphene oxide as an efficient electrocatalyst for oxygen reduction reaction,Electrochim. Acta.224 (2017) 561–570. https://doi.org/10.1016/j.electacta.2016.12.021.

[22] D. M. Nguyen, M. H. Nguyen, Q. B. Bui, Uniform WMoSx nanoparticles attached graphene nanosheets as a highly effective electrocatalyst for oxygen reduction reaction in alkaline medium,Mater. Chem. Phys.224 (2019) 186–195. https://doi.org/10.1016/j.matchemphys.2018.11.079.

[23] Z.S. Wu, S. Yang, Y. Sun, K. Parvez, X. Feng, K. Müllen, 3D Nitrogen-doped graphene aerogel-supported Fe 3 O 4 nanoparticles as efficient electrocatalysts for the oxygen reduction reaction,J. Am. Chem. Soc.134 (2012) 9082–9085. https://doi.org/10.1021/ja3030565.

[24] V. I. Zaikovskii, K. S. Nagabhushana, V. V. Kriventsov, K. N. Loponov, S. V. Cherepanova, R. I. Kvon, H. Bönnemann, D. I. Kochubey, E. R. Savinova, synthesis and structural characterization of Se-modified carbon-supported Ru nanoparticles for the oxygen reduction reaction,J. Phys. Chem. B.110 (2006) 6881–6890. https://doi.org/10.1021/jp056715b.

[25] J. Zhang, Z. Zhao, Z. Xia, L. Dai, A metal-free bifunctional electrocatalyst for oxygen reduction and oxygen evolution reactions,Nat. Nanotechnol.10 (2015) 444–452. https://doi.org/10.1038/nnano.2015.48.

[26] Z. Zhuang & W. Chen, Ultra-low loading of Pd^{5+} nanoclusters on carbon nanotubes as bifunctional electrocatalysts for the oxygen reduction reaction and the ethanol oxidation reaction,J. Colloid. Interface. Sci.In Press, (2018). https://doi.org/10.1016/j.jcis.2018.12.015.

[27] R. M. Félix-Navarro, M. Beltrán-Gastélum, E. A. Reynoso-Soto, F. Paraguay-Delgado, G. Alonso-Nuñez, J. R. Flores-Hernández, Bimetallic Pt–Au nanoparticles supported on multi-wall carbon nanotubes as electrocatalysts for oxygen reduction. Renew. Energ.87 (2016) 31–41. https://doi.org/10.1016/j.renene.2015.09.060.

[28] B. Lim, M. Jiang, P. H. C. Camargo, E. C. Cho, J. Tao, X. Lu, Y. Zhu, Y. Xia, Pd-Pt bimetallic nanodendrites with high activity for oxygen reduction,Science.324 (2009) 1302–1305. https://doi.org/10.1126/science.1170377.

[29] H. Yang, W. Vogel, C. Lamy, Structure and electrocatalytic activity of carbon-supported Pt - Ni Alloy nanoparticles toward the oxygen reduction reaction, J. Phys. Chem. B.108 (2004) 11024–11034. https://doi.org/10.1021/JP049034+.

[30] D. Raciti, J. Kubal, C. Ma, M. Barclay, M. Gonzalez, M. Chi, J. Greeley, K. L. More, C. Wang, Pt_3Re alloy nanoparticles as electrocatalysts for the oxygen reduction reaction, Nano Energy.20 (2016) 202–211. https://doi.org/10.1016/j.nanoen.2015.12.014.

[31] J. Kim, Y. Lee, S. Sun, Structurally ordered FePt nanoparticles and their enhanced catalysis for oxygen reduction reaction, J. Am. Chem. Soc.132 (2010) 4996–4997. https://doi.org/10.1021/ja1009629.

[32] C. Wang, H. Daimon, Y. Lee, J. Kim, S. Sun, Synthesis of monodisperse Pt nanocubes and their enhanced catalysis for oxygen reduction, J. Am. Chem. Soc.129 (2007) 6974–6975. https://doi.org/10.1021/ja070440r.

[33] J. Zhang, H. Yang, J. Fang, & S. Zou, Synthesis and oxygen reduction activity of shape-controlled Pt 3 Ni nanopolyhedra, Nano Lett.10 (2010) 638–644. https://doi.org/10.1021/nl903717z.

[34] D. Wang, H. L. Xin, R. Hovden, H. Wang, Y. Yu, D. A. Muller, F. J. DiSalvo, H. D. Abruña, Structurally ordered intermetallic platinum–cobalt core–shell nanoparticles with enhanced activity and stability as oxygen reduction electrocatalysts, Nat. Mater.12 (2013) 81–87. https://doi.org/10.1038/nmat3458.

[35] N. V. Long, T. Duy Hien, T. Asaka, M. Ohtaki, M. Nogami, Synthesis and characterization of Pt–Pd alloy and core-shell bimetallic nanoparticles for direct methanol fuel cells (DMFCs): Enhanced electrocatalytic properties of well-shaped core-shell morphologies and nanostructures, Int. J. Hydrogen Energ.36 (2011) 8478–8491. https://doi.org/10.1016/j.ijhydene.2011.03.140.

[36] V. Mazumder, M. Chi, K. L. More, S. Sun, Core/shell Pd/FePt nanoparticles as an active and durable catalyst for the oxygen reduction reaction, J. Am. Chem. Soc.132 (2010) 7848–7849. https://doi.org/10.1021/ja1024436.

[37] M. Shao, K. Sasaki, N. S. Marinkovic, L. Zhang, R. R. Adzic, Synthesis and characterization of platinum monolayer oxygen-reduction electrocatalysts with Co-Pd core-shell nanoparticle supports, Electrochem. Comm.9 (2007) 2848–2853. https://doi.org/10.1016/j.elecom.2007.10.009.

[38] C. Wang, H. Daimon, S. Sun, Dumbbell-like $Pt-Fe_3O_4$ Nanoparticles and their enhanced catalysis for oxygen reduction reaction, Nano Lett.9 (2009) 1493–1496. https://doi.org/10.1021/nl8034724.

[39] C. Cui, L. Gan, H.-H. Li, S.H. Yu, M. Heggen, P. Strasser, Octahedral PtNi nanoparticle catalysts: exceptional oxygen reduction activity by tuning the alloy particle surface composition,Nano Lett.12 (2012) 5885–5889. https://doi.org/10.1021/nl3032795.

[40] M. K. Carpenter, T. E. Moylan, R. S. Kukreja, M. H. Atwan, M. M. Tessema, solvothermal synthesis of platinum alloy nanoparticles for oxygen reduction electrocatalysis,J. Am. Chem. Soc.134 (2012) 8535–8542. https://doi.org/10.1021/ja300756y.

[41] S. Guo, S. Sun, FePt nanoparticles assembled on graphene as an enhanced catalyst for oxygen reduction reaction,J. Am. Chem. Soc.134 (2012) 2492–2495. https://doi.org/10.1021/ja2104334.

[42] Y. Mu, H. Liang, J. Hu, L. Jiang, L. Wan, Controllable Pt nanoparticle deposition on carbon nanotubes as an anode catalyst for direct methanol fuel cells,J. Phys. Chem. B.109 (2005) 22212–22216. https://doi.org/10.1021/jp0555448.

[43] L. Tao, S. Dou, Z. Ma, S. Wang, Platinum nanoparticles supported on nitrobenzene-functionalized multiwalled carbon nanotube as efficient electrocatalysts for the methanol oxidation reaction,Electrochim. Acta.157 (2015) 46–53. https://doi.org/10.1016/j.electacta.2015.01.054.

[44] Z. Peng, H. Yang, Synthesis and oxygen reduction electrocatalytic property of Pt-on-Pd bimetallic heteronanostructures,J. Am. Chem. Soc.131 (2009) 7542–7543. https://doi.org/10.1021/ja902256a.

[45] N. Rajalakshmi, H. Ryu, M. M. Shaijumon, S. Ramaprabhu, Performance of polymer electrolyte membrane fuel cells with carbon nanotubes as oxygen reduction catalyst support material,J. Power Sources.140 (2005) 250–257. https://doi.org/10.1016/j.jpowsour.2004.08.042.

[46] H.W. Ha, I.Y. Kim, S.J. Hwang, R. S. Ruoff, One-Pot Synthesis of platinum nanoparticles embedded on reduced graphene oxide for oxygen reduction in methanol fuel cells, Electrochem. Solid. State. Lett.14 (2011) B70–B73. https://doi.org/10.1149/1.3584092.

[47] W. Li, C. Liang, W. Zhou, J. Qiu, Zhou, G. Sun, .Q. Xin, Preparation and characterization of multiwalled carbon nanotube-supported platinum for cathode catalysts of direct methanol fuel cells,J. Phys. Chem. B.107 (2003) 6292–6299. https://doi.org/10.1021/jp022505c.

[48] Y. Qian, Wen, P. A. Adcock, Z. Jiang, N. Hakim, M. S. Saha, S. Mukerjee, PtM/C catalyst prepared using reverse micelle method for oxygen reduction reaction in PEM Fuel Cells,J. Phys. Chem. C.112 (2008) 1146–1157. https://doi.org/10.1021/jp074929i.

[49] K. Yamamoto, T. Imaoka, W.J. Chun, O. Enoki, H. Katoh, M. Takenaga, A. Sonoi, Size-specific catalytic activity of platinum clusters enhances oxygen reduction reactions,Nature Chem.1 (2009) 397–402. https://doi.org/10.1038/nchem.288.

[50] T. J. Schmidt, U. A. Paulus, H. A. Gasteiger, R. J. Behm, The oxygen reduction reaction on a Pt/carbon fuel cell catalyst in the presence of chloride anions,J. Electroanal. Chem. 508 (2001) 41–47. https://doi.org/10.1016/S0022-0728(01)00499-5.

[51] C. V. Rao, A. L. M. Reddy, Y. Ishikawa, P. M. Ajayan, Synthesis and electrocatalytic oxygen reduction activity of graphene-supported Pt3Co and Pt3Cr alloy nanoparticles,Carbon.49 (2011) 931–936. https://doi.org/10.1016/j.carbon.2010.10.056.

[52] R. Srivastava, P. Mani, N. Hahn, P. Strasser, Efficient oxygen reduction fuel cell electrocatalysis on voltammetrically dealloyed Pt–Cu–Co nanoparticles, Angew. Chem. Int. Ed.46 (2007) 8988–8991. https://doi.org/10.1002/anie.200703331.

[53] M. Oezaslan, P. Strasser, Activity of dealloyed PtCo$_3$ and PtCu$_3$ nanoparticle electrocatalyst for oxygen reduction reaction in polymer electrolyte membrane fuel cell,J. Power Sources.196 (2011) 5240–5249. https://doi.org/10.1016/j.jpowsour.2010.11.016.

[54] R. Sandström, E. Gracia-Espino, G. Hu, A. Shchukarev, J. Ma, T. Wågberg, Yttria stabilized,and surface activated platinum (Pt x YO y) nanoparticles through rapid microwave assisted synthesis for oxygen reduction reaction,Nano Energ.46(2018) 141–149. https://doi.org/10.1016/j.nanoen.2018.01.038.

[55] Z. Chen, M. Waje, W. Li, Y. Yan, Supportless Pt and PtPd nanotubes as electrocatalysts for oxygen-reduction reactions,Angew. Chem.119 (2007) 4138–4141. https://doi.org/10.1002/ange.200700894.

[56] Z. Zhou, S. Wang, W. Zhou, G. Wang, L. Jiang, W. Li, S. Song, J. Liu, G. Sun, Q. Xin, Novel synthesis of highly active Pt/C cathode electrocatalyst for the direct methanol fuel cell,Chem. Comm.0 (2003) 394–395. https://doi.org/10.1039/b211075j.

[57] Z. Wen, J. Liu, J. Li, Core/shell Pt/C nanoparticles embedded in mesoporous carbon as a methanol-tolerant cathode catalyst in direct methanol fuel cells, Adv. Mater.20 (2008) 743–747. https://doi.org/10.1002/adma.200701578.

[58] D. Zhao, J. Dai, N. Zhou, N. Wang, Xinwen Peng, Y. Qu, L. Li, Prussian blue analogues-derived carbon composite with cobalt nanoparticles as an efficient bifunctional electrocatalyst for oxygen reduction and hydrogen evolution,Carbon.142 (2019) 196–205. https://doi.org/10.1016/j.carbon.2018.10.057.

[59] K. K. Hazarika, C. Goswami, H. Saikia, B. J. Borah, P. Bharali, Cubic Mn2O3nanoparticles on carbon as a bifunctional electrocatalyst for oxygen reduction and oxygen evolution reactions,Molec. Catal.451 (2018) 153–160. https://doi.org/10.1016/j.mcat.2017.12.012.

[60] J. Zhang, Y. Sun, J. Zhu, Z. Kou, P. Hu, L. Liu, S. Li, S. Mu, Y. Huang, Defect and pyridinic nitrogen engineering of carbon-based metal-free nanomaterial toward oxygen reduction,Nano Energy.52 (2018) 307–314. https://doi.org/10.1016/j.nanoen.2018.08.003.

[61] Y. Liang, Y. Li, H. Wang, J. Zhou, J. Wang, T. Regier, H. Dai, Co_3O_4 nanocrystals on graphene as a synergistic catalyst for oxygen reduction reaction,Nature Mater.10 (2011) 780–786. https://doi.org/10.1038/nmat3087.

[62] L. Qu, Y. Liu, J.-B. Baek, L. Dai, Nitrogen-doped graphene as an efficient metal-free electrocatalyst for oxygen reduction in fuel cells,ACS Nano.4 (2010) 1321–1326. https://doi.org/10.1021/nn901850u.

[63] L. Zhang, Z. Xia, Mechanisms of oxygen reduction reaction on nitrogen-doped graphene for fuel cells,J. Phys. Chem. C. 115 (2011) 11170–11176. https://doi.org/10.1021/jp201991j.

[64] J. Shui, M. Wang, F. Du, L. Dai, N-doped carbon nanomaterials are durable catalysts for oxygen reduction reaction in acidic fuel cells,Sci. Adv.1 (2015) e1400129-37. https://doi.org/10.1126/sciadv.1400129.

[65] H. T. Chung, J. H. Won, P. Zelenay, Active and stable carbon nanotube/nanoparticle composite electrocatalyst for oxygen reduction,Nature Comm.4 (2013) 1922. https://doi.org/10.1038/ncomms2944.

[66] Z. Lin, G. Waller, Y. Liu, M. Liu, C.P. Wong, Facile synthesis of nitrogen-doped graphene via pyrolysis of graphene oxide and urea, and its electrocatalytic activity toward the oxygen-reduction reaction,Adv. Energy Mater.2 (2012) 884–888. https://doi.org/10.1002/aenm.201200038.

[67] Z. Chen, D. Higgins, H. Tao, R. S. Hsu, Z. Chen, Highly active nitrogen-doped carbon nanotubes for oxygen reduction reaction in fuel cell applications,J. Phys. Chem. C.113 (2009) 21008–21013. https://doi.org/10.1021/jp908067v.

[68] Y. Zhao, L. Yang, S. Chen, X. Wang, Y. Ma, Q. Wu, Y. Jiang, W. Qian, Z. Hu, Can boron and nitrogen Co-doping improve oxygen reduction reaction activity of carbon nanotubes,J. Am. Chem. Soc.135 (2013) 1201–1204. https://doi.org/10.1021/ja310566z.

[69] D.S. Yang, D. Bhattacharjya, S. Inamdar, J. Park, J.-S. Yu, Phosphorus-doped ordered mesoporous carbons with different lengths as efficient metal-free

electrocatalysts for oxygen reduction reaction in alkaline media,J. Am. Chem. Soc.134 (2012) 16127–16130. https://doi.org/10.1021/ja306376s.

[70] L. Zhang, J. Niu, M. Li, Z. Xia, Catalytic mechanisms of sulfur-doped graphene as efficient oxygen reduction reaction catalysts for fuel cells, J. Phys. Chem. C. 118 (2014) 3545–3553. https://doi.org/10.1021/jp410501u.

[71] Y. Liang, H. Wang, J. Zhou, Y. Li, J. Wang, T. Regier, H. Dai, Covalent hybrid of spinel manganese–cobalt oxide and graphene as advanced oxygen reduction electrocatalysts,J. Am. Chem. Soc.134 (2012) 3517–3523. https://doi.org/10.1021/ja210924t.

[72] U.A. do Rêgo, T. Lopes, J.L. Bott-Neto, A.A. Tanaka, E.A. Ticianelli, Oxygen reduction electrocatalysis on transition metal-nitrogen modified tungsten carbide nanomaterials,J. Electroanal. Chem. 810 (2018) 222–231. https://doi.org/10.1016/j.jelechem.2018.01.013.

[73] D. Wang, H. L. Xin, Y. Yu, H. Wang, E. Rus, D. A. Muller, H. D. Abruña, Pt-Decorated PdCo@Pd/C Core−shell nanoparticles with enhanced stability and electrocatalytic activity for the oxygen reduction reaction,J. Am. Chem. Soc.132 (2010) 17664–17666. https://doi.org/10.1021/ja107874u.

[74] H. Yin, H. Tang, D. Wang, Y. Gao, Z. Tang, Facile synthesis of surfactant-free Au cluster/graphene hybrids for high-performance oxygen reduction reaction,ACS Nano.6 (2012) 8288–8297. https://doi.org/10.1021/nn302984x.

[75] C. Zhang, R. Hao, H. Liao, Y. Hou, Synthesis of amino-functionalized graphene as metal-free catalyst and exploration of the roles of various nitrogen states in the oxygen reduction reaction,Nano Energy.2 (2013) 88–97. https://doi.org/10.1016/j.nanoen.2012.07.021.

[76] W. Niu, L. Li, X. Liu, N. Wang, J. Liu, W. Zhou, Z. Tang, S. Chen, Mesoporous N-doped carbons prepared with thermally removable nanoparticle templates: an efficient electrocatalyst for oxygen reduction reaction,J. Am. Chem. Soc.137 (2015) 5555–5562. https://doi.org/10.1021/jacs.5b02027.

[77] D. Deng, L. Yu, X. Chen, G. Wang, L. Jin, X. Pan, J. Deng, G. Sun, X. Bao, Iron Encapsulated within Pod-like carbon nanotubes for oxygen reduction reaction,*Angewandte Chemie International Edition*,Angew. Chem. Int. Ed.52 (2013) 371–375. https://doi.org/10.1002/anie.201204958.

[78] X. Chen, G. Wu, J. Chen, X. Chen, Z. Xie, X. Wang, Synthesis of "Clean" and well-dispersive Pd nanoparticles with excellent electrocatalytic property on graphene oxide. J. Am. Chem. Soc.133 (2011) 3693–3695. https://doi.org/10.1021/ja110313d.

[79] A. Sarkar, A. V. Murugan, A. Manthiram, Synthesis and characterization of nanostructured Pd–Mo electrocatalysts for oxygen reduction reaction in fuel cells,J. Phys. Chem. C.112 (2008) 12037–12043. https://doi.org/10.1021/jp801824g.

[80] I.Y. Jeon, H.J. Choi, M. Choi, J.M. Seo, S.M. Jung, M.J. Kim, S. Zhang, L. Zhang, Z. Xia, L. Dai, N. Park, J.B. Baek, Facile, scalable synthesis of edge-halogenated graphene nanoplatelets as efficient metal-free eletrocatalysts for oxygen reduction reaction,Sci. Rep.3 (2013) 1810. https://doi.org/10.1038/srep01810.

[81] W. Yang, X. Liu, X. Yue, J. Jia, S. Guo, Bamboo-like carbon nanotube/Fe 3 C nanoparticle hybrids and their highly efficient catalysis for oxygen reduction,J. Am. Chem. Soc.137 (2015) 1436–1439. https://doi.org/10.1021/ja5129132.

[82] N. Alonso-Vante, Y. Feng, T. He, Carbon-supported CoSe$_2$ nanoparticles for oxygen reduction and hydrogen evolution in acidic environments, 2010.

[83] H. Zhu, S. Zhang, Y.X. Huang, L. Wu, S. Sun, Monodisperse M x Fe 3– x O 4 (M = Fe, Cu, Co, Mn) Nanoparticles and their electrocatalysis for oxygen reduction reaction, Nano Lett.13 (2013) 2947–2951. https://doi.org/10.1021/nl401325u.

[84] S. Han, Y. Yun, K.W. Park, Y.E. Sung, T. Hyeon, Simple solid-phase synthesis of hollow graphitic nanoparticles and their application to direct methanol fuel cell Electrodes,Adv. Mater.15 (2003) 1922–1925. https://doi.org/10.1002/adma.200305697.

[85] D. Hu, X. Wang, H. Yang, D. Liu, Y. Wang, J. Guo, T. Wu, Host-guest electrocatalyst with cage-confined cuprous sulfide nanoparticles in etched chalcogenide semiconductor zeolite for highly efficient oxygen reduction reaction,Electrochim. Acta.282 (2018) 877–885. https://doi.org/10.1016/j.electacta.2018.06.106.

[86] T. Li, Z. Tang, K. Wang, W. Wu, S. Chen, C. Wang, Palladium nanoparticles grown on β-Mo 2 C nanotubes as dual functional electrocatalysts for both oxygen reduction reaction and hydrogen evolution reaction,Int. J. Hydrogen Energ.43 (2018) 4932–4941. https://doi.org/10.1016/j.ijhydene.2018.01.107.

[87] Y. Gorlin, T. F. Jaramillo, A bifunctional nonprecious metal catalyst for oxygen reduction and water oxidation,J. Am. Chem. Soc.132 (2010) 13612–13614. https://doi.org/10.1021/ja104587v.

[88] V. Andrey,S. P.Nomoev, Structure and mechanism of the formation of core–shell nanoparticles obtained through a one-step gas-phase synthesis by electron beam evaporation, Beilstein J. Nanotechnol. 6(2015) 874-880.

Nanomaterials for Alcohol Fuel Cells
Materials Research Foundations **49** (2019) 271-292

Materials Research Forum LLC
doi: https://doi.org/10.21741/9781644900192-9

Chapter 9

Polymer-based Nanocomposites for Direct Alcohol Fuel Cells

S. Ghosh, R.N. Basu

Fuel Cell & Battery Division, CSIR-Central Glass & Ceramic Research Institute, Kolkata – 700032, India

ghosh.srabanti@gmail.com

Abstract

Direct alcohol fuel cells (DAFCs) have been widely considered as an energy conversion device used in portable electronic devices and stationary power systems. The fabrication of inexpensive, high performance and durable electrocatalysts is the major challenge in DAFCs application. Particularly, platinum and other low-cost metal such as palladium and transition metal-based polymer nanocomposites are known as efficient electrocatalysts for electrooxidation of alcohol. The chapter describes some of the recent developments of polymer composites catalysts as anode material for fuel cell applications. The present chapter includes a deeper understanding of the composition-tunable metal based polymer nanocomposites as electrocatalysts and their effectiveness of catalytic activity and energy conversion in DAFCs.

Keywords

Direct Alcohol Fuel Cell (DAFCs), Noble Metal, Conducting Polymer Nanostructure, Polymer Composites, Methanol, Ethanol, Electrooxidation

Contents

1.   Introduction

Direct alcohol fuel cells (DAFCs) technology have been deliberated as a suitable electrochemical energy conversion device for portable electronic and stationary power systems with high efficiency, zero pollutant emissions, and operating at low temperature [1,2]. Efforts have been made to explore the direct electrochemical oxidation of different alcohol such as methanol, ethanol, propanol, ethylene glycol, and glycerol, etc. [3] Among the liquid organic fuels, methanol or ethanol has shown promising characteristics regarding reactivity at low temperature [4]. In DAFC, the anode side is provided with alcohol, for example methanol (CH_3OH) and the cathode side is fed with oxygen or air. In the presence of catalysts, methanol is split into protons and electrons through an oxidation reaction at the anode. Then protons travel across the membrane, and the electrons transfer in an external circuit to the cathode. On the other hand, oxygen combines with protons and electrons to form water through a reduction reaction at the cathode (Figure 1).

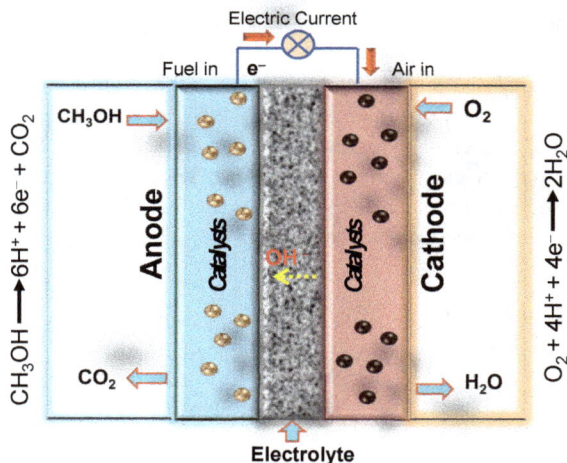

Figure 1. Schematic presentation and components of direct alcohol fuel cells using methanol as fuel.

Hence, various small alcohol molecules oxidation reactions (e.g., methanol, ethanol) take place on the anode side and the sluggish alcohol reaction kinetics can be improved by using the active surface of electrocatalysts, typically Pt-based catalysts for widespread applications of DAFCs. Moreover, complete oxidation of alcohol molecule in the presence of anode catalysts leads to enhance fuel utilization and energy conversion efficiency of fuel cells. The broad scale application of DAFC relies on the development of active, stable and low-cost catalyst materials. However, high cost originated from the use of Pt-based noble metal catalysts as anode materials. Palladium and transition metals-based nanoalloys are also known as efficient and low-cost electrocatalysts for DAFCs applications [5–7]. Electrocatalytic activity and effective utilization of the metal catalysts can be improved through incorporation of support materials which possess high surface area, good electronic conductivity and provide stability under critical experimental condition [8–10]. Various carbon-based support such as carbon black, carbon nanotubes (CNTs), carbon fibers, mesoporous carbon, graphene, etc. are widely used in order to provide large surface area, enhance electrical conductivity, and good resistance to corrosion and allow stability to metal catalysts. There are several advantages of using carbon as a catalyst-support material which originated from its unique advantages, low-cost, high stability in both acidic and basic media, and possibility of easy recovery of the precious metal catalysts. However, high hydrophobic surface of carbon needs functionalization which prevents interaction between supporting materials with the catalyst as well as make metal catalysts prone to oxidation [11–13]. Polymers can be used as a matrix to produce noble metal catalysts and also supporting material for electrooxidation of alcohol molecules [14–16]. The commonly used conducting polymers include polypyrrole (Ppy), polyaniline (PANI), polyacetylene (PAc), polythiophene (PTh) (Figure 2) while widely incorporated nanofillers such as carbon nanotubes, metal oxides, metals nanostructures, etc.The polymer nanocomposites are of particular interest due to their smart functionalities such as enhanced physical, chemical, thermal,electrical conductivity, where polymer matrix reinforced with inorganic metal nanoparticles (NPs) [17]. Recently, various polymer nanocomposite using conducting polymers such as polypyrrole and polyaniline having exceptional high stability, π-conjugated structures with high electrical conductivity as supporting materials and metal catalysts deposited on such polymer-based composites have been tested in fuel cell applications [18–20]. The chapter provides an overview of preparation and catalytic activity of polymer nanocomposites as anode materials for alcohol fuel cells.The present chapter also focuses on the development of stable, low cost, polymer nanocomposites based electrocatalyst, understanding their effectiveness of catalytic activity with the composition-tunable nanocomposites.

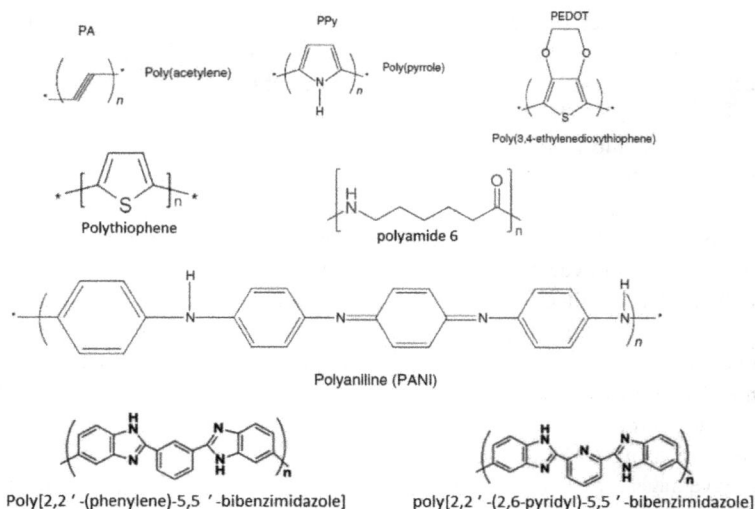

Figure 2. Chemical structures of polymer used as support for metal nanoparticles.

2.   Application of polymer nanocomposite for alcohol fuel cells

Combing inorganic particles into the inner part of the conducting polymers (CP) provides an interesting aspect of nanocomposite synthesis [21–23]. The most commonly used method which includes direct mixing via *ex-situ* where the polymer is dissolved in an appropriate solvent, then nanofiller is added to the polymer solution under vigorous stirring *in situ* methods through *in situ* polymerization in the presence of nanoparticles or synthesis of polymer network with nanostructures [24–26]. The as-prepared composite materials differ from the pure polymers and at the same time differ from individual components. For example, Ghosh *et al.* [27] reported an alternative strategy using poly(diphenyl butadiyne) (PDPB) polymer nanofiber in combination with Nafion (an ion-exchange membrane) as support for 2D Pd nanoplates which significantly improved the electrocatalytic activity of Pd nanostructures for ethanol oxidation reaction (EOR) compared to Nafion supported Pd catalysts. The strong interaction between Pd nanostructures and polymer-based composite support may enhance the availability of nanostructures during catalysis and make them promising anode catalyst in DAFCs. On the other hand, *in situ* strategy also provided uniform nanostructures in various morphologies within the polymer matrix [28]. Selvaraj et al. [29] developed a novel supporting material based on carbon nanotubes (CNTs) with *in situ* polymerizations of thiophene (Th) and further Pt and Pt–Ru NPs deposited on polythiophene/CNT composites (PTh-CNTs) for ethylene glycol oxidation reaction. Conducting polymer modified carbon nanotubes provides higher electrochemically active surface area, enhanced electronic conductivity and facile charge-transfer at the interfaces. Moreover, polymer composites based supporting material permits higher dispersion of Pt–Ru NPs which in turn improve the electrocatalytic activity of Pt–Ru/PTh–CNT electrodes for ethylene glycol electro-oxidation than the Pt/PTh electrodes. In another example, Prasanna et al. [30] reported high catalytic activity of platinum (Pt) and platinum-gold (Pt-Au) NPs using multifunctional carbon nanotube (amine terminated cyclophosphazene, ATCP/cyclophosphazene, CP/hexafluoroisopropylidenedianiline, HFPA-CNT) composite as support for methanol electrooxidation reaction (MOR) under alkaline medium. Compared to Pt/CNT and Pt/C catalysts, ATCP/CP/HFPA-CNT composite supported Pt catalysts display high anodic current density and significant CO tolerance. Shi et al. [31] also prepared polyaniline wrapped carbon nanotubes (CNT–PANI) core-shell composites via interfacial polymerization and used as a support for Pt NPs as shown in Figure 3a. The interfacial polymerization at the interface between CNT and PANI formed due to the presence of π–π interactions.

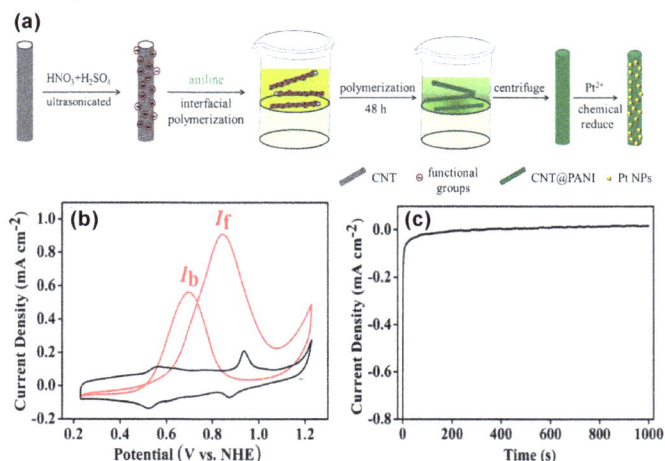

Figure 3. (a) Synthesis of CNT–PANI composites supported Pt NPs. (b) CV profiles for 1M methanol oxidation using CNT–PANI and Pt/CNT–PANI catalysts in 0.5 M H₂SO₄ solution. (c) The chronoamperometric curve of Pt/CNT–PANI electrode in the presence of 1 M CH₃OH +0.5M H₂SO₄. Reproduced from Ref. 31 with permission from American Chemical Society, 2012.

Then, Pt NPs are deposited with controllable loading density on the CNT–PANI composites and consequently, form a bridge between the Pt NPs and CNT walls through PANI via platinum–nitride (Pt–N) and π–π bonding which provided an exceptional electrochemical activity for MOR. Figure 3b shows the typical cyclic voltammetry (CV) profile of MOR using the Pt/CNT–PANI as an electrode with lower anodic overpotential suggests high catalytic activity. The polarization current displayed a slow decay and reached steady state within 300 s revealed excellent electrochemical stability towards methanol oxidation as shown in Figure 3c. Wei et al. [32] reported a similar trend of high catalytic activity were highly dispersed Pt NPs deposited on multi-walled carbon nanotubes (MCNTs) modified with transition metal oxides such manganese oxide and poly(3,4-ethylenedioxythiophene) polymer for enhanced MOR as compared with the polymer modified MWCNTs or MWCNTs supported Pt catalysts. Recently, Wang et al. [33] deposited Pt NPs on a polyindole-functionalized MWCNT and the as-prepared composite materials used as anode materials for the MOR with high electrocatalytic activity and durability than the commercial Pt/C catalysts.

Furthermore, synthesis of metal nanoparticles (MNPs)/conducting polymer nanocomposites can be achieved through *in situ* methods, for example, electrochemical deposition or electropolymerization route, redox reaction etc. which are low cost and environmentally benign without using strong reducing agents and allow to control over the size, and structure of the MNPs [34]. The noble metal catalysts such as Pt, Pd or Au NPs supported on polymer represent an important class of heterogeneous catalysts for electrocatalytic reactions [35,36]. Patra and Munichandraiah [37] described the high catalytic activity of electrodeposited Pt NPs of about 5 nm on 3,4-polyethylenedioxythiophene (PEDOT) coated carbon paper compared to Pt deposited on carbonpaper alone. Figure 4a shows an FESEM image of Pt-PEDOT/C having a uniform distribution of Pt NPs on PEDOT modified carbon paper.

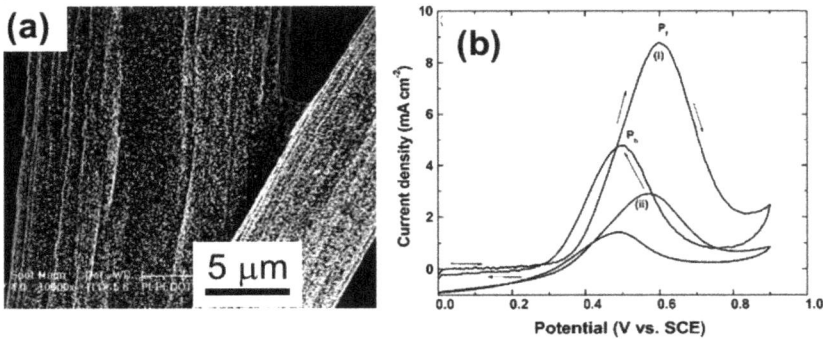

Figure 4. (a)SEM image of Pt-PEDOT/C. (b) Cyclic voltammograms of (A) Pt-PEDOT/C electrode (i) and Pt/C electrode (ii) with a Pt loading of 10 μg cm⁻². Reproduced from Ref. 37 with permission from American Chemical Society, 2009.

CV profile for the Pt-PEDOT/C electrode illustrates a high peak current density for MOR compared to Pt/C electrode. This indicates that PtNPs on PEDOT possesses a high electrochemical activity and beneficial role of PEDOT to speeding up the kinetics of MOR (Figure 4b). Dash *et al.* [38] studied electrodeposition of Pd nanodendritic on PEDOT modified carbon paper (Pd–PEDOT/C) and used as an electrodefor the electrooxidation of a series of alcohol such as propanol, 1, 2- propanediol, 1, 3-propanediol, and glycerol under alkaline medium. The chronoamperometry and repeated cyclic voltammetry data suggest superior catalytic activity toward electrooxidation of glycerol with high electrochemical stability compared to Pd/C electrode. Pandey et al. [39,40] developed a generalized method for the polymer-based composites by using electrochemical deposition of Pd NPs on PANI nanofibers and PEDOT film and the as-prepared composites based electrode shows superior catalytic activity for both ethanol and methanol electrooxidation. There are many reports on electrodeposition of metal NPs on polymer and using them as active electrode material for fuel cell application such as, Kost et al. [41] showed methanol electrooxidation in acidic media by Pt particles dispersed in a PANI film electrode. Rajesh et al. [42] fabricated Pt NPs supported on well aligned Ppy nanotubules which show high catalytic activity for MOR. A series of mixed supporting materials also employed for electrodeposition of metal NPs and used a composite electrode for alcohol fuel cell applications, such as Wu et al. [43] electrodeposited Pt into PANI modified single wall CNTs which exhibited superior catalytic activity. Selvaraj and Alager [44] dispersed Pt, and Pt-Ru NPs in Ppy/multiwall CNTs and the synthesized composites catalysts reveals high catalytic activity for MOR due to the large electrochemically accessible surface areas, high conductivity contributed from CP as well as easier charge-transfer. Pt particles are distributed into a PEDOT–poly(styrene sulfonic acid) (PSS) which displayed high current density for MOR (2.51 mA cm^{-2}) in comparison to bulk Pt electrode (0.45 mA cm^{-2}) as reported by Kuo et al. [45].

Further, Kuo et al. [46] also designed a complex system comprising by electrochemical deposition of Pt and ruthenium oxide (RuO$_2$) particles on PANI-poly(acrylic acid-co-maleic acid) (PAMA) film to obtain composite electrodes which showed superior catalytic activity toward MOR. Gharibi et al. [47] prepared Vulcan XC-72, and PANI-doped trifluoromethane sulfonic acid supports based composites catalyst which demonstrates significant high catalytic activity and stability toward MOR. Yang et al. [48] fabricated a vertically aligned polyaniline-reduced graphene oxide nanosheets by single-step electrodeposition technique which anchoring Pd NPs bya redox reaction between PANI and palladium precursorsand the composites employed as anactive and stable electrocatalyst for MOR and EOR.

Alternatively, conducting polymer can be utilized as both the template and reducing agent for the synthesis of metal composites where metal ions with a higher reduction potential, reduced to form zero-valent metal NPs. For example, Ghosh et al. [49] prepared Pd/Ppy nanocomposites by a one-pot redox process using a surfactant based soft temple which exhibited high catalytic activity towards EOR. TEM image shows the presence of very small Pd NPs (2.2 – 4 nm) dispersed on Ppy nanofibers with high crystallinity as shown in Figure 5a. Moreover, Pd NP-based polymer nanocomposites using three different types of a polymer such as Ppy, PEDOT, and PANI show superior electrocatalytic activity for EOR. Figure 5b-d displays superposition of the 1st and the 100th CV profiles of polymer nanocomposites. The specific catalytic activity of Pd/Ppy is higher in comparison to the other two composites. The high ratio between the forward (I$_f$) and the reverse (I$_b$) anodic peak current density indicates efficient alcohol oxidation with low accumulation of carbonaceous residues. The I$_f$/I$_b$ ratio for Pd/Ppy composites (1.37) is lower in comparison to Pd/PEDOT (1.59), and Pd/PANI (1.48) respectively, which show the nature of polymer at the surface of the Pd NPs may influence the adsorption of organic species. The mass activity of Pd/Ppy nanocomposites is 7.5 times higher compared to the Pd/C catalysts which suggest the effective dispersion of the Pd NPs within the CP nanostructures.

Materials Research Foundations **49** (2019) 271-292   doi: https://doi.org/10.21741/9781644900192-9

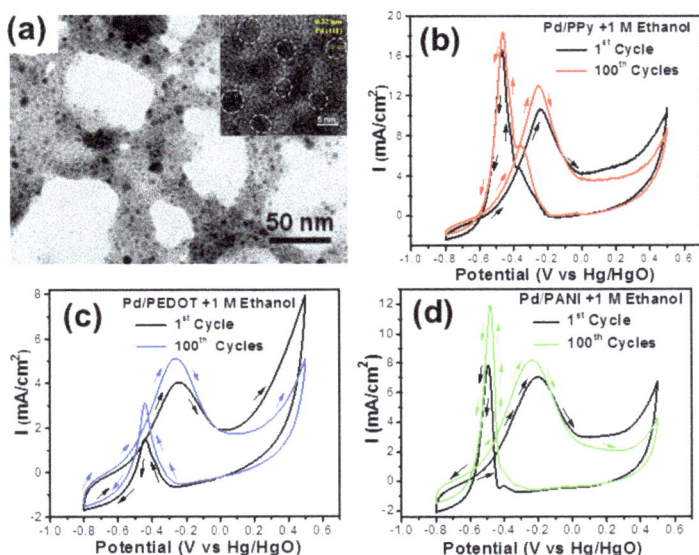

Figure 5. (a) Transmission electron micrograph of Pd/Ppy polymer nanocomposites.Inset: HRTEM image of Pd/Ppy.(b-d)Cyclic voltammograms of the 1st and the 100thcyclic for electrooxidation of ethanol using polymer nanocompositesas the working electrode.Reproduced from Ref. 49 with permission from Springer Publishers, 2017.

Su et al. [50] invented a freestanding and conductive Pd/PA6 nanofibers by the electrospun method. Fig.5a and b show FESEM images of Pd/PA6 nanofibers with a Pd layer (thickness ~85 nm) and an average diameter of nanofibers of 322 nm (Inset of Figure 6a). The as-prepared Pd/PA6 nanofibers have been used as electrodes for a series of alcohol oxidation such as methanol, ethanol, and isopropanol without using any conductive support.

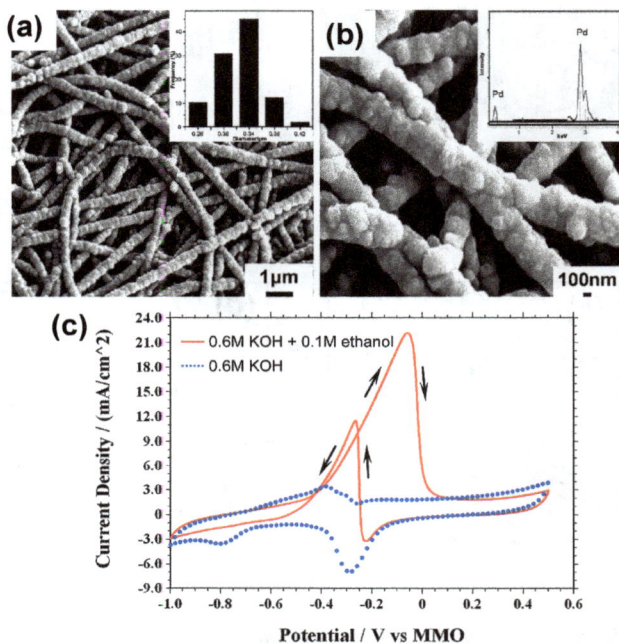

Figure 6. (a-b)SEM images of Pd/PA6 nanofibers at two different magnification. The insets in Fig.5a the diameter size distribution of nanofibers, Fig.5b EDX analysis for Pd/PA6 nanofibers. (c) CVs using Pd/PA6 nanofibrous as a working electrode (red line forKOH and ethanol, blue dotted line for KOH without ethanol). Reproduced from Ref. 50 with permission from American Chemical Society, 2009.

EDX spectrum analysis shows the presence of Pd on PA6 nanofibers (inset of Figure 6b). The Pd/PA6 nanofibers greatly promote the EOR as shown in CV profile indicating the formation of the absorbed intermediate products such as hydroxyl group on the active sites of Pd and removal of the carbonaceous species from ethanol oxidation (Figure 6c) [51]. In contrast, no oxidation peak can be observed in the absence of ethanol suggests a catalytic activity of Pd/PA6 nanofibers. Kim et al. [52] reported a new approach of using ice template for the synthesis of monodispersed Pt nanocrystals supported by two-dimensional PANI nanosheets with an active surface area of 94.57 m^2 g^{-1} and demonstrated high catalytic activity for MOR with high current densities, good

Materials Research Foundations **49** (2019) 271-292   doi: https://doi.org/10.21741/9781644900192-9

durability, and excellent carbon monoxide tolerance. Xu et al. [53] fabricated Ppy functionalized multilayered PtPd/Ppy/PtPd nanotube arrays (TNTAs) for the electrooxidation of alcohol. Fig.7a illustrates PtPd/Ppy/PtPd TNTAs have a sandwich-like structure with a wall thickness of ~ 60 nm and an inner diameter of ~200 nm as shown in Figure 7b. The PtPd/Ppy/PtPd TNTAs showed ~ 6.5 times higher specific electroactivity than that of the Pt/C catalysts, respectively (Figure 7c). Further for long cycling test, the lowering of the peak current density~ 4 % after 500 cycles indicating excellent cycle stability of the PtPd/Ppy/PtPd TNTAs compared to PtPd NTAs (~ 29%). The enhanced catalytic activity and improved stability may be associated with the hollow, array sandwich-like structures of PtPd/Ppy/PtPd electrocatalysts which allow rapid transport of electroactive species and high surface area. Moreover, electron transfer from Ppy polymer to the metal Pt and Pd atoms due to electron delocalization among the π-conjugated ligands of Ppy with Pt 4f orbitals and Pd 3d orbitals which may contribute to enhance the catalytic activity.

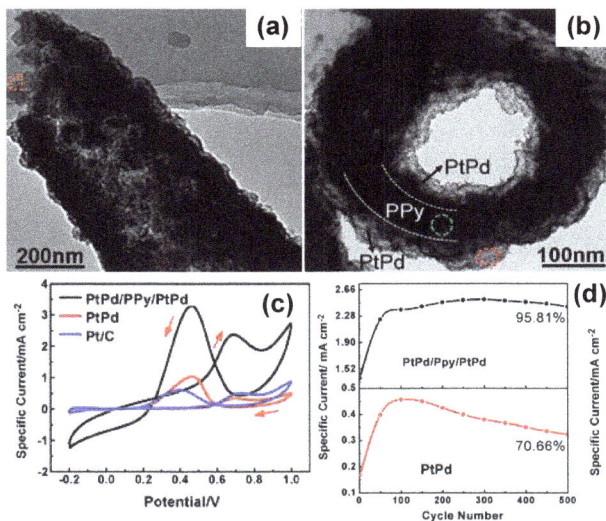

Figure 7. (a-b) TEM images of a PtPd/Ppy/PtPd nanotube arrays (TNTAs), (c) Cyclic voltammograms of PtPd/Ppy/PtPd TNTAs, PtPd NTAs and Pt/C catalysts for electrooxidation of ethanol in 0.5 M H_2SO_4. (d) Cycling study for EOR using nanotube arrays based electrocatalysts. Reproduced from Ref. 53 with permission from Nature Publishers, 2013.

Lee and Co-worker [54] used amine-terminated poly(N-isopropylacrylamide) functionalized ultrathin $CuPt_3$ wavy nanowires as catalysts having excellent catalytic activity for the MOR. Ghosh et al. [55] developed an alternative approach for the synthesis of Pt NPs based bimetallic alloys on Ppy nanofibers via radiolys is.Figure 8a shows $Pt_{66}Pd_{34}$ nanoparticles are uniformly deposited on Ppy nanofibers having a remarkably high electrocatalytic activity for MOR compared with the Pd/C catalyst.

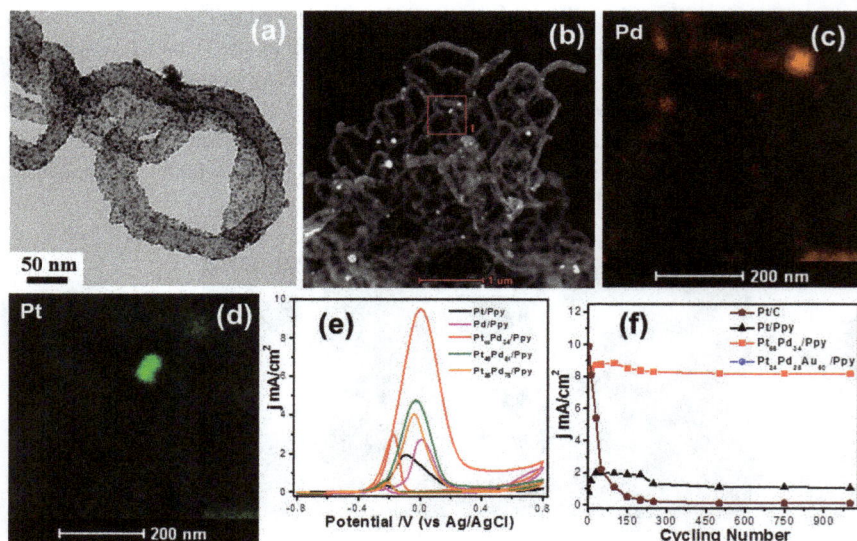

Figure 8. Transmission electron micrograph and high-angle annular dark-field (HAADF) images of (a,b)and (c,d) corresponding elemental mapping of $Pt_{66}Pd_{34}$/Ppy composites. (e)CV profiles for the electrooxidation of methanol using Pt/Ppy, Pd/Ppy and other bimetallic composites, $Pt_{66}Pd_{34}$/Ppy, $Pt_{49}Pd_{51}$/Ppy, $Pt_{25}Pd_{75}$/Ppy as working electrodes. (f) Long cycling study of Pt/C, polymer composite based electrodes for methanol oxidation at scan rate 20 mV Sec^{-1}. Reproduced from Ref. 55 with permission from Royal Society of Chemistry, 2017.

Figure 8b-d shows scanning transmission electron microscopy (STEM) of $Pt_{66}Pd_{34}$/Ppy composites which reveal small NPs composed of Pt-Pd alloys deposited to the Ppy nanofiber. Cyclic voltammograms (Figure 8e) shows $Pt_{66}Pd_{34}$/Ppy electrocatalysts have superior catalytic activity for MOR with high current density and better tolerance to CO or carbonaceous intermediates under alkaline medium. Due to a large number of active sites at Pt−Pd heterojunctions and cooperative action of the metals in the alloy may enhance catalytic activity. A minimal change in current density of the $Pt_{66}Pd_{34}$/Ppy electrode material (5.7%) obtained compared to commercial Pt/C (98%) decay after 1000 potential cycle during MOR as shown in Fig.8f. Additionally, Pt loading in bimetallic catalysts is significantly less, which lower the overall cost of the catalyst. Hence, polypyrrole nanofibers supported metal NPs potential anode catalyst for direct methanol fuel cells application.

Wang et al. [56] fabricated Pd/PANI/Pd sandwich-structured nanotube array (SNTA) which display high catalytic activity for EOR and demonstrates significant improvement in catalytic activity and durability compared with Pd/C catalysts. Xu et al. [57] also reported multilayered Pt/CeO_2/PANI with three-layered hollow nanorod arrays (THNRAs) and $ZnO/Pt/CeO_2$/PANI hybrid by combining CP with metal oxides as multilayered supporting material which illustrates elevated electrocatalytic activity toward MOR. Multimetallic nanoalloys have been explored as active catalysts for electrooxidation of alcohol [58,59]. Recently, Ghosh et al. [60] reported a radiolytic synthetic route for the synthesis of Pd based multimetallic nanoalloys decorated onPpy nanofibers which can be used as anode material for EOR. For example, bimetallic, Pd-Pt/Ppy, Pd-Au/Ppy, and trimetallic Pd-Pt-Au/Ppy composites have been prepared by pre-calculated mass ratio. CV profiles suggest that the trimetallic Pd alloy based polymer composite catalysts demonstrate dramatically enhanced electrocatalytic activity for EOR in the alkaline medium as shown in Figure 9a.

The peak current density for $Pd_{30}Pt_{29}Au_{41}$/Ppy shows exceptionally high catalytic activity which is ~5.5 times higher compared to Pd/C catalyst (Figure 9b). Chronoamperometric measurement shows bimetallic and trimetallic electrode materials reach a stable current after 500 seconds, whereas the current nearly decreases to zero for Pd/Ppy composite electrode as shown in Figure 9c. Moreover, FESEM image of trimetallic electrode shows nanoparticles are well distributed on polymer nanofibers after1000 cycling of EOR (Figure 9d). The superior catalytic activity and durability of nanocomposites obtained due to electronic structural modification of Pd metals and increased active sites at the Pt−Pd−Au heterojunctions and the strong interactions between Ppy nanofibers support with metallic NPs.The electrochemical performance of the trimetallic catalysts have been compared with other Pd-based catalysts from the literature in Table 1. This clearly indicates that bimetallic and tri metallic alloy based composites performed as stable electrocatalyst compared to other monometallic Pd catalysts for alcohol oxidation.

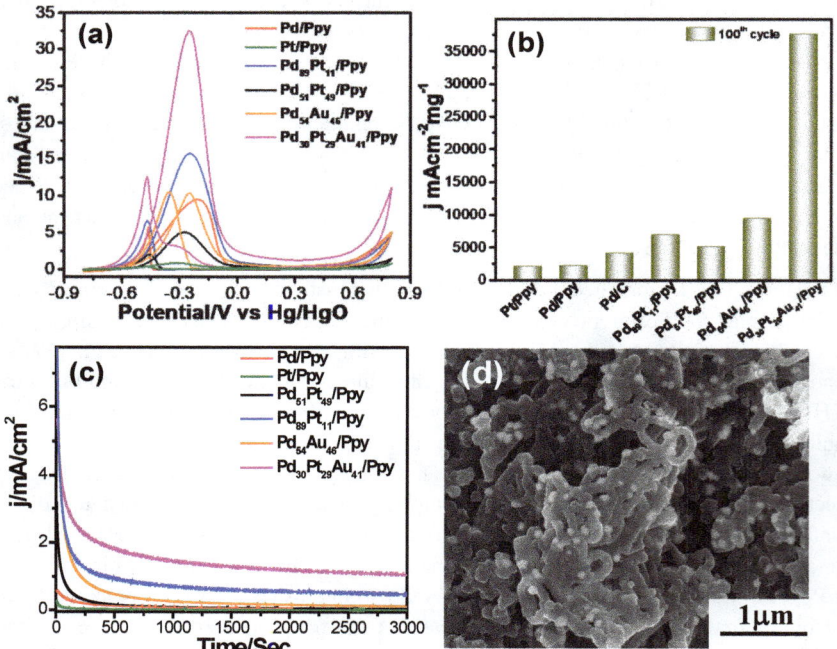

Figure 9. (a)CV profiles for ethanol oxidation using Pd based multimetallic nanoalloys decorated on Ppy nanofibers as working electrodes. (b) The mass activity of Pd/C, Pd-polymer composite electrodes for ethanol oxidation. (c) Chronoamperometric curves for the ethanol oxidation at constant potential $-0.25V$ vs. Hg/HgO. (d) FESEM image of $Pd_{30}Pt_{29}Au_{41}$/Ppy composites electrodes after 1000 cycling of ethanol electrooxidation. Reproduced from Ref. 60 with permission from American Chemical Society, 2017.

Table 1 *Comparison of the electrochemical performance of Pd electrocatalysts for the ethanol oxidation.*

Electrode	Fuel	E_{onset}, mV/SCE	I_f, mA.cm^{-2}	I_f, mA.cm^{-2}.mg^{-1}	Specific Current, I_f, mA.mg^{-1}	Reference
Pd/Ppy	Ethanol	−628	7.05	4147	248.70	49
Pd/PANI/Pd		-	-	-	310	56
Pd/PEDOT		-	-	-	458.5	40
PtPd/PPy/PtPd nanotube		-	3.1	-	-	53
Pd/Ppy		−640	9.50	2168	325	66
Pd$_{89}$Pt$_{11}$/Ppy		−630	15.8	5197	782	66
Pd$_{54}$Au$_{46}$/Ppy		−680	10.35	5280	792	66
Pd$_{30}$Pt$_{29}$Au$_{41}$/Ppy		−630	32.45	12018	1802	66
Pt/PEDOT	Methanol	-	-	-	614	37
Pt/CNT–PANI		−430	1	-	-	31
Pt/Ppy		−265	0.40	-	396	55
Pd/Ppy		−81	0.27	-	4578	55
Pt$_{66}$Pd$_{34}$/Ppy		−256	2.98	-	4522	55
Pt$_{49}$Pd$_{51}$/Ppy		−215	0.594	-	2982	55
Pt$_{25}$Pd$_{75}$/Ppy		−200	0.627	-	3620	55
Pt$_{24}$Pd$_{26}$Au$_{50}$/Ppy		−227	1.73	-	8259	55
Pd-PEDOT/C	Propanol	-	1.5	-	-	38
Pd-PEDOT/C	Glycerol	-	8.4	-	-	38
Pd-PEDOT/C	1, 3-propanediol	-	6	-	-	38

Conclusion

In summary, metal/polymer composite materials recognized as a new generation electrocatalysts for direct alcohol fuel cells. Different kinds of metal, as well as nanoalloys, can be deposited on polymer via *ex situ* method where the polymer can be dissolved in an appropriate solvent, then NPs may combine to the polymer solution and *in situ* method through *in situ* polymerization in the presence of NPs or synthesis of polymer network with nanostructures. Experimental results suggest that conducting polymer significantly contributed to improving the catalytic activity of metal nanostructures for alcohol oxidation regarding high specific current density and durability compared with Pd/C catalysts. The mass activity has been successful increasing via the synergetic combination of polymer with carbon nanotubes or graphene or metal oxides as supporting materials and using multimetallic nanoalloys as catalysts and demonstrated better tolerance towards intermediate poisoning. Development of polymer nanocomposite and its modifications, recent trend towards the anode materials for alcohol oxidation have been reviewed. Detailed understanding of structure-property-processing relationships between metal catalysts and the polymers are useful for further development of active electrocatalysts for fuel cell application. The simple, low-cost and environmentally friendly approach must be developed for large scale production of metal NPs/polymer nanocomposites based electrocatalysts with controlled structure and properties, which make them the potential for applications in fuel cells.

References

[1] U. Lucia, Overview on fuel cells, Renew. Sustain. Energy Rev. 30 (2014) 164–169. https://doi.org/10.1016/j.rser.2013.09.025

[2] C. Lamy, A. Lima, V. LeRhun, F. Delime, C. Coutanceau, J.M. Léger, Recent advances in the development of direct alcohol fuel cells (DAFC), J. Power Sources. 105 (2002) 283–296. https://doi.org/10.1016/S0378-7753(01)00954-5

[3] S. Ghosh, T. Maiyalagan, R.N. Basu, Recent Advances in Nanostructured Electrocatalysts for Low-temperature Direct Alcohol Fuel Cells, in: Electrocatal. Low Temp. Fuel Cells, Wiley-VCH Verlag GmbH & Co. KGaA, Weinheim, Germany, 2017, pp. 347–371. https://doi.org/10.1002/9783527803873.ch11

[4] S.P.S. Badwal, S. Giddey, A. Kulkarni, J. Goel, S. Basu, Direct ethanol fuel cells for transport and stationary applications – A comprehensive review, Appl. Energy. 145 (2015) 80–103. https://doi.org/10.1016/j.apenergy.2015.02.002

[5] E. Antolini, Palladium in fuel cell catalysis, Energy Environ. Sci. 2 (2009) 915–931. https://doi.org/10.1039/b820837a

[6] C. Bianchini, P.K. Shen, Palladium-Based Electrocatalysts for Alcohol Oxidation in Half Cells and in Direct Alcohol Fuel Cells, Chem. Rev. 109 (2009) 4183–4206. https://doi.org/10.1021/cr9000995

[7] S. Ghosh, H. Remita, P. Kar, S. Choudhury, S. Sardar, P. Beaunier, P.S. Roy, S.K. Bhattacharya, S.K. Pal, Facile synthesis of Pd nanostructures in hexagonal mesophases as a promising electrocatalyst for ethanol oxidation, J. Mater. Chem. A. 3 (2015) 9517–9527. https://doi.org/10.1039/C5TA00923E

[8] P.T. Yu, W. Gu, J. Zhang, R. Makharia, F.T. Wagner, H.A. Gasteiger, Carbon-Support Requirements for Highly Durable Fuel Cell Operation, in: Polym. Electrolyte Fuel Cell Durab., Springer New York, New York, 2009,pp. 29–53. https://doi.org/10.1007/978-0-387-85536-3_3

[9] Y.C. Park, H. Tokiwa, K. Kakinuma, M. Watanabe, M. Uchida, Effects of carbon supports on Pt distribution, ionomer coverage and cathode performance for polymer electrolyte fuel cells, J. Power Sources. 315 (2016) 179–191. https://doi.org/10.1016/j.jpowsour.2016.02.091

[10] J. Goel, S. Basu, Effect of support materials on the performance of direct ethanol fuel cell anode catalyst, Int. J. Hydrogen Energy. 39 (2014) 15956–15966. https://doi.org/10.1016/j.ijhydene.2014.01.203

[11] P. Serp, J.L. Figueiredo, Carbon Materials for Catalysis, John Wiley & Sons, Inc., Hoboken, NJ, USA, 2008. https://doi.org/10.1002/9780470403709

[12] N. Mackiewicz, G. Surendran, H. Remita, B. Keita, G. Zhang, L. Nadjo, A. Hagège, E. Doris, C. Mioskowski, Supramolecular self-assembly of amphiphiles on carbon nanotubes: A versatile strategy for the construction of CNT/metal nanohybrids, application to electrocatalysis, J. Am. Chem. Soc. 130 (2008) 8110–8111. https://doi.org/10.1021/ja8026373

[13] S. Ghosh, Y. Holade, H. Remita, K. Servat, P. Beaunier, A. Hagège, K.B. Kokoh, T.W. Napporn, One-pot synthesis of reduced graphene oxide supported gold-based nanomaterials as robust nanocatalysts for glucose electrooxidation, Electrochim. Acta. 212 (2016) 864–875. https://doi.org/10.1016/j.electacta.2016.06.169

[14] C.T. Hable, M.S. Wrighton, Electrocatalytic oxidation of methanol and ethanol: a comparison of platinum-tin and platinum-ruthenium catalyst particles in a conducting polyaniline matrix, Langmuir. 9 (1993) 3284–3290. https://doi.org/10.1021/la00035a085

[15] B. Rajesh, K.R. Thampi, J.M. Bonard, A.J. McEvoy, N. Xanthopoulos, H.J. Mathieu, B. Viswanathan, Pt particles supported on conducting polymeric nanocones as electro-catalysts for methanol oxidation, J. Power Sources. 133 (2004) 155–161. https://doi.org/10.1016/j.jpowsour.2004.02.008

[16] K. Dutta, S. Das, D. Rana, P.P. Kundu, Enhancements of catalyst distribution and functioning upon utilization of conducting polymers as supporting matrices in DMFCs: A review, Polym. Rev. 55 (2015) 1–56. https://doi.org/10.1080/15583724.2014.958771

[17] C. Zhan, G. Yu, Y. Lu, L. Wang, E. Wujcik, S. Wei, Conductive polymer nanocomposites: a critical review of modern advanced devices, J. Mater. Chem. C. 5 (2017) 1569–1585. https://doi.org/10.1039/C6TC04269D

[18] Y. Shi, L. Peng, Y. Ding, Y. Zhao, G. Yu, Nanostructured conductive polymers for advanced energy storage, Chem. Soc. Rev. 44 (2015) 6684–6696. https://doi.org/10.1039/C5CS00362H

[19] S. Ghosh, T. Maiyalagan, R.N. Basu, Nanostructured conducting polymers for energy applications: towards a sustainable platform, Nanoscale. 8 (2016) 6921–6947. https://doi.org/10.1039/C5NR08803H

[20] S. Ghosh, N.A. Kouamé, L. Ramos, S. Remita, A. Dazzi, A. Deniset-Besseau, P. Beaunier, F. Goubard, P.-H. Aubert, H. Remita, Conducting polymer nanostructures for photocatalysis under visible light, Nat. Mater. 14 (2015) 505–511. https://doi.org/10.1038/nmat4220

[21] X. Chen, S. Wei, A. Yadav, R. Patil, J. Zhu, R. Ximenes, L. Sun, Z. Guo, Poly(propylene)/carbon nanofiber nanocomposites: ex situ solvent-assisted preparation and analysis of electrical and electronic properties, Macromol. Mater. Eng. 296 (2011) 434–443. https://doi.org/10.1002/mame.201000341

[22] S. Sardar, P. Kar, H. Remita, B. Liu, P. Lemmens, S.K. Pal, S. Ghosh, Enhanced charge separation and fret at heterojunctions between semiconductor nanoparticles and conducting polymer nanofibers for efficient solar light harvesting, Sci. Rep. 5 (2015) 17313. https://doi.org/10.1038/srep17313

[23] G. Wang, A. Morrin, M. Li, N. Liu, X. Luo, Nanomaterial-doped conducting polymers for electrochemical sensors and biosensors, J. Mater. Chem. B. 6 (2018) 4173–4190. https://doi.org/10.1039/C8TB00817E

[24] K. Mallick, M.J. Witcomb, A. Dinsmore, M.S. Scurrell, Fabrication of a metal nanoparticles and polymer nanofibers composite material by an in situ chemical synthetic route, Langmuir. 21 (2005) 7964–7967. https://doi.org/10.1021/la050534j

[25] H. Wei, D. Ding, S. Wei, Z. Guo, Anticorrosive conductive polyurethane multiwalled carbon nanotube nanocomposites, J. Mater. Chem. A. 1 (2013) 10805. https://doi.org/10.1039/c3ta11966a

[26] H. Wu, G. Yu, L. Pan, N. Liu, M.T. McDowell, Z. Bao, Y. Cui, Stable Li-ion battery anodes by in-situ polymerization of conducting hydrogel to conformally coat silicon nanoparticles, Nat. Commun. 4 (2013) 1943. https://doi.org/10.1038/ncomms2941

[27] S. Ghosh, A.L. Teillout, D. Floresyona, P. de Oliveira, A. Hagège, H. Remita, Conducting polymer-supported palladium nanoplates for applications in direct alcohol oxidation, Int. J. Hydrogen Energy. 40 (2015) 4951–4959. https://doi.org/10.1016/j.ijhydene.2015.01.101

[28] P.K. Rastogi, V. Ganesan, S. Krishnamoorthi, A promising electrochemical sensing platform based on a silver nanoparticles decorated copolymer for sensitive nitrite determination, J. Mater. Chem. A. 2 (2014) 933–943. https://doi.org/10.1039/C3TA13794E

[29] V. Selvaraj, M. Alagar, Ethylene glycol oxidation on Pt and Pt–Ru nanoparticle decorated polythiophene/multiwalled carbon nanotube composites for fuel cell applications, Nanotechnology. 19 (2008) 45504. https://doi.org/10.1088/0957-4484/19/04/045504

[30] D. Prasanna, V. Selvaraj, Cyclophosphazene based conductive polymer-carbon nanotube composite as novel supporting material for methanol fuel cell applications, J. Colloid Interface Sci. 472 (2016) 116–125. https://doi.org/10.1016/j.jcis.2016.03.032

[31] L. Shi, R.P. Liang, J.D. Qiu, Controllable deposition of platinum nanoparticles on polyaniline-functionalized carbon nanotubes, J. Mater. Chem. 22 (2012) 17196–17203. https://doi.org/10.1039/c2jm31859h

[32] L. Wei, Y.J. Fan, J.H. Ma, L.H. Tao, R.X. Wang, J.P. Zhong, H. Wang, Highly dispersed Pt nanoparticles supported on manganese oxide–poly(3,4-ethylenedioxythiophene)–carbon nanotubes composite for enhanced methanol electrooxidation, J. Power Sources. 238 (2013) 157–164. https://doi.org/10.1016/j.jpowsour.2013.03.051

[33] R.X. Wang, Y.J. Fan, L. Wang, L.N. Wu, S.N. Sun, S.G. Sun, Pt nanocatalysts on a polyindole-functionalized carbon nanotube composite with high performance for methanol electrooxidation, J. Power Sources. 287 (2015) 341–348. https://doi.org/10.1016/j.jpowsour.2015.03.181

[34] P. Xu, X. Han, B. Zhang, Y. Du, H.-L. Wang, Multifunctional polymer–metal nanocomposites via direct chemical reduction by conjugated polymers, Chem. Soc. Rev. 43 (2014) 1349–1360. https://doi.org/10.1039/C3CS60380F

[35] V. Armel, O. Winther-Jensen, R. Kerr, D.R. MacFarlane, B. Winther-Jensen, Designed electrodeposition of nanoparticles inside conducting polymers, J. Mater. Chem. 22 (2012) 19767–19773. https://doi.org/10.1039/c2jm34214f

[36] T. Thirugnanasambandan, Polymers-metal nanocomposites, in: Environ. Nanotechnol., Springer, Cham, 2019: pp. 213–254. https://doi.org/10.1007/978-3-319-98708-8_8

[37] S. Patra, N. Munichandraiah, Electrooxidation of methanol on pt-modified conductive polymer PEDOT, Langmuir. 25 (2009) 1732–1738. https://doi.org/10.1021/la803099w

[38] S. Dash, N. Munichandraiah, Electrocatalytic oxidation of C 3 -aliphatic alcohols on electrodeposited pd-pedot nanodendrites in alkaline medium, J. Electrochem. Soc. 160 (2013) H197–H202. https://doi.org/10.1149/2.007304jes

[39] R.K. Pandey, V. Lakshminarayanan, Electro-Oxidation of Formic Acid, Methanol, and ethanol on electrodeposited pd-polyaniline nanofiber films in acidic and alkaline medium, J. Phys. Chem. C. 113 (2009) 21596–21603. https://doi.org/10.1021/jp908239m

[40] R.K. Pandey, V. Lakshminarayanan, Enhanced electrocatalytic activity of Pd-dispersed 3,4-polyethylenedioxythiophene film in hydrogen evolution and ethanol electro-oxidation reactions, J. Phys. Chem. C. 114 (2010) 8507–8514. https://doi.org/10.1021/jp1014687

[41] K.M. Kost, D.E. Bartak, B. Kazee, T. Kuwana, Electrodeposition of platinum microparticles into polyaniline films with electrocatalytic applications, Anal. Chem. 60 (1988) 2379–2384. https://doi.org/10.1021/ac00172a012

[42] B. Rajesh, K.R. Thampi, J.M. Bonard, H.J. Mathieu, N. Xanthopoulos, B. Viswanathan, Conducting polymeric nanotubules as high performance methanol oxidation catalyst support, Chem. Commun. 0 (2003) 2022. https://doi.org/10.1039/b305591d

[43] G. Wu, L. Li, J.H. Li, B.Q. Xu, Methanol electrooxidation on Pt particles dispersed into PANI/SWNT composite films, J. Power Sources. 155 (2006) 118–127. https://doi.org/10.1016/j.jpowsour.2005.04.035

[44] V. Selvaraj, M. Alagar, Pt and Pt–Ru nanoparticles decorated polypyrrole/multiwalled carbon nanotubes and their catalytic activity towards methanol oxidation, Electrochem. Commun. 9 (2007) 1145–1153. https://doi.org/10.1016/j.elecom.2007.01.011

[45] C.W. Kuo, L.M. Huang, T.C. Wen, A. Gopalan, Enhanced electrocatalytic performance for methanol oxidation of a novel Pt-dispersed poly(3,4-ethylenedioxythiophene)–poly(styrene sulfonic acid) electrode, J. Power Sources. 160 (2006) 65–72. https://doi.org/10.1016/j.jpowsour.2006.01.100

[46] C.W. Kuo, Z.Y. Kuo, T.Y. Wu, J.Y. Chen, W.B. Li, Enhanced Electrocatalytic Performance for Methanol Oxidation via Insertion of Ruthenium Oxide Particles into Pt and Polyaniline-Poly(Acrylic Acid-co-Maleic Acid) Composite Electrode, ECS Trans. 50 (2013) 1997–2000. https://doi.org/10.1149/05002.1997ecst

[47] H. Gharibi, K. Kakaei, M. Zhiani, Platinum nanoparticles supported by a vulcan XC-72 and PANI doped with trifluoromethane sulfonic acid substrate as a new electrocatalyst for direct methanol fuel cells, J. Phys. Chem. C. 114 (2010) 5233–5240. https://doi.org/10.1021/jp9119414

[48] L. Yang, Y. Tang, D. Yan, T. Liu, C. Liu, S. Luo, Polyaniline-reduced graphene oxide hybrid nanosheets with nearly vertical orientation anchoring palladium nanoparticles for highly active and stable electrocatalysis, ACS Appl. Mater. Interfaces. 8 (2016) 169–176. https://doi.org/10.1021/acsami.5b08022

[49] S. Ghosh, N. Bhandary, S. Basu, R.N. Basu, Synergistic effects of polypyrrole nanofibers and pd nanoparticles for improved electrocatalytic performance of Pd/PPy nanocomposites for ethanol oxidation, Electrocatalysis. 8 (2017) 329–339. https://doi.org/10.1007/s12678-017-0374-x

[50] L. Su, W. Jia, A. Schempf, Y. Ding, Y. Lei, Free-standing palladium/polyamide 6 nanofibers for electrooxidation of alcohols in alkaline medium, J. Phys. Chem. C. 113 (2009) 16174–16180. https://doi.org/10.1021/jp905606s

[51] Z.X. Liang, T.S. Zhao, J.B. Xu, L.D. Zhu, Mechanism study of the ethanol oxidation reaction on palladium in alkaline media, Electrochim. Acta. 54 (2009) 2203–2208. https://doi.org/10.1016/j.electacta.2008.10.034

[52] K. Kim, H. Ahn, M.J. Park, Highly Catalytic Pt Nanoparticles Grown in Two-Dimensional Conducting Polymers at the Air–Water Interface, ACS Appl. Mater. Interfaces. 9 (2017) 30278–30282. https://doi.org/10.1021/acsami.7b10821

[53] H. Xu, L.-X. Ding, C.-L. Liang, Y.-X. Tong, G.-R. Li, High-performance polypyrrole functionalized PtPd electrocatalysts based on PtPd/PPy/PtPd three-layered nanotube arrays for the electrooxidation of small organic molecules, NPG Asia Mater. 5 (2013) e69–e69. https://doi.org/10.1038/am.2013.54

[54] G. Fu, X. Yan, Z. Cui, D. Sun, L. Xu, Y. Tang, J.B. Goodenough, J.-M. Lee, Catalytic activities for methanol oxidation on ultrathin CuPt3 wavy nanowires with/without smart polymer, Chem. Sci. 7 (2016) 5414–5420. https://doi.org/10.1039/C6SC01501H

[55] S. Ghosh, S. Bera, S. Bysakh, R.N. Basu, Conducting polymer nanofiber-supported Pt alloys: unprecedented materials for methanol oxidation with enhanced electrocatalytic performance and stability, Sustain. Energy Fuels. 1 (2017) 1148–1161. https://doi.org/10.1039/C7SE00126F

[56] A.L. Wang, H. Xu, J.X. Feng, L.X. Ding, Y.X. Tong, G.R. Li, Design of Pd/PANI/Pd sandwich-structured nanotube array catalysts with special shape effects and synergistic effects for ethanol electrooxidation, J. Am. Chem. Soc. 135 (2013) 10703–10709. https://doi.org/10.1021/ja403101r

[57] H. Xu, A.L. Wang, Y.X. Tong, G.R. Li, Enhanced catalytic activity and stability of Pt/CeO$_2$/PANI hybrid hollow nanorod arrays for methanol electro-oxidation, ACS Catal. 6 (2016) 5198–5206. https://doi.org/10.1021/acscatal.6b01010

[58] Y. Liu, S. Liu, Z. Che, S. Zhao, X. Sheng, M. Han, J. Bao, Concave octahedral Pd@PdPt electrocatalysts integrating core–shell, alloy and concave structures for high-efficiency oxygen reduction and hydrogen evolution reactions, J. Mater. Chem. A. 4 (2016) 16690–16697. https://doi.org/10.1039/C6TA07124D

[59] J. Datta, A. Dutta, S. Mukherjee, The beneficial role of the cometals Pd and Au in the carbon-supported PtPdAu catalyst toward promoting ethanol oxidation kinetics in alkaline fuel cells: Temperature effect and reaction mechanism, J. Phys. Chem. C. 115 (2011) 15324–15334. https://doi.org/10.1021/jp200318m

[60] S. Ghosh, S. Bera, S. Bysakh, R.N. Basu, Highly active multimetallic palladium nanoalloys embedded in conducting polymer as anode catalyst for electrooxidation of ethanol, ACS Appl. Mater. Interfaces. 9 (2017) 33775–33790. https://doi.org/10.1021/acsami.7b08327

Chapter 10

Advances in Electrocatalyst for Ethanol Electro-Oxidation

R. Chauhan[1], V.C. Srivastava[1,*]

[1]Department of Chemical Engineering, Indian Institute of Technology Roorkee, Roorkee 247667, Uttarakhand, India

*Corresponding Author: Phone: +91–1332–285889; fax: +91–1332–276535;

vimalcsr@yahoo.co.in, vimalfch@iitr.ac.in (VCS), rohitchn18@gmail.com (RC)

Abstract

Ethanol is used in a fuel cell as a source of fuel due to its low-cost and low-toxicity towards the environment. Direct ethanol fuel cells (DEFC) are of two types, i.e., acid and alkaline type, in which the alkaline type DEFC shows better performance. Different types of catalysts like metal (noble or non-noble) and non-metal (carbon) based are being used as a catalyst for ethanol oxidation. Ethanol electro-oxidation has complex reaction mechanism having different types of intermediates and final products. This chapter aims to present the advances in the analysis of electrocatalytic activity, reaction mechanism and synthesis of the electrocatalyst for electro-oxidation of ethanol.

Keywords

Direct Ethanol Fuel Cell, Advanced Electro-Oxidation, Nanocatalyst, Reaction Mechanism, Pathways

List of Abbreviations

AEM   Anion exchange membrane
CND   Carbon nanodots
DAFC   Direct alcohol fuel cell
DEFC   Direct ethanol fuel cell
DEMS   Differential electrochemical mass spectrometry
DMFC   Direct methanol fuel cell
GO   Graphene oxide
MEA   Membrane electrode assembly
PEM   Proton exchange membrane
r-GO   Reduced graphene oxide

Contents

1.   Introduction

Energy plays a very vital role in the economic and social growth of any society. In the 20th century, coal and fossil fuels were extensively used for energy needs. In the 21st century, fossil fuel remains the primary source of energy. Its usage is creating problems for the future generations as its generates many types of harmful emissions. Future human generations need are energy sources which produce zero or very low harmful emissions to the environment. The demand for sustainable and clean energy source can be fulfilled by the fuel cell. A fuel cell is an electrochemical device which could be used for conversion of chemical energy into electrical energy. In the fuel cell, H_2 and alcohol are used as a fuel having chemical energy, and are converted into electrical energy. The fuel cell produces zero or very low harmful emission. The fuel cell can be used as a portable electrochemical device for energy production [1,2].

Different types of fuels can be used in fuel cells; however, hydrogen and alcohols (mainly methanol and ethanol) are the most frequently used fuels. Hydrogen is one of the most popular fuel because of its high electro-kinetics, and since oxidation of hydrogen produces only water [3]. However, hydrogen gas is too expensive to be used as fuel. Its storage is also complicated. Therefore, liquid alcohols are used as an alternative fuel in fuel cells which is so-called direct alcohol fuel cells (DAFCs). Methanol is an elementary alcohol, therefore, it is frequently used in fuel cells, however, it is toxic in nature and not easy to produce whereas ethanol can be produced in large scale from biomass and/or from many agricultural products and has higher specific energy (8.01 kWh/kg) than methanol (6.09 kWh/kg) which is closer to gasoline (11 kWh/kg) and hydrocarbons (10 kWh/kg) [4,5]. During electro-oxidation of ethanol in direct ethanol fuel cells (DEFCs), 12 e$^-$ are produced in a single complete reaction of oxidation forming CO_2, whereas in DMFC (direct methanol fuel cells), six e$^-$ are produced for a single complete oxidation reaction of methanol to CO_2 [6].

Worldwide production of ethanol was ~27000 million gallons in 2017 (Fig. 1). The United States of America (U.S.A.) and Brazil are major producers of ethanol acounting for ~84% of world ethanol production. The use of ethanol as a fuel reportedly enhanced from 155 million gallons in 2014 to 280 million gallons in 2017 in India, and in the USA, it increased from 14.3 billion gallons in 2014 to 15.8 billion gallons in 2017 [7].

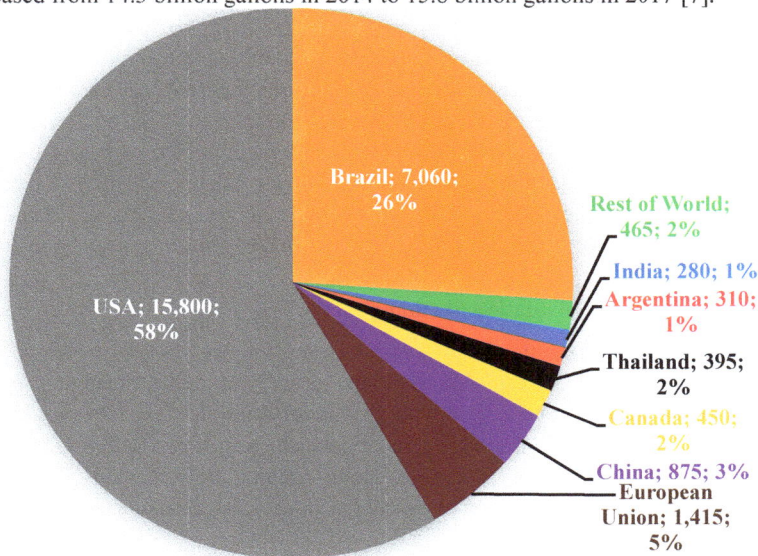

Figure1. Production of ethanol in millions of gallons by worldwide countries in 2017[6].

Electro-oxidation reaction of ethanol occurs through different paths and gives different types of products i.e., methane (CH_4), acetic acid (CH_3COOH), acetaldehyde (CH_3CHO) and carbon dioxide (CO_2), and also small quantities of formic acid, ethylene glycol, ethane and ethyl acetate [8–11]. The incomplete oxidation reaction of ethanol in DEFC happens due to blockage of the electrode surface resulting in less C-C bond cleavage [7,8–12].

Improvement in the electro-oxidation of ethanol is necessary for higher utilization of DEFCs. Higher electrocatalytic activity for ethanol oxidation reaction can be achieved by providing sufficient active surface area for cleaving of C-C bond and achieving simultaneous removal of –CH_x and CO species with an appropriate surface structure for enhanced production of CO_2. Furthermore, simultaneous adsorption of water and ethanol molecules is also essential for bringing them together. Hence, the catalyst used in the fuel cells must be capable of adsorbing and activating the water and ethanol molecules with the right combination of textural properties. Several types of metal and non-metal catalysts have been used for electro-oxidation of ethanol in DEFCs. Pt, Pd, Ti, Zn, etc. are reported as base metals which have been used as electrocatalysts [13–16]. Many methods are being researched for the advancement of metal-based electro-catalysts. In one method, the main metal is doped by another noble metal (Au, Ru, or Ir) or non-noble metal (Ni, Sn, or Cu) which enhances the electro-oxidation of ethanol in DEFCs [14,17]. In another method, carbon-based support material (carbon nanosheets, carbon nanotubes, carbon nanofiber, graphene oxide and/or reduced graphene oxide, etc.) [18–20] and conducting polymers (polythiophene, polypyrrole, polyaniline, etc.) are used for electro-oxidation of ethanol [21]. Many challenges exist during electro-oxidation of ethanol for commercialization of the DEFCs. This chapter highlights the contexts and challenges in the advancement of electro-oxidation of ethanol in the DEFCs.

2.   Theory of electro-oxidation of ethanol

The ethanol electro-oxidation produces electrons in the DEFC which generates electricity. All the electro-oxidation reactions occur within the membrane electrode assembly (MEA) which is connected with the current controller that generates electricity over the system. The electrons generated on anode which is shown by Eq. 1, travel to the cathode through the solution. Electro-oxidation reaction flow diagram for DEFCs is shown in Fig. 2. Following types of chemical reactions occur on anode and cathode in the DEFC [22]:

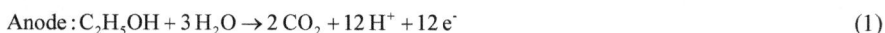

$$\text{Anode}: C_2H_5OH + 3\,H_2O \rightarrow 2\,CO_2 + 12\,H^+ + 12\,e^- \tag{1}$$

$$\text{Cathode}: 3\,O_2 + 12\,H^+ + 12\,e^- \rightarrow 6\,H_2O \tag{2}$$

$$\text{Overall}: C_2H_5OH + 3\,O_2 \rightarrow 2\,CO_2 + 3\,H_2O \tag{3}$$

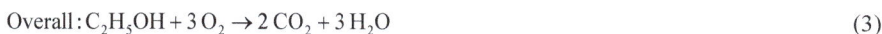

Two types of medium i.e., acidic and alkaline can be used for the oxidation of ethanol in DEFC. Therefore, DEFC is two types: i) acidic and ii) alkaline type DEFC. Mechanism of oxidation is different in these media, but the aim is the same i.e., to produce the enhanced number of electrons by oxidizing ethanol.

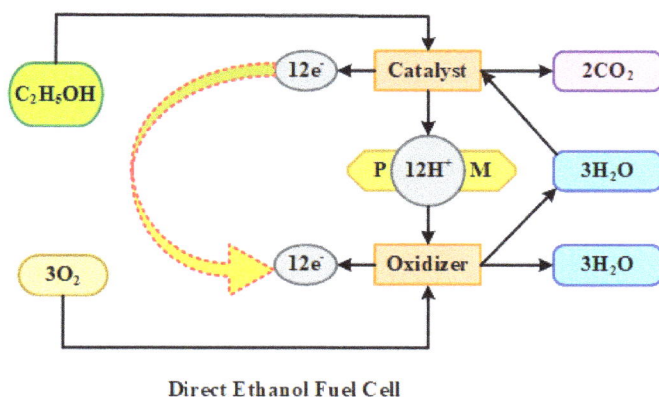

Direct Ethanol Fuel Cell

Figure 2. Electro-oxidation reaction flow diagram for direct ethanol fuel cell (DEFC) [22].

2.1 Acidic medium DEFC (Proton exchange membrane/PEM)

Acidic medium DEFC oxidizes ethanol at a low pH (pH<5). Proton exchange membrane (PEM) is utilized in acidic medium for migration of cations into the electrolyte. The circuit is completed (at all over phase) due to the weakly bonded cations of membrane structures. H^+ ion has a high concentration in the acidic medium, therefore, it shows better transfer as membrane cations in the alkaline medium. Hence, low pH favors the PEM. Schematic diagram of PEM is shown in Fig. 3(a). It illustrates that ethanol (as fuel) flows all over the anodic cell whereas O_2/air flows through the cathodic cell. As the reaction proceeds, ethanol concentration declines in the anodic cell and requires re-supply of ethanol. At the same time, water concentration increases in the cathodic cell. Hence, there is a requirement of a rigorous fuel cell system which can evacuate the water form the cathodic cell or maintain water concentration in both the electrode cell.

There are many challenges for designing of PEM such as (i) partial oxidation of ethanol into CO_2 which alters the fuel cell Faradic efficiency, (ii) high cost of acidic electrolyte membranes, (iii) parasitic current generation with the transfer of ethanol from anode to cathode within the PEM, (iv) oxidation of ethanol has stagnant reaction kinetics in acidic medium which results in loss of huge amount of activation energy, (v) durability of used catalysts like Pt-Ru/C and Pt-Sn/C, (vi) need to apply cathode catalyst for enhancement of ethanol oxidation rate [7,23].

It has been reported that Pt catalyst is a very good promoter of oxidation (methanol oxidation) reaction at low pH [24]. Hence, the combination of Pt catalyst with PEM is a suitable option for advanced oxidation of ethanol in the acidic medium; therefore, there is very low physical interference of contamination of in-situ generated salt on the active surface area of the electrode within the reactor. These salts are generated from acid, which is soluble. Hence it does not create as much problem [25]. Application of low pH fuel cell requires more attention because of the deterioration of the cell during its operation.

2.2   Alkaline medium DEFC (Anion exchange membrane/AEM)

Alkaline medium DEFC oxidizes ethanol at a high pH (8<pH<12). It differs from acidic medium DEFC; therefore, PEM is not suitable for this type of DEFC as at high value of pH, the concentration of H^+ ions is very low which results in very low diffusion kinetics all over the cell. To overcome this problem, the anion exchange membrane (AEM) is used. This type of membrane allows diffusion of anions from the anode to cathode [25]. Alkaline electrolyte is the best medium for AEM because this membrane is based on anion exchange migration Fig. 3(b) shows the schematic diagram of alkaline medium DEFC consisting of AEM. It works opposite to that of PEM/DEFC as water consumption occurs at the cathode whereas reformation of water is done on the anode cell. Therefore, the concentration of water decreases in the cathodic cell which is opposite of PEM/DEFC. AEM reduces the problem of cathode flooding as in the AEM, and ions move in the opposite direction in comparison with PEM. However, the problem of fuel crossover occurs when the component concentration is neglected in both the electrode cells.

Several types of challenges exist in designing AEM. The partial oxidation of ethanol into CO_2 is to be minimized. There is a need to enhance the catalytic activity of non-Pt catalyst at the cathode and Pd-based catalyst for enhanced oxidation of ethanol. The chemical and mechanical stability, and ionic conductivity of the ionomers in the catalyst layers also needs improvement. There is a need to improve ionic conductivity, mechanical, chemical and thermal stability of OH$^-$ ions for AEM/DEFC.

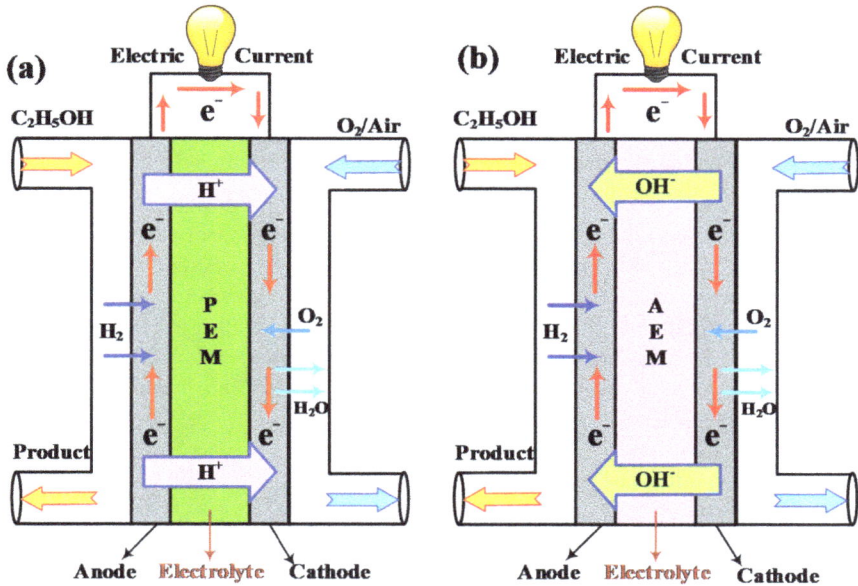

Figure3. Schematic diagram for direct ethanol fuel cell (DEFC) for (a) Proton Exchange Membrane (PEM), (b) Anion Exchange Membrane (AEM) [23].

Several differences exist between acidic and alkaline medium DEFC; however, few principles declare alkaline medium DEFC better than acidic type DEFC. High pH environment shows better reaction kinetics than low pH environment for oxidation of ethanol. High pH environment provides high OH$^-$ ion concentration than low pH environment which results in an additional amount of OH$^-$ adsorbed on the surface of the catalyst, and it enhances the oxidation of ethanol. However, the catalyst needs to be focused oxidizing ethanol rather than dissociating H$_2$O into OH$^-$ in the anodic cell for AEM/DEFC whereas PEM/DEFC needs an additional catalyst for dissociating water into OH$^-$ ions. However, a high concentration of OH$^-$ ions is a significant setback when CO$_2$ reacts with OH$^-$ ions to produce bi-carbonates and carbonates ions. These ions can further react with added primary material like KOH and NaOH which can produce K_2CO_3 and Na_2CO_3, respectively. These in-situ produced carbonate salts can alter the diffusion activity of the ions in an alkaline medium that results in poor performance of DEFC [6]. These carbonates salts affect the active surface area of the electrodes which destroys the active layer of the catalyst. Therefore, the selectivity of electrodes and catalyst becomes important, and these should have high resistivity towards the ions and salts. These drawbacks need to be minimized for the advancement of alkaline medium DEFC over acidic medium DEFC. It can be done by optimizing all the parameters which control the OH^- ion concentration as well as decreasing CO_2 concentration from the electrode cell.

Also, non-Pt catalyst shows much more catalytic activity in the alkaline medium when compared to the conventional Pt catalyst. It was previously reported that Pd catalyst shows approximately four times better catalytic activity in the alkaline medium than Pt catalyst for the oxidation of ethanol [26]. Au and Ag also show better catalytic activity than Pt catalyst in an alkaline medium [27–29]. This suggests that expensive Pt can be replaced by less costly Pd, Au, and Ag in alkaline medium DEFC. We can use a wide variety of metals as a catalyst for an alkaline medium because it has a less corrosive environment than the acidic medium DEFC.

3. The mechanism for electro-oxidation of ethanol

The mechanism for the electro-oxidation of ethanol is essential for the analytical design of catalysts as well as optimization of reactions in the DEFCs. This objective can be achieved by making the proper relation of synergy between experimental methods and theoretical simulations. The reaction intermediates and adsorbed species can be monitored by different techniques like liquid chromatography, ion chromatography, and infrared spectroscopy and online differential electrochemical mass spectrometry (DEMS) [30–33]. There is a general consent in the literature that mechanism of electro-oxidation of ethanol has a dual-pathway mechanism in either alkaline or acidic medium DEFC, on different types Pt-based or non-Pt based catalysts. Fig. 4 shows a dual pathway mechanism for the oxidation of ethanol. The main reactions are as follows:

The path I (C_l):

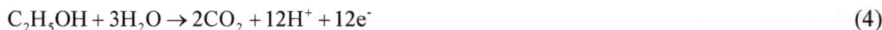

$$C_2H_5OH + 3H_2O \rightarrow 2CO_2 + 12H^+ + 12e^- \tag{4}$$

$$C_2H_5OH + 5H_2O \rightarrow 2HCO_3^- + 14H^+ + 12e^- \tag{5}$$

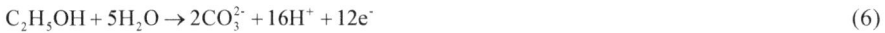

Materials Research Foundations **49** (2019) 293-320-   doi: https://doi.org/10.21741/9781644900192-10

$$C_2H_5OH + 5H_2O \rightarrow 2CO_3^{2-} + 16H^+ + 12e^- \tag{6}$$

Path II (C$_{II}$):

$$C_2H_5OH + H_2O \rightarrow CH_3COOH + 4H^+ + 4e^- \tag{7}$$

$$C_2H_5OH \rightarrow CH_3CHO + 2H^+ + 2e^- \tag{8}$$

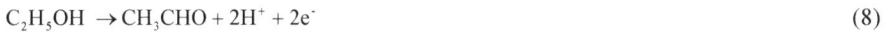

First pathway (a path I/C$_I$) is the total oxidation of ethanol into carbon dioxide (CO$_2$) or carbonates (CO$_3^{2-}$) with breakage of C-C bonds, followed by CO$_{ads}$ intermediate as a final product with the delivery of 12 e$^-$. Second pathway (path II/C$_{II}$) is the incomplete oxidation of ethanol into acetic acid, acetate or acetaldehyde as final products with the delivery of two electrons, and C-C bond does not break in this pathway.

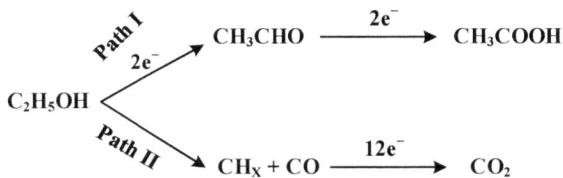

Figure4. Possible parallel pathway (dual pathway mechanism) of ethanol oxidation reactions on Pt electrodes in acidic/alkaline medium DEFCs [30].

From different types of intermediates and species identified by several techniques, it was reported that electrochemical oxidation of ethanol involves a series of many complex reactions which involves some parallels and sequential reaction steps. These results suggest that there are more than 40 possible highly oxidative or adsorbed intermediates like hydroxyl free radicals ($^\bullet$OH) in both acidic as well as alkaline medium DEFCs [34–36]. Fig. 5 shows extensive schematic presentation of electro-oxidation of ethanol. This schematic is universal in such a manner that it can explain the oxidation mechanism and pathway in both alkaline and acidic medium DEFCs. Some reaction involves (OH)$_{ads}$ which results in the regeneration of catalysts. It should also note that detected intermediates and final products of electrochemical oxidation may be affected by the nature and type of electrolyte, catalyst, and applied potential. Therefore, it is not possible to show all the intermediates and products for each oxidation reactions. There are three possible pathways for electrochemical oxidation of ethanol in DEFCs shown schematically in Fig.5. In the two possible pathways, dehydrogenation reaction leads to the formation of either C_2H_5O or CH_3CHOH in the very first step. Acetaldehyde becomes the central molecule for the second pathway. There are some other possibilities for these two pathways which are shown in Fig.5. In the third possible pathway, water molecule losses and gives $(C_2H_5)_{ads}$ which further converts into ethane. It was observed that for the third path, ethane could be the final product in an acidic medium at low cathodic potential. There are few key steps which are essential to highlight (from Fig.5)like (i) identification of germinal diol $(CH_3CH(OH)_2)$, ethyl acetate $(CH_3COOCH_2CH_3)$ and $CH_3CH(OH)OCH_2CH_3$ is very critical and it can be identified by NMR technique, (ii) CO_2 forms by breaking of C-C bond in the acetyl (CH_3CO) pathway but it can be altered by the generation of ethyl acetate $(CH_3COOCH_2CH_3)$ and/or acetic acid (CH_3COOH), (iii) there are some other products like $CH_3CH(OH)_2$-CH_3COOH which can be linked by some of the steps shown in Fig. 5. This schematic accomplishes the problem of breaking the C-C bond in electro-oxidation of ethanol in alkaline/acidic medium DEFCs.

Very limited number of research articles are available which provide molecular studies on electro-oxidation of ethanol in alkaline medium DEFCs [37,38]. The formation of soluble carbonates from CO_2 due to the aqueous alkaline electrolyte is the main reason for lack of studies on electro-oxidation mechanism as it creates hindrance for for studies using techniques like FTIR. This problem was solved by using alkaline polymer electrolyte membranes which helped in identfying produced CO_2 during electro-oxidation of ethanol in the effluents of the fuel cell which was coupled with the DEMS system [39]. DEMS is a critical technique to explain that C-C bond breakage which is more efficient in the alkaline medium than in acidic medium DEFCs. Rao et al. [39] observed that current efficiency for CO_2 was ≈55% at 0.8 V/RHE at 60 °C for alkaline media whereas it was only ≈2% in the acidic media. This observation was confirmed by Cremers et al. [40] reported that oxidation reaction kinetics for alkaline medium was higher than for acidic media with the same conditions. They also reported that: (i) only CO_2 and CH_4 desorbs from adsorbates of ethanol in alkaline medium; (ii) there are only two regions of potential i.e., above and below of 0.9 V/RHE for oxidation of ethanol adsorbates in alkaline medium; (iii) they described that formation of CH_4 occurs from adsorbed species like CH_x and/or CO_xH_y which comes from ethanol adsorbate; (iv) they calculated two e^- per molecule of carbon dioxide (CO_2) produced over the potential region below 0.9 V/RHE which was independent of the adsorption potential. They also observed that formation of CO_2 occurs only from adsorbed ethanol in the alkaline environment not from bulk solution of ethanol; (v) as the concentration of ethanol increases the efficiency of cleaving C-C bond decreases, and when it was compared between Pt and Pd electrode,

Nanomaterials for Alcohol Fuel Cells   Materials Research Forum LLC
Materials Research Foundations **49** (2019) 293-320-   doi: https://doi.org/10.21741/9781644900192-10

it was observed that oxidation of ethanol in alkaline environment showed selectivity towards the formation of acetate on Pd electrode. Also, the Au electrode showed almost 100% selectivity towards ethyl acetate and acetate with no breaking of C-C bonds [41]. In the last few decades, the difficulty of the cleaving of C-C bond with the identification of intermediates during the electro-oxidation of ethanol has been solved by many authors [42]. At potential below than 0.56 V/RHE, it is possible to cleave C-C bond into CO_2 in the alkaline environment on Pd/C electrode. However, the selectivity of CO_2 was not so good due to the formation of acetate and other by-products formation. As the potential goes higher than 0.72 V/RHE, the selectivity becomes worst to break the C-C bond.

In summary, the electro-oxidation of ethanol follows different pathways due to its complex reaction steps. There are different types of operating conditions like electrode material, temperature, catalyst material and its structure, initial concentration of ethanol, applied potential, reaction medium, etc. which can be dominating factors for reaction pathway and product distribution [36,43]. The adsorption of hydroxyl radicals is very critical for alkaline as well as the acidic environment. In the acidic environment, cleavage of C-C bond occurs (easily into CO) at low applied potential [44]. Limitation of CO_2 production is due to the poisoned surface which occurrs by lack of oxidants to eliminate CO_{ads}. There is a lack of oxidants at high applied potential; however lower selectivity of CO_2/CO products can also be due to the inhibition of C-C bond rupture by a high concentration of oxidants [36]. The dissociation of adsorbed ethanol in alkaline environment happens more quickly, and elimination of the adsorbed compounds becomes rate-determining step by the adsorbed hydroxyl. This observation varies among several kinds of literature. The reaction kinetics is well affected by the evolution of the inactive layer of surface oxides as well as electrochemical adsorption of hydroxyl ions (OH^-) at higher applied potential. Accordingly, electro-oxidation on the Pd and Pt-based catalyst generally follow the C_{II} pathway with acetate and acetic acid as the final product, and CO_2/carbonate shows relatively low distribution as final products in this scheme. In the acidic environment, dehydrogenation reaction at α–carbon is the initial step of bond breaking, whereas in a highly alkaline environment, it is O-H bond breaking [36]. Various oxidation reaction mechanisms of ethanol reported in the literature help in elucidating desirable properties of a highly efficient catalyst for total oxidation of ethanol into carbon dioxide (CO_2). This rationally synthesized catalyst should have following properties like: (i) a proper surface structure and composition to enhance selectivity towards production of CO_2, (ii) relevant surface area for breaking of C-C bond, (iii) it should have bi-functional characteristics to promote the adsorption as well as activation of water molecules to produce OH_{ads} for removing –CH_x and CO species [36]. These guidelines can be very useful to the researchers for designing and developing new and the advance catalyst having higher activity and selectivity with more extended durability for the advancement of electro-oxidation of ethanol [36–44].

4.   Advances in electro-oxidation of ethanol

The MEA plays the most vital role in determining the performance of the DEFCs system. The MEA has three major cells, i.e., electrolyte cell, support for the metal electrode, and the catalyst for the electrode. Catalyst acts as a site that provides an active surface for electro-oxidation of ethanol which takes place at low activation energy with an enhanced rate of reaction. As the activation energy gets lowered due to the catalyst surface, the energy consumption for the cleaving of the bonds of reactants and formation of products gets affected. Also, the time required for the process decreases due to the enhanced rate of reaction. Anode and cathode both require catalyst in the MEA of DEFCs. Electro-oxidation of ethanol is influenced by the catalyst at the anode, whereas the electro-reduction rate of oxygen is influenced by the catalyst at the cathode [45]. Both compartment reactions are most important and influence the performance of the system. If any of the reaction slows, the performance of the whole system goes down. Both the electrodes have different half reactions; therefore, advanced design of catalyst and cell is mandatory to provide excellent performance. There are many types of advanced nano-catalyst available like metal and non-metal (carbon-based) based which can be used for the electro-oxidation of ethanol in DEFCs system.

4.1   Advanced nano electro-catalyst

In the last few decades, the researchers have developed many types of electro-catalyst on nano-scale for the enhanced oxidation of ethanol. Pt and Pd are the most common catalysts which are used as a monometallic catalysts for ethanol oxidation. There are other metals which were used to synthesize bimetallic and trimetallicnano electro-catalysts like Pt-Sn/C [46] and Pt-Ru-Ni [47], respectively. Non-metal based or carbon-based electrocatalysts like Pd/CNT (carbon nanotube) [48] and Pd/r-GO (reduced graphene oxide) [49] have also been used.

4.1.1   Metal base electro-catalyst

As discussed earlier, noble metals like Pt, Pd, Ru and Ag and non-noble metal like Sn, Co and Ni have been used in different ways to synthesize various types of electro-catalyst with different properties for ethanol oxidation in DEFCs [6]. Table 1 consist of details of different types and structures of an advanced metal catalyst for the enhanced electro-oxidation of ethanol in alkaline medium [46–55].

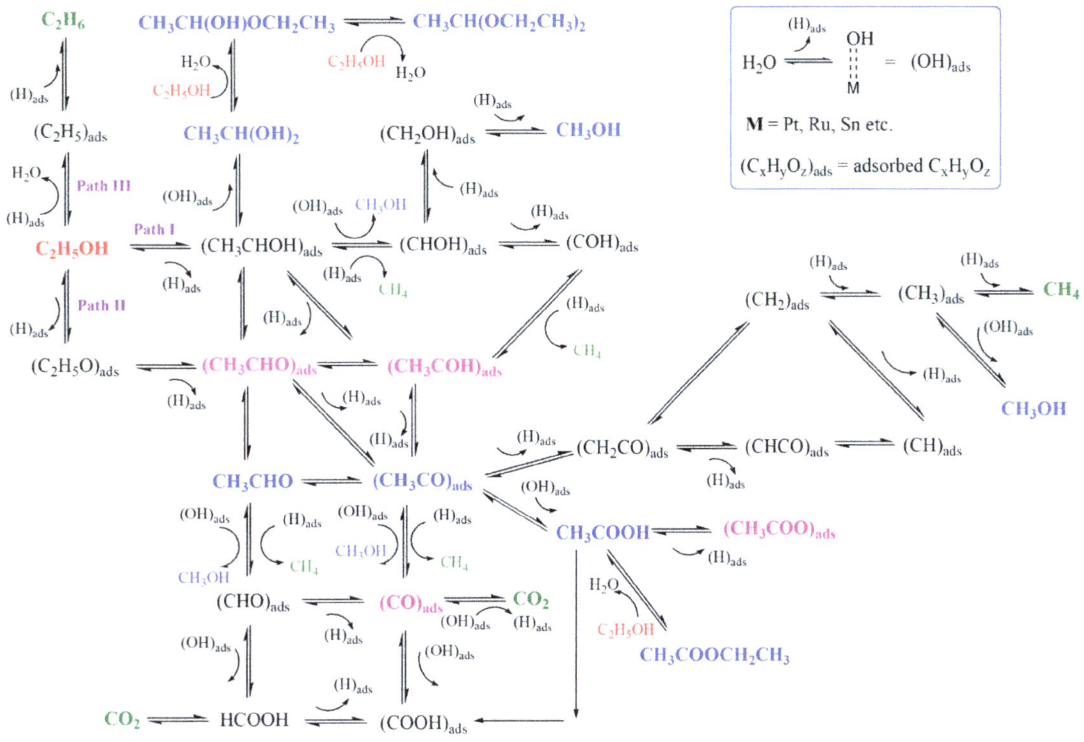

Figure 5. Schematic pathway for electrochemical oxidation of ethanol from literature available [23].

305

Table 1. Metal-based advanced catalyst for electro-oxidation of ethanol in alkaline medium.

Catalyst	Type	Co-catalyst/ Support	Media	Active surface area/ particle size	Operating condition	Catalytic activity	Ref.
Pt-Sn-Co	Colloidal nano particles	Sn & Co	Alkaline	31 $[m^2g^{-1}]$	1.0 M NaOH; 1.0 M ethanol; T= 80 °C; electrolyte: Nafion® 117 membrane	47.9 $[mA/cm^2]$	[46]
Pt-Sn-Ni	Colloidal nano particles	Sn & Ni	Alkaline	33 $[m^2g^{-1}]$	1.0 M NaOH; 1.0 M ethanol; T= 80 °C; electrolyte: Nafion® 117 membrane	47.9 $[mA/cm^2]$	[46]
Pt	Nanoparticles	Au	Alkaline	4 [nm]	0.1 M NaOH; 0.2 M ethanol	0.23 $[mCcm^{-2}]$	[49]
Pt/C-TiO$_2$	Nanoparticles	TiO$_2$	Alkaline	30 $[m^2g_{Pt}^{-1}]$	0.1 M NaOH; 0.5 M ethanol	-	[50]
Pt/C-CeO$_2$	Nanoparticles	CeO$_2$	Alkaline	30 $[m^2g_{Pt}^{-1}]$	0.1 M NaOH; 0.5 M ethanol	-	[50]
Pd	Nano structures (NB, NP, NR)	C	Alkaline	NB=5.1; NP= 6.8; NR= 6.7 [nm]	1 M KOH; 1 M & 3M ethanol	NB=8.1; NP= 6.8; NR= 6 $[mWcm^{-2}]$	[51]
Pd	Nanosheet (aerogel)	-	Alkaline	56.1 $[m^2g^{-1}]$	1M NaOH + 1M ethanol	-	[52]
Pd-TiO$_2$/ SnO$_2$	Nanotubes	TiO$_2$ & SnO$_2$	Alkaline	SnO$_2$=16 [nm]; TiO$_2$=10 [nm]	1 M KOH; 1 M ethanol	-	[53]
NiAl-LDH-NS	Nanosheets	Al	Alkaline		1.0 M NaOH; 1.0 M ethanol;	45.8 $[mA/cm^2]$	[54]
NiCo$_2$O	Mesoporous fibers	-	Alkaline	54.47 $[m^2g^{-1}]$	0.1 M NaOH; 0.5 M ethanol	-	[55]

Table 2. Carbon-based advanced nano-catalyst for electro-oxidation of ethanol in alkaline medium.

Catalyst	Type	Metal	Base	Media	EC active surface area	Mass Ratio for max. oxidation	Catalytic activity [mA/cm^2]	Ref.
3D Pd/rGO	Nano structure	Pd	r-GO	Alkaline	-	GO:Pd=5:20	-	[49]
Pd/CNTs	Nano tube	Pd	Carbon nanotube	Alkaline	33.1 [m^{2}g^{-1}]	-	-	[48]
Pd/3D NCNTs	Nano tube	Pd	Nitrogen-doped carbon nanotubes	Alkaline	20.1 [m^{2}g^{-1}]	NCNT:Pd=2:29	-	[71]
Pd-NiO/ MWCNTs/rGO	Nano tube	Pd, Ni	Multiwalled carbon nanotubes and r-GO	Alkaline	42.65 [m^{2}g^{-1}]	-	10.2 Pd	[72]
Pd-FP-rGONs	Nano sheet	Pd	Reduced graphene oxide	Alkaline	0.207 [cm^{2}g^{-1}] of Pd	-	10.2	[73]
Pd/B-N-G	Nano structure	Pd	B-N co-doped GO	Alkaline	55.3 [m^{2}g^{-1}]	Pd%=16.81	108.8	[74]
NiCo/C-N/CNT	Nano tube	Ni & Co	Carbon nanotube	Alkaline	-	NiCo=35%	181	[75]
Pt-SrCoO$_{3-\delta}$/ MV-RGO-CH	Nano composite	Pt	Methyl viologemr-GO	Acidic	74 [m^{2}g^{-1}]	-	3.22	[76]

4.1.1.1 Noble metals

Pt, Pd, Ru, Ag, Ir, and Rd are the noble metals which have been used to synthesize many types of catalysts. Among these noble metals, Pt and Pd were used extensively for synthesizing electro-catalysts. These two noble metals are default catalysts for electro-oxidation of ethanol. Role of OH^- ion was studied for ethanol electro-oxidation on the Pt and CeO_2 modified Pt catalyst [56]. It was observed that adsorbed OH^- ion changed its behavior in blank KOH solution, and it was applied for enhanced activity of ethanol oxidation reaction. This phenomenon was not observed in the acidic environment. Therefore, pH plays a vital role in reaction kinetics as well as equilibrium properties of surface species and solutions. In the alkaline environment, the consumption of OH^- ions influences the pH at the surface of the electrode which decreases the reaction kinetics. In the literature, Pt nanoparticles show enhanced electro-oxidation of ethanol in the alkaline environment [57–62]. Buso-Rogero et al. [57] reported Pt nanoparticles having different types of shades and loads by using spectroscopic and electrochemical techniques. They observed that the acetate was the final product with a very low amount of CO_2 production. Godoi et al. [50] used several metal oxides like ZrO_2/C, SnO_2/C, TiO_2/C, CeO_2/C, $WO_{3,}$ and MoO_3 was used on the Pt nanoparticles for enhanced ethanol electro-oxidation in an alkaline environment. They reported that the lowest and highest activities for Pt nanoparticles were for CeO_2/C and TiO_2/C, respectively. They also obtained acetate as the final product and CO_2 was only in traceable amount as observed by infrared reflection absorption spectroscopy (IRRAS) technique.

Pt-based electrodes are considered best for alcohol oxidation, however, Pd catalyst is good alternative for ethanol oxidation reaction in the alkaline environment. As the Pt metal is too costly and precious, researchers have tried Pd-based electrodes as well [23]. Pt-based electrodes exhibit CO poisoning which limits its application towards ethanol oxidation reaction [62]. Ma et al. [58] reported that the activity of carbon-supported Pd (Pd/C) towards ethanol oxidation was better than the carbon-supported Pt (Pt/C) due to the higher oxyphilic characteristics of the Pd/C and the relative inertness of the Pd/C towards C-C bond cleavage. Kumar and Buttry [59] synthesized 8% Pd nanoparticles from palladium halide precursors. Another study reported the different nanostructures of Pd like nano-rod, nanobar and nano polyhedral with enhanced stability and performance in DEFCs system [60]. Microwave-assisted Pd nano electro-catalyst showed more enhanced active electrochemical surface area, uniform dispersion, great amounts of palladium oxides and showed exceptional activity towards ethanol oxidation. The morphological study was carried out by Cerritos et al. [60] with three different types of nanostructures like nanorods (ND), nanoparticles (NP) and nanobars (NB). They reported that NB was the best-nanostructured catalyst of Pd in these for ethanol oxidation. A comparative study between Pd and Pt catalyst in the alkaline environment has been reported in the literature [61,62]. Pd-based catalysts were found to show better activity than the Pt-based catalysts.

Several researchers like Sun et al. [62] and Ma et al. [58] concluded that both Pd and Pt catalysts have unique and vital characteristics. Pt was found to possess great catalytic activity for ethanol oxidation not only in alkaline medium but also in acidic medium whereas Pd showed no activity in acidic medium. However, Pd showed exceptionally enhanced catalytic activity higher than Pt catalyst in an alkaline environment. Researchers prefer Pt catalyst rather than Pd catalyst concerning performance.

Other noble metals have also been used for electro-oxidation of ethanol in DEFCs system. Chen et al. [27] used gold (Au) for electro-oxidation of ethanol as the base catalyst. They reported that Au showed the same activity and characteristics as Pd i.e., inert in acidic condition (low pH) but comparability active in alkaline condition (high pH). They observed that Au has slow oxidizing capacity than Pt catalyst for ethanol. Despite adding Ru as co-catalyst, the oxidation reaction of ethanol does not accelerate because of sluggish adsorption/desorption taking place on the active surface area of Au. Cao et al. [63] used the Ir catalyst and compared it with the Pt catalyst and obtained similar results.

4.1.1.2 Non-noble metals

Several authors used non-noble metals (non-platinum group) for the electro-oxidation of ethanol in DEFCs. Zhan et al. [55] used Ni-based catalyst, i.e. $NiCo_2O_4$ having specific surface area and average pore size of 54.47 m^2g^{-1} and 13.5 nm, respectively, for the ethanol oxidation. This catalyst shows significant ethanol electro-oxidation with better current density and inferior onset potential than NiO and Co_3O_4. Xu et al. [54] synthesized ultrathin layered double hydroxide nanosheets consisting Ni(III) for enhanced electro-oxidation of ethanol. They observed that this catalyst has a high current density and extended durability. Sharma et al. [64] reported the eco-friendly and straightforward synthesis of non-noble based metal catalysts could play a vital role in the production of sustainable energy from alcohol. They fabricated porous Ni phosphate foams and used them for ethanol electro-oxidation and reported great cyclability as well as good catalytic activity and stability. The nanoparticles of nickel nitride (Ni_3N)were synthesized by Mazloum-Ardakani et al. [65] and used for ethanol oxidation in an alkaline environment. Hassan and Hamid et al. [66] synthesized nanocomposites of Ni/Cr_2O_3 and electrodeposited on the carbon electrode for the electro-oxidation of ethanol. They reported that the fabrication of catalyst on the electrode increases, the electrocatalytic activity of the constructed electrodes. Oh et al. [67] applied tungsten carbide (WC) as the anode for the electro-oxidation of ethanol. They reported that WC shows comparably similar results to Pd by the cyclic voltammetry technique. They also suggest that if WC is added with Pd as co-catalyst, it enhances the electrochemical surface area. Therefore, WC can provide better active surface sites for the highly active oxygen species to be adsorbed. Barakat et al. [68] synthesized Cd-doped Co (Cd/Co) electro-catalyst and used as the anode in the DEFCs. They found that Cd plays a vital role in the catalyst for the adsorption/desorption as well as oxidation of CO_{ads} species. They reported that the addition of Co in the Cd enhances the active surface area of the doped catalyst when compared to the alone Cd catalyst. They further observed that Co acts as co-catalyst in the same manner like Ni and Ru, and this synergistic effect enhances the catalytic reactivity by lowering the bond energy of the in-situ generated intermediates on the active surface area sites. They also compared cyclic voltammetry study between Cd/Co and Pt catalyst and found that Cd/Co control the performance due to the lower onset potential with a higher ultimate peak.

4.1.2 Carbon-based advanced electro-catalyst

Supported electrocatalysts have enhanced surface area, better electrical conductivity, and easiness of catalyst recovery. A better interface between the support and the catalysts enhances the catalyst efficiency and decreases the catalyst loss, however, it governs the transfer of charge. Better catalyst support reduces catalyst poisoning (e.g. S, CO, etc.) which enhances the catalyst activity. It may be noted that corrosion of electrode happens sometimes due to the catalyst support which essentially leads to additional issues like loss of electrocatalysts [23,69]. Hence, the selection of proper support material is of utmost importance in determining the overall performance of the fuel cell. Carbon as support has characteristics that can enhance the catalytic activity, morphology, and dispersion of any catalyst. It has better electrical conductivity which attracts researchers to use carbon as support material for synthesizing advanced electrocatalysts [69]. The use of carbon has increased extensively in DEFCs in the last few decades for the synthesis of the bipolar electrode plate, a supporting agent in the electrocatalyst layer and fabrication of the gas diffusion layer. Antolini et al. [70] reported that carbon enhanced the catalytic activity as well as the durability and stability of the catalyst for electro-oxidation of ethanol. Nowadays, graphene, graphene oxide (GO) and reduced graphene oxide (r-GO) are being used by the researchers as a better alternative support for stabilization and dispersion of the electro-catalyst nanoparticles. Table 2 shows properties of different types of carbon-based advanced catalyst for electro-oxidation of ethanol in alkaline medium.

Carbon nanodots (CND) was used for enhancing the catalytic activity of Pt-Sn nanoparticles in acidic medium [71]. Zhang et al. [49] reported an easy and green electrochemical method for the synthesis of rGO supported 3D Pd nanoparticles without adding another reagent. They observed that synthesized Pd/rGO nano electro-catalyst expresses better catalytic activity than available 10% Pd/C electrocatalyst and Pd nanoparticles towards the ethanol electro-oxidation in an alkaline environment. They reported that this is due to the synergistic effect in between r-GO and Pd and 3D coated nanostructure. Karuppasamy et al. [72] fabricated Au nano decorative and r-GO supported electro-catalyst for the ethanol oxidation in an alkaline environment and observed that r-GO nanosheets are the essential component for enhanced catalytic activity. Wang et al. [73] synthesized 3D series rGO/CNT supported advanced electro-catalyst by the electrochemical depletion of copper. They reported that synthesized catalyst have a sandwich-like structure, and it advances the catalytic activity towards electro-oxidation of ethanol by transferring electrons through the fast channel. Kumar et al. [74] reported controlled action of deformity in graphene-based materials is a bright approach to enhance the mechanical, electrical as well as electrochemical properties towards the innovative and advanced applications. They synthesized on a large scale, GO nanosheets with Pd hybrids which can be a promising electro-catalytic element by in-situ production and infusion of Pd-NPs by continuously applying microwave irradiation. They used this catalyst and reported improved electro-oxidation of ethanol in an alkaline medium having current density of 10 mAcm^{-2}. Kabir et al. [75] investigated ethanol electro-oxidation by Pd-NPs supported on the 3D-graphene nanosheets with different types of physiochemical as well as morphological properties. They observed that 3D-Graphene nanosheets have a better degree of C-C sp^2 hybridization which enhanced the diffusion and lowered the average size of the Pd nanoparticle's crystals. They also reported that this synthesized advanced electro-catalyst expressed better resistance against poisoning elements and showed better catalytic activity than commercial Pt/C electrocatalyst. Li et al. [76] synthesized graphite carbon supported g-C$_3$N$_4$ and Pd (Pd/g-C$_3$N$_4$/graphite carbon) electro-catalyst having high electrical conductivity for the electro-oxidation of ethanol in an alkaline environment. From obtained results, they suggested that Pd/g-C$_3$N$_4$/graphite carbon is one of the auspicious electro-catalyst for the enhanced electro-oxidation of ethanol.

5. Applications of electro-oxidation of ethanol

Despite new at the global market, DEFC shows better achievement and validation of it's potential to play a vital role as the green and sustainable energy source for the next generation. DEFC is one of the promising product of NDC (Nanomaterials Discovery

Corporation) power that adopted non-Pt electro-catalyst for cathode and anode. Air and ethanol are used as fuel as a power source for this device. The application of fuel cells is increasing in the automobile industries. Toyota's fuel cell-based bus named 'Sora' is ready for the production. This corporation is going to sell around 100 buses to shuttle the passengers all around Tokyo city for the Paralympic and Olympic games in 2020 [77]. The Fuel Cell and Hydrogen Energy Association (FCHEA, U.S.A.) is working on the portable use of fuel cell. These portable cells are mobile electricity generation units which are to be used as the power source of electronic equipment or to recharge batteries. They can also be used as the power back up for households and offices [78]. Shell Eco-marathon Asia, 2018 was held at Singapore, and team BIT's from India used an ethanol-based fuel cell. They applied ethanol as fuel very significantly in the engine which gives a very tight competition to hydrogen fuel cell [79].

Conclusions

Ethanol is a potential source of fuel. DEFC (direct ethanol fuel cell) has the possibility of use in green vehicles. Advanced catalyst to be used for the enhanced performance of ethanol electro-oxidation reaction should have the following properties: (i) highly active surface area for the cleaving of C-C bonds; (ii) a well-established morphological structures to enhance the selectivity towards CO_2 production; (iii) the capability to promote the adsorption as well as activation of H_2O for the elimination of $-CH_x$ and CO species. Application of these guidelines can help in the synthesis of advanced nanocatalyst having enhanced catalytic activity, selectivity, and high durability. Therefore, the main focus of advancement of DEFC is substantially towards the fabrication of catalysts to increase the ethanol electro-oxidation which can overcome the problem of the cleaving of ethanol C-C bond and crossover of the fuel. Also, alkaline medium DEFC shows quite better performance than acidic medium and the use of the non-noble metal-based catalyst in DEFC can reduce the total cost of the system. Analytical technique help to identify several types of intermediates and final products and help in understanding the reaction mechanism of electro-oxidation of ethanol. Metal-oxide supported nano-catalysts like TiO_2, CeO_2, SnO_2, and NiO/foams show enhanced catalytic activity for breaking the C-C bond. In all of these nano-catalysts, CeO_2 seems to be the best metal oxide support. Despite enhancing the reaction kinetics for ethanol electro-oxidation, SnO_2 has poor selectivity towards breaking the C-C bond. Now it is feasible to synthesize nanoparticles with favored composition (mono- or multi-metallic), dimensions (1D, 2D, and 3D nano-spheres, nano-sheets, nano-tubes, nano-wires, nano-boxes, nano-cages, etc.) and surface (111, 100, etc.). Graphene oxide (GO) and reduced graphene oxide (r-GO) are widely used as the base of electro-catalyst for electro-oxidation of ethanol. All these feasible possibilities support great opportunities for studying the structure-activity relationship. Direct ethanol fuel cells (DEFC) have the capability for use in high-performance lightweight vehicles. The armed market has a valuable aspect for the implementation of micro-type fuel cells as soldiers have to carry large batteries to power their equipment for a long time span. With further research, these fuel cells could be used for generation of power and in the transportation sector using H_2 as fuel. The use of DEFC is continuously increasing as the portable mode of transport which is convincing the people towards the reliability of ethanol as fuel in the fuel cell. In the last, the advancement of DEFCs looks promising for future generations.

References

[1] B.M. Daas, S. Ghosh, Fuel cell applications of chemically synthesized zeolite modified electrode (ZME) as catalyst for alcohol electro-oxidation - A review, J. Electroanal. Chem. 783 (2016) 308–315. https://doi.org/10.1016/j.jelechem.2016.11.004

[2] Y. Li, System Design and performance in alkaline direct ethanol fuel cells, in: Lect. Notes Energy, 2018, pp. 217–247. https://doi.org/10.1007/978-3-319-71371-7_7

[3] G.K. Dinesh, R. Chauhan, S. Chakma, Influence and strategies for enhanced biohydrogen production from food waste, Renew. Sustain. Energy Rev. 92 (2018) 807–822. https://doi.org/10.1016/j.rser.2018.05.009

[4] T.Y. Chen, G.W. Lee, Y.T. Liu, Y.F. Liao, C.C. Huang, D.S. Lin, T.L. Lin, Heterojunction confinement on the atomic structure evolution of near monolayer core–shell nanocatalysts in redox reactions of a direct methanol fuel cell, J. Mater. Chem. A. 3 (2015) 1518–1529. https://doi.org/10.1039/C4TA04640D

[5] H. Huang, J. Zhu, W. Zhang, C.S. Tiwary, J. Zhang, X. Zhang, Q. Jiang, H. He, Y. Wu, W. Huang, P.M. Ajayan, Q. Yan, Controllable codoping of nitrogen and sulfur in graphene for highly efficient Li-oxygen batteries and direct methanol fuel cells, Chem. Mater. 28 (2016) 1737–1745. https://doi.org/10.1021/acs.chemmater.5b04654

[6] M.A.F. Akhairi, S.K. Kamarudin, Catalysts in direct ethanol fuel cell (DEFC): An overview, Int. J. Hydrogen Energy. 41 (2016) 4214–4228. https://doi.org/10.1016/j.ijhydene.2015.12.145

[7] R.F. Association, World ethanol production, (2017)

[8] J. Sánchez-Monreal, P.A. García-Salaberri, M. Vera, A genetically optimized kinetic model for ethanol electro-oxidation on Pt-based binary catalysts used in direct ethanol fuel cells, J. Power Sources. 363 (2017) 341–355. https://doi.org/10.1016/j.jpowsour.2017.07.069

[9]A. Brouzgou, A. Podias, P. Tsiakaras, PEMFCs and AEMFCs directly fed with ethanol: a current status comparative review, J. Appl. Electrochem. 43 (2013) 119–136. https://doi.org/10.1007/s10800-012-0513-2

[10]M. Meyer, J. Melke, D. Gerteisen, Modelling and simulation of a direct ethanol fuel cell considering multistep electrochemical reactions, transport processes and mixed potentials, Electrochim. Acta. 56 (2011) 4299–4307. https://doi.org/10.1016/j.electacta.2011.01.070

[11]R.M. Antoniassi, A. Oliveira Neto, M. Linardi, E.V. Spinacé, The effect of acetaldehyde and acetic acid on the direct ethanol fuel cell performance using $PtSnO_2/C$ electrocatalysts, Int. J. Hydrogen Energy. 38 (2013) 12069–12077. https://doi.org/10.1016/j.ijhydene.2013.06.139

[12]R. Kavanagh, X.M. Cao, W.F. Lin, C. Hardacre, P. Hu, Origin of low CO_2 selectivity on platinum in the direct ethanol fuel cell, Angew. Chemie Int. Ed. 51 (2012) 1572–1575. https://doi.org/10.1002/anie.201104990

[13]J.E. Sulaiman, S. Zhu, Z. Xing, Q. Chang, M. Shao, Pt–Ni Octahedra as electrocatalysts for the ethanol electro-oxidation reaction, ACS Catal. 7 (2017) 5134–5141. https://doi.org/10.1021/acscatal.7b01435

[14]T. Wu, J. Fan, Q. Li, P. Shi, Q. Xu, Y. Min, Palladium nanoparticles anchored on anatase titanium dioxide-black phosphorus hybrids with heterointerfaces: Highly electroactive and durable catalysts for ethanol electrooxidation, Adv. Energy Mater. 8 (2018) 1701799. https://doi.org/10.1002/aenm.201701799

[15]K.M. Hassan, A.A. Hathoot, R. Maher, M. Abdel Azzem, Electrocatalytic oxidation of ethanol at Pd, Pt, Pd/Pt and Pt/Pd nano particles supported on poly 1,8-diaminonaphthalene film in alkaline medium, RSC Adv. 8 (2018) 15417–15426. https://doi.org/10.1039/C7RA13694C

[16]W.H. Yang, Q.H. Zhang, H.H. Wang, Z.Y. Zhou, S.G. Sun, Preparation and utilization of a sub-5 nm PbO_2 colloid as an excellent co-catalyst for Pt-based catalysts toward ethanol electro-oxidation, New J. Chem. 41 (2017) 12123–12130. https://doi.org/10.1039/C7NJ02620J

[17]L. Chen, L. Lu, H. Zhu, Y. Chen, Y. Huang, Y. Li, L. Wang, Improved ethanol electrooxidation performance by shortening Pd–Ni active site distance in Pd–Ni–P nanocatalysts, Nat. Commun. 8 (2017) 14136. https://doi.org/10.1038/ncomms14136

[18]O.Y. Podyacheva, A.S. Lisitsyn, L.S. Kibis, A.I. Stadnichenko, A.I. Boronin, E.M. Slavinskaya, O.A. Stonkus, S.A. Yashnik, Z.R. Ismagilov, Influence of the nitrogen-doped carbon nanofibers on the catalytic properties of supported metal and oxide nanoparticles, Catal. Today. 301 (2018) 125–133. https://doi.org/10.1016/j.cattod.2017.01.004

[19]A. Perry, S. Kabir, I. Matanovic, M.S. Chavez, K. Artyushkova, A. Serov, P. Atanassov, Novel hybrid catalyst for the oxidation of organic acids: Pd nanoparticles supported on Mn-N-3D-graphene nanosheets, ChemElectroChem. 4 (2017) 2336–2344. https://doi.org/10.1002/celc.201700285

[20]S. Kabir, A. Serov, A. Zadick, K. Artyushkova, P. Atanassov, Palladium nanoparticles supported on three-dimensional graphene nanosheets: Superior cathode electrocatalysts, ChemElectroChem. 3 (2016) 1655–1666. https://doi.org/10.1002/celc.201600245

[21]W. Zheng, H.W. Man, L. Ye, S.C.E. Tsang, Electroreduction of carbon dioxide to formic acid and methanol over a palladium/polyaniline catalyst in acidic solution: a study of the palladium size effect, Energy Technol. 5 (2017) 937–944. https://doi.org/10.1002/ente.201600659

[22]M.Z.F. Kamarudin, S.K. Kamarudin, M.S. Masdar, W.R.W. Daud, Review: Direct ethanol fuel cells, Int. J. Hydrogen Energy. 38 (2013) 9438–9453. https://doi.org/10.1016/j.ijhydene.2012.07.059

[23]E.A. Monyoncho, T.K. Woo, E.A. Baranova, Ethanol electrooxidation reaction in alkaline media for direct ethanol fuel cells, in: SPR Electrochem. 2019, pp. 1–57. https://doi.org/10.1039/9781788013895-00001

[24]A.M. Zainoodin, S.K. Kamarudin, W.R.W. Daud, Electrode in direct methanol fuel cells, Int. J. Hydrogen Energy. 35 (2010) 4606–4621. https://doi.org/10.1016/j.ijhydene.2010.02.036

[25]E. Antolini, E.R. Gonzalez, Alkaline direct alcohol fuel cells, J. Power Sources. 195 (2010) 3431–3450. https://doi.org/10.1016/j.jpowsour.2009.11.145

[26]Z. Zhang, L. Xin, K. Sun, W. Li, Pd–Ni electrocatalysts for efficient ethanol oxidation reaction in alkaline electrolyte, Int. J. Hydrogen Energy. 36 (2011) 12686–12697. https://doi.org/10.1016/j.ijhydene.2011.06.141

[27]Y. Chen, L. Zhuang, J. Lu, Non-Pt anode catalysts for alkaline direct alcohol fuel cells, Chinese J. Catal. 28 (2007) 870–874. https://doi.org/10.1016/S1872-2067(07)60073-4

[28]Y.Y. Feng, Z.H. Liu, W.Q. Kong, Q.Y. Yin, L.X. Du, Promotion of palladium catalysis by silver for ethanol electro-oxidation in alkaline electrolyte, Int. J. Hydrogen Energy. 39 (2014) 2497–2504. https://doi.org/10.1016/j.ijhydene.2013.12.004

[29]J.B. Xu, T.S. Zhao, Y.S. Li, W.W. Yang, Synthesis and characterization of the Au-modified Pd cathode catalyst for alkaline direct ethanol fuel cells, Int. J. Hydrogen Energy. 35 (2010) 9693–9700. https://doi.org/10.1016/j.ijhydene.2010.06.074

[30]Z.X. Liang, T.S. Zhao, J.B. Xu, L.D. Zhu, Mechanism study of the ethanol oxidation reaction on palladium in alkaline media, Electrochim. Acta. 54 (2009) 2203–2208. https://doi.org/10.1016/j.electacta.2008.10.034

[31]H. Hitmi, E.M. Belgsir, J.-M. Léger, C. Lamy, R.O. Lezna, A kinetic analysis of the electro-oxidation of ethanol at a platinum electrode in acid medium, Electrochim. Acta. 39 (1994) 407–415. https://doi.org/10.1016/0013-4686(94)80080-4

[32]D.J. Tarnowski, C. Korzeniewski, Effects of surface step density on the electrochemical oxidation of ethanol to acetic acid, J. Phys. Chem. B. 101 (1997) 253–258. https://doi.org/10.1021/jp962450c

[33]Y. Wang, K. Jiang, W.B. Cai, Enhanced electrocatalysis of ethanol on dealloyed Pd-Ni-P film in alkaline media: an infrared spectroelectrochemical investigation, Electrochim. Acta. 162 (2015) 100–107. https://doi.org/10.1016/j.electacta.2014.11.182

[34]T. Sheng, W.F. Lin, C. Hardacre, P. Hu, Significance of β-dehydrogenation in ethanol electro-oxidation on platinum doped with Ru, Rh, Pd, Os and Ir, Phys. Chem. Chem. Phys. 16 (2014) 13248–13254. https://doi.org/10.1039/C4CP00737A

[35]A. Dutta, A. Mondal, J. Datta, Tuning of platinum nano-particles by Au usage in their binary alloy for direct ethanol fuel cell: Controlled synthesis, electrode kinetics and mechanistic interpretation, J. Power Sources. 283 (2015) 104–114. https://doi.org/10.1016/j.jpowsour.2015.01.113

[36]H.A. Asiri, A.B. Anderson, Mechanisms for ethanol electrooxidation on Pt(111) and adsorption bond strengths defining an ideal catalyst, J. Electrochem. Soc. 162 (2015) F115–F122. https://doi.org/10.1149/2.0781501jes

[37]A.N. Geraldes, D. Furtunato da Silva, J.C. Martins da Silva, O. Antonio de Sá, E.V. Spinacé, A.O. Neto, M. Coelho dos Santos, Palladium and palladium–tin supported on multi wall carbon nanotubes or carbon for alkaline direct ethanol fuel cell, J. Power Sources. 275 (2015) 189–199. https://doi.org/10.1016/j.jpowsour.2014.11.024

[38]Y.Y. Yang, J. Ren, Q.X. Li, Z.Y. Zhou, S.G. Sun, W. Bin Cai, Electrocatalysis of ethanol on a pd electrode in alkaline media: An in situ attenuated total reflection surface-enhanced infrared absorption spectroscopy study, ACS Catal. 4 (2014) 798–803. https://doi.org/10.1021/cs401198t

[39]V. Rao, C. Cremers, U. Stimming, L. Cao, S. Sun, S. Yan, G. Sun, Q. Xin, Electro-oxidation of ethanol at gas diffusion electrodes a DEMS study, J. Electrochem. Soc. 154 (2007) B1138. https://doi.org/10.1149/1.2777108

[40] C. Cremers, D. Bayer, B. Kintzel, M. Joos, M. Krausa, D. Martin, J. Bernard, Investigation on denaturing agents for use with ethanol in direct ethanol fuel cells (DEFC), in: ECS Trans., ECS, 2009, pp. 517–524. https://doi.org/10.1149/1.3142783

[41] D. Bayer, C. Cremers, H. Baltruschat, J. Tübke, The electro-oxidation of ethanol in alkaline medium at different catalyst metals, in: 2011, pp. 1669–1680. https://doi.org/10.1149/1.3635698

[42] E.A. Monyoncho, S.N. Steinmann, C. Michel, E.A. Baranova, T.K. Woo, P. Sautet, Ethanol electro-oxidation on palladium revisited using polarization modulation infrared reflection absorption spectroscopy (PM-IRRAS) and density functional theory (DFT): Why is it difficult to break the C–C bond, ACS Catal. 6 (2016) 4894–4906. https://doi.org/10.1021/acscatal.6b00289

[43] M. Zhiani, S. Majidi, H. Rostami, M.M. Taghiabadi, Comparative study of aliphatic alcohols electrooxidation on zero-valent palladium complex for direct alcohol fuel cells, Int. J. Hydrogen Energy. 40 (2015) 568–576. https://doi.org/10.1016/j.ijhydene.2014.10.144

[44] T. Sheng, W.F. Lin, C. Hardacre, P. Hu, Role of water and adsorbed hydroxyls on ethanol electrochemistry on Pd: new mechanism, active centers, and energetics for direct ethanol fuel cell running in alkaline medium, J. Phys. Chem. C. 118 (2014) 5762–5772. https://doi.org/10.1021/jp407978h

[45] J.N. Tiwari, R.N. Tiwari, G. Singh, K.S. Kim, Recent progress in the development of anode and cathode catalysts for direct methanol fuel cells, Nano Energy. 2 (2013) 553–578. https://doi.org/10.1016/j.nanoen.2013.06.009

[46] S. Beyhan, C. Coutanceau, J.M. Léger, T.W. Napporn, F. Kadırgan, Promising anode candidates for direct ethanol fuel cell: Carbon supported PtSn-based trimetallic catalysts prepared by Bönnemann method, Int. J. Hydrogen Energy. 38 (2013) 6830–6841. https://doi.org/10.1016/j.ijhydene.2013.03.058

[47] E. Ribadeneira, B.A. Hoyos, Evaluation of Pt–Ru–Ni and Pt–Sn–Ni catalysts as anodes in direct ethanol fuel cells, J. Power Sources. 180 (2008) 238–242. https://doi.org/10.1016/j.jpowsour.2008.01.084

[48] M.Z. Yazdan-Abad, M. Noroozifar, N. Alfi, A.R. Modarresi-Alam, H. Saravani, A simple and fast method for the preparation of super active Pd/CNTs catalyst toward ethanol electrooxidation, Int. J. Hydrogen Energy. 43 (2018) 12103–12109. https://doi.org/10.1016/j.ijhydene.2018.04.179

[49] Q. Zhang, X. Wu, M. Gao, H. Qiu, J. Hu, K. Huang, S. Feng, Y. Yang, T. Wang, B. Zhao, Z. Liu, Highly active electrocatalyst of 3D Pd/reduced graphene oxide nanostructure for electro-oxidation of methanol and ethanol, Inorg. Chem. Commun. 94 (2018) 43–47. https://doi.org/10.1016/j.inoche.2018.05.028

[50]D.R.M. Godoi, H.M. Villullas, F.-C. Zhu, Y.-X. Jiang, S.-G. Sun, J. Guo, L. Sun, R. Chen, A comparative investigation of metal-support interactions on the catalytic activity of Pt nanoparticles for ethanol oxidation in alkaline medium, J. Power Sources. 311 (2016) 81–90. https://doi.org/10.1016/j.jpowsour.2016.02.011

[51]R. Carrera-Cerritos, R. Fuentes-Ramírez, F.M. Cuevas-Muñiz, J. Ledesma-García, L.G. Arriaga, Performance and stability of Pd nanostructures in an alkaline direct ethanol fuel cell, J. Power Sources. 269 (2014) 370–378. https://doi.org/10.1016/j.jpowsour.2014.06.161

[52]M. Zareie Yazdan-Abad, M. Noroozifar, A.R. Modaresi Alam, H. Saravani, Palladium aerogel as a high-performance electrocatalyst for ethanol electro-oxidation in alkaline media, J. Mater. Chem. A. 5 (2017) 10244–10249. https://doi.org/10.1039/C7TA03208K

[53]M.K.S. Barr, L. Assaud, N. Brazeau, M. Hanbücken, S. Ntais, L. Santinacci, E.A. Baranova, Enhancement of Pd catalytic activity toward ethanol electrooxidation by atomic layer deposition of SnO_2 onto TiO_2 nanotubes, J. Phys. Chem. C. 121 (2017) 17727–17736. https://doi.org/10.1021/acs.jpcc.7b05799

[54]L. Xu, Z. Wang, X. Chen, Z. Qu, F. Li, W. Yang, Ultrathin layered double hydroxide nanosheets with Ni(III) active species obtained by exfoliation for highly efficient ethanol electrooxidation, Electrochim. Acta. 260 (2018) 898–904. https://doi.org/10.1016/j.electacta.2017.12.065

[55]J. Zhan, M. Cai, C. Zhang, C. Wang, Synthesis of mesoporous $NiCo_2O_4$ fibers and their electrocatalytic activity on direct oxidation of ethanol in alkaline media, Electrochim. Acta. 154 (2015) 70–76. https://doi.org/10.1016/j.electacta.2014.12.078

[56]Y. Katayama, T. Okanishi, H. Muroyama, T. Matsui, K. Eguchi, Enhanced supply of hydroxyl species in CeO_2 -modified platinum catalyst studied by in situ ATR-FTIR spectroscopy, ACS Catal. 6 (2016) 2026–2034. https://doi.org/10.1021/acscatal.6b00108

[57]C. Busó-Rogero, J. Solla-Gullón, F.J. Vidal-Iglesias, E. Herrero, J.M. Feliu, Oxidation of ethanol on platinum nanoparticles: surface structure and aggregation effects in alkaline medium, J. Solid State Electrochem. 20 (2016) 1095–1106. https://doi.org/10.1007/s10008-015-2970-0

[58]L. Ma, D. Chu, R. Chen, Comparison of ethanol electro-oxidation on Pt/C and Pd/C catalysts in alkaline media, Int. J. Hydrogen Energy. 37 (2012) 11185–11194. https://doi.org/10.1016/j.ijhydene.2012.04.132

[59]A. Kumar, D.A. Buttry, Influence of halide ions on anodic oxidation of ethanol on palladium, Electrocatalysis. 7 (2016) 201–206. https://doi.org/10.1007/s12678-015-0298-2

[60]R.C. Cerritos, M. Guerra-Balcázar, R.F. Ramírez, J. Ledesma-García, L.G. Arriaga, Morphological effect of Pd catalyst on ethanol electro-oxidation reaction, materials (basel). 5 (2012) 1686–1697. https://doi.org/10.3390/ma5091686

[61]S. Cherevko, X. Xing, C.H. Chung, Pt and Pd decorated Au nanowires: Extremely high activity of ethanol oxidation in alkaline media, Electrochim. Acta. 56 (2011) 5771–5775. https://doi.org/10.1016/j.electacta.2011.04.052

[62]C.L. Sun, J.S. Tang, N. Brazeau, J.J. Wu, S. Ntais, C.W. Yin, H.L. Chou, E.A. Baranova, Particle size effects of sulfonated graphene supported Pt nanoparticles on ethanol electrooxidation, Electrochim. Acta. 162 (2015) 282–289. https://doi.org/10.1016/j.electacta.2014.12.099

[63]L. Cao, G. Sun, H. Li, Q. Xin, Carbon-supported IrSn catalysts for direct ethanol fuel cell, Fuel Cells Bull. 2007 (2007) 12–16. https://doi.org/10.1016/S1464-2859(08)70142-1

[64]P. Sharma, S. Radhakrishnan, M.S. Khil, H.Y. Kim, B.S. Kim, Simple room temperature synthesis of porous nickel phosphate foams for electrocatalytic ethanol oxidation, J. Electroanal. Chem. 808 (2018) 236–244. https://doi.org/10.1016/j.jelechem.2017.12.025

[65]M. Mazloum-Ardakani, V. Eslami, A. Khoshroo, Nickel nitride nanoparticles as efficient electrocatalyst for effective electro-oxidation of ethanol and methanol in alkaline media, Mater. Sci. Eng. B. 229 (2018) 201–205. https://doi.org/10.1016/j.mseb.2017.12.038

[66]H.B. Hassan, Z.A. Hamid, Electrodeposited $Ni–Cr_2O_3$ nanocomposite anodes for ethanol electrooxidation, Int. J. Hydrogen Energy. 36 (2011) 5117–5127. https://doi.org/10.1016/j.ijhydene.2011.01.024

[67]Y. Oh, S.K. Kim, D.H. Peck, J. Jang, J. Kim, D.H. Jung, Improved performance using tungsten carbide/carbon nanofiber based anode catalysts for alkaline direct ethanol fuel cells, Int. J. Hydrogen Energy. 39 (2014) 15907–15912. https://doi.org/10.1016/j.ijhydene.2014.02.010

[68]N.A.M. Barakat, M.A. Abdelkareem, H.Y. Kim, Ethanol electro-oxidation using cadmium-doped cobalt/carbon nanoparticles as novel non precious electrocatalyst, Appl. Catal. A Gen. 455 (2013) 193–198. https://doi.org/10.1016/j.apcata.2013.02.004

[69]S. Sharma, B.G. Pollet, Support materials for PEMFC and DMFC electrocatalysts - A review, J. Power Sources. 208 (2012) 96–119. https://doi.org/10.1016/j.jpowsour.2012.02.011

[70]E. Antolini, E.R. Gonzalez, Polymer supports for low-temperature fuel cell catalysts, Appl. Catal. A Gen. 365 (2009) 1–19. https://doi.org/10.1016/j.apcata.2009.05.045

[71]S.S. Gwebu, P.N. Nomngongo, N.W. Maxakato, Pt-Sn nanoparticles supported on carbon nanodots as anode catalysts for alcohol electro-oxidation in acidic conditions, Electroanalysis. 30 (2018) 1125–1132. https://doi.org/10.1002/elan.201800098

[72]L. Karuppasamy, C.Y. Chen, S. Anandan, J.J. Wu, Sonochemical fabrication of reduced graphene oxide supported Au nano dendrites for ethanol electrooxidation in alkaline medium, Catal. Today. 307 (2018) 308–317. https://doi.org/10.1016/j.cattod.2017.06.032

[73]M. Wang, Z. Ma, R. Li, B. Tang, X.Q. Bao, Z. Zhang, X. Wang, Novel flower-like PdAu(Cu) Anchoring on a 3D rGO-CNT sandwich-stacked framework for highly efficient methanol and ethanol electro-oxidation, Electrochim. Acta. 227 (2017) 330–344. https://doi.org/10.1016/j.electacta.2017.01.046

[74]R. Kumar, R. Savu, R.K. Singh, E. Joanni, D.P. Singh, V.S. Tiwari, A.R. Vaz, E.T.S.G. da Silva, J.R. Maluta, L.T. Kubota, S.A. Moshkalev, Controlled density of defects assisted perforated structure in reduced graphene oxide nanosheets-palladium hybrids for enhanced ethanol electro-oxidation, Carbon N. Y. 117 (2017) 137–146. https://doi.org/10.1016/j.carbon.2017.02.065

[75]S. Kabir, A. Serov, K. Artyushkova, P. Atanassov, Design of novel graphene materials as a support for palladium nanoparticles: highly active catalysts towards ethanol electrooxidation, Electrochim. Acta. 203 (2016) 144–153. https://doi.org/10.1016/j.electacta.2016.04.026

[76]Z. Li, R. Lin, Z. Liu, D. Li, H. Wang, Q. Li, Novel graphitic carbon nitride/graphite carbon/palladium nanocomposite as a high-performance electrocatalyst for the ethanol oxidation reaction, Electrochim. Acta. 191 (2016) 606–615. https://doi.org/10.1016/j.electacta.2016.01.124

[77]N. Manthey, Toyota's fuel cell bus Sora ready for production - electrive.com, (2018)

[78]Fuel Cell & Hydrogen Energy Association, Portable Power — Fuel Cell & Hydrogen Energy Association, (2015)

[79]Shell eco-marathon Asia, news and highlights, (2018)

Nanomaterials for Alcohol Fuel Cells
Materials Research Foundations **49** (2019) 321--350

Materials Research Forum LLC
doi: https://doi.org/10.21741/9781644900192-11

Chapter 11

Proton Transport and Design of Proton Electrolyte Membranes for Methanol Oxidation

P.S. Kumar[1], S.K. Pal[1], R. Rajasekar[2*], M.H. Kumar[2], A.M. Kumar[2]

[1]Department of Mining Engineering, Indian Institute of Technology Kharagpur,
West Bengal – 721 302, India

[2]Department of Mechanical Engineering, Kongu Engineering College,
Tamil Nadu – 638 060, India

rajasekar.cr@gmail.com

Abstract

The fuel cell has been widely used in automobiles and has a bright future in our country. Fuel Cell development is making adequate progress in the direct methanol fuel cell (DMFC) discipline. This chapter explains the extent and trends of theoretical developments in the DMFC. This chapter identifies the scope for proton transport and design of PEM for methanol oxidation in DMFCs. It also highlights predominant theories, frameworks, and constructs that can be utilized by practitioners to improve their understanding of DMFCs, their ability to predict future scenarios and solve practical problems. This chapter will also play a significant role in the further development of DMFC discipline.

Keywords

Proton Transport, Methanol Oxidation, Proton Electrolyte Membranes, Direct Methanol Fuel Cell, Methanol Crossover

Contents

1.   Introduction

Fuel cell technology is finding global importance due to less greenhouse gas (GHG) emissions associated with energy production and influence on climate change. An electrochemical device converts 90% of chemical energy with an oxidant directly into electrical energy is known as fuel cell which has some similarities with batteries and engines. At present, there is a rapid increase in involvement in improving the design of fuel cells. Hydrogen and oxygen in its pure form are the two typical reactants for fuel cells. Normally, H_2 available in a mixture with other gases and oxygen obtained from ambient air conditions will be used as oxidants in fuel cells. The different types of fuel cells based on their field of applications are listed below:

For space exploration:

   (i) direct methanol, (ii) proton exchange membrane, and (iii) alkaline

For stationary applications:

   (i) phosphoric acid, (ii) solid oxide, and (iii) molten carbonate

Compared to all current available types of fuel cells, PEMFC produce electricity directly from the reactants and finds its most attention in the automotive industry because of the alternative energy-converting devices which also has following advantages like (i) generate high current density, (ii) low operating temperature, which is less than 100 °C, (iii) high efficiency, and (iv) low pollution levels. Mainly, the polymer membrane is sandwiched between two gas-diffusion electrodes. However, oxygen is fed into the inlet of the cathode channel and hydrogen to the anode channel. Still, researchers are hardly working to advance the performance, cost-effectiveness, and durability of PEMFCs [1, 2]. The DMFC technology is mostly used for small portable power applications like Mobiles, Laptops, and Tablets etc. DMFC is having more scope among other fuel cells and functions similar to PEMFC except for the type of fuel used (fuel for PEMFC is Hydrogen & fuel for DMFC is Methanol). In the past few decades, the development of DMFCs finds its interest in fuel cells and applied as a promising alternative clean power generator [3-5] because of their desired properties [6-8] like:

1. Stable low-temperature operation

2. Improved efficiency of the fuel cell

3. The improved power density at increased content of methanol

4. Less noise

5. Low weight

6. Ease of transport, storage, and production of liquid methanol

7. Less fossil fuel consumption [9] and

8. Reduced environmental pollution [10]

DMFCs are also considered as heart of fuel cell [11, 12] because of its ionic pathways for protons transfer. It also prevents direct combustion by acting as a barrier between reactant gases. The two drawbacks which hindered the commercialization of DMFCs and affect its overall cell performances are (i) methanol crossover phenomenon [13], which results in 20% fuel loss and low cell voltage during the utilization of Nafion membranes [14] and (ii) slow kinetics of the oxidation reaction. The fuel cells have high proton conductivity, intermediate temperature [15], high permeability [16] and excellent stability in chemical and oxidation reaction, and desired mechanical properties. The application of Nafion membranes on DMFCs doesn't allow it for commercialization because of various limitations. To overcome those, new proton conductive membranes has been developed based on two main stratagems. They are (i) modification of perfluoro sulfonic acid membranes i.e., blocking Nafion surface by metallic layers [17] and (ii) replace Nafion membranes with cheaper, eco-friendly and other acceptable non-fluorinated materials. Researchers proposed various other polymers with diverse chemical structures and functionalized with sulfonic acid families based on their preferred advantages [18-20]. Those developed polymers areSPEEK [21, 22], SPSf [23], SPI [24-26], SPAES [27], PPSS [28], PBI [29, 30], PSS [31], PPENK [32], SPEK [33-35], PAE copolymers [36], PES [37] and SPAEK [38].

1.1   Design of flow-transport

During electrochemical reactions, fast transport of reactants and products limits the concentration overpotentials. Most common designs of bipolar plates with grooved channels are

(i)   parallel channels which run in co-flow and counterflow,

(ii)   interdigitated channels,

(iii)   flow distributors with 'dead' zones between the channels, and

(iv)   serpentine flow channels with many passes.

2.   Direct methanol fuel cell (DMFC)

The variant of PEM fuel cell is the direct alcohol fuel cell (DAFC), whereas methanol as a fuel is called DMFC and ethanol as a fuel is called direct ethanol fuel cell (DEFC). Similarly, DAFC means fuel cell with PEM as an electrolyte, direct alcohol alkaline fuel cell (DAAFC) ways anion exchange membrane as the electrolyte. It shows quicker fuel reduction kinetics of oxidation and oxygen is faster in alkaline medium [39].

In the recent past, DAFCs have become popular as fuel crossover is absent, which is predominantly present in PEM-based DAFCs. As mentioned earlier, DMFC uses methanol as an alternative instead of hydrogen gas. The anode, cathode and overall reactions involved in DMFCs are shown below [39]:

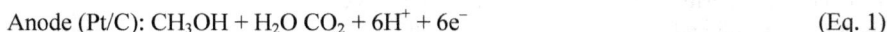

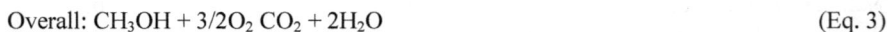

Anode (Pt/C): $CH_3OH + H_2O \; CO_2 + 6H^+ + 6e^-$ (Eq. 1)

Cathode (Pt/C): $3/2O_2 + 6H^+ + 6e^- \; 3H_2O$ (Eq. 2)

Overall: $CH_3OH + 3/2O_2 \; CO_2 + 2H_2O$ (Eq. 3)

The application of methanol in DMFC solves the major problem of storing hydrogen for fuel cells. The utilization of methanol, which is readily available and a low-cost liquid fuel, on the fuel cell system makes it very easy to refill and simple in its usage. These are the main advantages of DMFCs. However, the main problem associated with DMFC is that it has very slow methanol electro-oxidation reaction as compared to hydrogen. Two major problems which affect the overall performances are lower power generation and methanol crossover from the anode compartment.

Six electrons were generated by electrooxidation (a complex multi-step process, slow in reaction and may process through several ways) of methanol, which was clearly explained in Eq. 1 [40]. The reversible cell potential is around 1.199 V for DMFC. Generally, CO performs as a poison with Pt catalyst, but ruthenium (Ru) was also added to reduce Pt catalyst poisoning. Momentous actions have been made towards the escalation of PEM for DMFCs in order to reduce the methanol crossover and increase its operating temperature [41]. The few typical approaches for reducing the methanol crossover will be explained in later sections. Researchers modified the Nafion by incorporating organic or inorganic materials and by proposing the alternate PEM for DMFC membranes [42-44]. Yang et al. [45] obtained reduced methanol crossover even at elevated temperatures while utilizing composite Nafion/Zr(HPO$_4$)$_2$ membrane. Hygroscopic inorganic oxides such as SiO$_2$, Zirconium phosphate, and TiO$_2$ have also been used for Nafion membrane modifications [46]. Furthermore, several other composite membranes like PVA/MMT, ErTfO/Nafion, SPEEK, etc., are under research for DMFC and DEFC.

Recently, Barbora et al. [47] compared pure Nafion membrane with developed ErTfO/Nafion composite membranes. The developed membranes showed decreased fuel permeability, decreased membrane swelling, higher proton conductivity, and increase in tensile strength, higher oxidation stability and reduction of around 80% in ethanol and methanol permeability only by adding 1% ErTfO/Nafion composite membrane. These results confirm that the developed composite membrane may be a suitable alternative for direct alcohol fuel cell applications.

DMFCs can be operated at very low temperatures [48] and acts as a portable power source and in transport applications. The ionic membranes commonly used in DMFCs were basically developed for PEMFC applications [49-51]. The similarities in methanol properties and water make the way for methanol crossover to occur [52, 53]. In general, Nafion exhibit high proton conductivity, but 40% of methanol has been lost due to membrane swelling [54]. These swelling facilitate the penetration of both methanol and water as well as that of protons to some level [54, 55]. Manipulating block copolymer method was used for controlling the structure of phase-separated microdomains of the matrix. Partially sulfonated SEBS triblock copolymers and SBR copolymers have been chosen for that study, because of their low cost and ease of morphology control [56-61]. Hydrogenation of SBR avoids the possible cross-linking and improved the chemical stability [60, 62-64]. Researchers have done detailed analysis on morphology and physical properties of sSEBS membrane [57, 65, 66]. After a thorough analysis on disentanglement behavior on ionic channels, the author concluded that it is very much necessary to reduce swellability by fixing the micro-domain structure of the membrane.

UV irradiation method was used for cross-linking hydrophobic matrix to obtain membrane morphology [67-69]. Poly (styrene-b-butadiene-b-styrene) (SBS) triblock (microphase-separated structure) and SBR copolymers (homogeneous random structure) are the two matrixes which have double bonds for cross-linking [70-73] and the same has been later sulfonated for various extent to enhance the proton conductivity. The membranes performance and structural characterizations have been analyzed to understand the effects of structural variations on proton and methanol permeabilities. The knowledge of proton transport mechanism is very much important to develop low-cost fuel cell membranes and membrane-electrode assemblies (MEAs).

3.   Research on developments of parameters, design, and materials

A number of frameworks for parameter, design, and material in the DMFC discipline have been presented. The flow channel designs and materials have been changed by improving the performance of DMFC. Serpentine flow channel design with porous material is the optimal performance of DMFC. The nature of research and the outcome of the research are discussed in the following sections.

Aarne Halme et al. [74] proposed the substitute for DMFC combined electrolyzer and H_2 PEMFC. This paper presents a novel idea for producing electricity from methanol by using a combining PEMFC and electrolyzer. In some situations, the efficiency in making electricity from methanol could be fairly higher. This is owing to the reality that production of hydrogen can be done using very low electrolysis energy on the one case, and in the other way, a PEM H_2 fuel cell is more competent than a DMFC. In a PEM electrolyzer, the cathode electrode is in a worse potential and no air is participated, which leads to preventing effectively forming of CO from the crossover methanol. According to the author experience, the methanol content at the electrolyzer anode can be improved without the risk of poisoning the Pt catalyst to many advanced concentrations than in DMFC. From a practical overview, the possibility to work the system using higher methanol concentrations without water dilution is a vital aspect, because it shortens the technical construction. Andrea Calabriso et al. [75] discussed in his study about the CO_2 bubble generation influence on DMFC performance. The occurrence of gaseous CO_2 hindrances the free layer of Gas Diffusion Layer (GDL), which deteriorates the energetic area. The author also explored some operating factors and manipulates the connection between gas phase fraction and the same. The results depicted a 40% reduction in power during bubbles generation. Balaiah Kuppan et al. [76] prepared Pt-supported ordered mesoporous carbon catalysts along with colloidal Pt by using four different reducing agents. Paraformaldehyde reduction method is employed to establish Pt nanoparticles over mesoporous CMK-3 carbon. This method is more suited for the preparation of high dispersed uniform seized Pt nanoparticles with an average size of 4 nm. Hence, as a result, the optimized reduction method is the suitable choice for carbon supports, which can offer major cost savings by lowering the catalyst loading.

Chunguang Suo et al. [77] defined the design of MEMS-based micro DMFC stack. The silicon-based DMFC stack was proposed as a "flip-flop" style and was manufactured using MEMS technology. Compared to the SS DMFC stack in planar connection, the flip flop connection way can significantly decrease the space of the connection resistance electric and connection. When fed 2M methanol solution and air as oxygen both the stainless steel stack and the silicon stack can power the LED with a comparable output of 6.77mW and 6.75mW, correspondingly. Chun-Chen Yang et al. [78] explained the preparation of novel composite membrane and PtRu/Hollow carbon sphere (HCS) anode catalyst for alkaline DMFC. The solution casting method was employed to develop, examine and assemble PVA/QASP/TAMPFS-PET composite electrolyte based alkaline composite anionic-exchange membrane. Maximum power density is acquired for the developed alkaline DMFCs. Halim F.A et al. [79] enhanced outline on vapour feed DMFC. Methanol can be fed in liquid or vapor phase. Methanol crossover problem occurred due to the addition of methanol in the liquid phase at a high concentration level, this can be overcome by means of adding the same in vapor phase. Because of the vapor phase, the formation of methanol obstacle layer increases the mass transfer resistance in the fuel cell, which also decreases the major problem of methanol crossover. Water management layer was added to push back the water and to hydrate the membrane for the methanol oxidation reaction. Jean Marcel R. Gallo et al. [80] prepared novel mesoporous carbon ceramics composites which take the merits of fine structured mesoporous silica and graphite. In this study, fuel can be diffused in the developed MCC composites than in the Vulcan XC-72R structure.

Joan M. Ogden et al. [81] discussed the study about an assessment of different fuels like methanol, hydrogen, and gasoline for fuel cell vehicles. The design of the vehicle is very easy in the case of direct hydrogen storage, but still, it needs improvement in the refueling infrastructure. The three factors compared or evaluated using modeling studies are (a) on board partial oxidation (POX) of hydrocarbon fuels, (b) compressed gas hydrogen storage for reforming of methanol, and (c) considering both vehicle and infrastructure issues. A strategy for commercializing feasible fuel for fuel cell vehicles is dignified. The following reasons like cost, vehicle design and efficiency proves that hydrogen is the opt fuel for fuel cell vehicles. In order to increase the methanol oxidation reaction, the co-sputtering technique was used to prepare novel Pt-C and Pt-Ru-C electrodes, which was explained by Minoru Umeda et al. [82]. O_2-enhancing methanol oxidation was conducted for Pt0.56 C0.44 and Pt 0.61Ru 0.34C0.05, whereas the results explained that the addition of Ru enhances the methanol oxidation current and tends to negative shift. In addition to the above analysis, DMFC power generation efficiency was also evaluated by shifting the counter-electrode reactant from H_2 to O_2. Based on all the above results, the degraded DMFC power generation was resolved due to methanol crossover.

Mullai Sudaroli.B et al. [83] explained about heat and mass transfer characteristics of DMFC. Temperature distribution and methanol in the anode side are expected. The double channel flow field is used to explore the methanol distribution and its consequence on cell performance. Water crossovers and methanol in the cell are the main controlling factors which control the overall cell efficiency. Effect of methanol crossover on cell efficiency and Fuel Utilization Efficiency (FUE) can be calculated by using this method. The cell efficiency had an increased range from 7 to 13% with decreasing methanol concentration of 1 to 0.25 M. The cell efficiency also increases from 7 to 13% due to high FUE. Phuttachart. T et al. [84] fabricate bipolar plates for DMFC to increase creep behavior and compressive strength. The composites consist of poly methyl methacrylate (PMMA), carbon black (CB) and polyurethane (PU) and were manufactured via bulk polymerization in a casting process. The composites were prepared in various weight ratios PMMA/PU/CB. The potential of using PMMA/PU/CB composites as original polymeric bipolar plates for DMFCs. PMMA/PU/CB composites were fabricated by bulk polymerization via casting process were victorious in further improving mechanical strengths of the composites like brittleness, creep behavior and compression strength. Vasco S. Silva et al. [85] examined the membranes for DMFC applications and an outline of the research development regarding DMFC. Explicit hard works are being given over to research feature related to membrane modeling, membrane preparation, DMFC tests and membrane characterization.

3.1   Methanol oxidation and oxygen reduction kinetics

Arico et al. [86] and Ren et al. [87] already reported about unsupported catalyst and thinner layers as support which directed to superior performances. The researcher also discussed that the particles which are of high surface areas and smaller in size are useful for the process. Perversely, the atomic ratio between ruthenium and platinum appears to have a minimum impact. The optimum performance has been discovered with an atomic ratio in the region of 1:1 at greater operating temperatures. For minimum temperatures, a maximum relative platinum substance seems to be beneficial. In DMFC, anode two-phase flow pattern technique was studied by Argyropoulos et al. [88, 89]. Inside the DMFC operating environment, high-speed video camera and acrylic cells have been used to visually examine the CO_2 gas evolution procedures. The influence of operating circumstances on the gas management using various gas diffusion layers (carbon paper and carbon cloth), cell sizes, flow channel designs, and exhaust manifold configurations were examined. The visual imagining studies revealed that carbon paper has a comparatively lower ability to the gas removal process. Further, elevating the methanol inlet rate of flow was favourable for gas removal. Upon escalating the density of current which guides to a superior production of gas and in the gas slug pattern, mainly for small flow rate environment, which tends to block the channels and in addition the cell performance is lowered. The authors introduced the design of new flow channels based on the concept of the heat exchanger which tends to be more effective in gas management. This can also be performed under low cell performance. Heat exchanger and picture analysis methodology were implemented in a transparent anode and cathode end plate for developing a visual DMFC [90] to acquire smart visual data with high quality and time effective. The authors analyzed the combination of visual cells with picture analysis and immediately, two-phase flow in the anode of DMFC was examined through video recordings.

Bewer et al. [91] demonstrated about the identification of new method for analyzing the bubble generation in aqueous medium and also interaction of the flow dispersal. They have also insisted that the introduced method is based on the disintegration of hydrogen peroxide solution with water and oxygen in the presence of catalyst. On employing suitable level of hydrogen peroxide, the rate of gas evolution influences the magnitude which will be revealed in real DMFC. The bubble formation can be simulated in the anode cubicle of DMFC without using current. The density of current that is to be simulated can be accustomed by setting the hydrogen peroxide content. The whole cell can be assembled without any current conducting parts by using transparent material namely perspex in order to ensure the complete flow visibility. The developed cell is in such a way that it is simple and also the flow fields can also be tested. Mostly all the researchers claim that the reaction can be progressed based on different mechanisms. Also, it is generally accepted that more considerable reactions are the adsorption of methanol and carbon monoxide oxidation. So the reaction mechanism related to the one used by Meyers et al. [92] was followed by many researchers.

3.2   Gaseous carbon-dioxide in the anode

Lu et al. [93] have discussed obvious DMFC to imagine the flow of bubbles in the DMFC anode. The transparent cells utilized for studies were constructed with stainless steel plates coupled with the polycarbonate plate, which allows the direct observation of flow. Also, the PC plate which is in concave drawing and also the steel plate had a convex model that avoids leakage inflow. These produced channels are machined through

SS plates and the surface contacts with MEA coated with chromium and gold for minimizing the contact resistance. Two basic types of MEAs were adopted by them for investigating the effects of structure and wettability on cells, namely hydrophilic carbon cloth and the other is hydrophobic carbon paper. For the first type, they informed the formation of uniform bubbles which is in a smaller size and the latter one indicated a formation of isolated and large bubbles in the channel. Yang et al. [94, 95] developed a transparent DMFC and investigated about the effect of parallel flow fields (PFF) and single serpentine (SFF) on the cell performance and also behaviour studies on carbon dioxide bubble. It results that DMFCs furnished with SFFs indicates superior performance when compared with PFFs. Also, they have found that the flow channels in PFF at a great current density and less methanol flow have been blocked by the gas bubbles. As the cells with PFFs showed poor performance the authors focused on SFF fields, which includes design parameters, channel length and open ratio.

Liao et al. [96] studied about transparent DMFC that was constructed for visualizing the methanol solution in two-phase and also carbon dioxide bubbles employing high-resolution cameras. However, the dynamic behavior studies of carbon dioxide were recorded and understanding the behaviour of cell performance and carbon dioxide gas bubbles were studied using polarization curves. The effects of parameters on carbon dioxide gas bubble behaviour and cell performance were evaluated based on the methanol flow rate, cell pressure, temperature and the concentration between cathode and anode. They have noticed that the gas bubbles were first emerged around the corner of porous diffusion and the ribs of the channel which produces gas slugs and amalgamation in the channel. The performance of the cell was enhanced on increasing the flow rate, feed concentration, feed temperature and pressure gradient between cathode and anode. Lundin et al. [97] have done an experiment to identify the rate of bubble formation and indicated the reduced bubble formation on raising the flow of fuel as more fuel is obtained for carbon dioxide to dissolve. They have also found that the addition of potassium hydroxide and lithium hydroxide to fuel eliminates CO_2 gas formation at less concentration due to higher solubility in nature.

3.3   Liquid water transport in the cathode

Mench et al. [98] enlightened that maintaining the appropriate water level on the cathode is essential to hydrate the polymer membrane to increase the proton conductivity. Nevertheless, a huge amount of water in the cathode leads to water flood on the pores, which declines the cathode performance. To achieve the precise critical operation parameters to avoid flooding, picturing of the cathode side is important to get the basic physics after the flooding rate. Lu et al. [99] designed a transparent DMFC to visualize the bubble flows in the cathode flooding as well as DMFC anode. Many researchers across the world applied two kinds of GDLs, carbon paper, and ELAT carbon cloth. Visualization of cathode overflowing shows that the more water dews arise upon on the carbon paper GDL compared to single side ELAT GDL, which may be due to the greater hydrophobicity of the latter substantial at raised temperatures. Chen et al. [100] depict the simplified simulations, which are based on macroscopic strength equilibriums and droplet-geometry estimates for calculating the beginning of instability leads to removal of water precipitations at the diffusion layer/flow channel interface. They also carried out visualization tests to detect the growth, formation, and instability of the water condensations on the specific interface of a simulated polymer electrolyte fuel cell cathode.

3.4   Methanol crossover and water management

Tricoli et al. [101] utilized fluorinated membranes to examine the methanol crossover rate (by measuring steady state current at the cathode) and proton conductivity. Hikita et al. [102] measured the content of methanol, CO and CO_2 (common method) in the exhaust to understand the methanol crossover rates. In DMFC operation, a huge content of CO_2 is generated during the anodic reaction and some can disperse onto the cathode side. A method to divide two contributions under real DMFC operating atmosphere and to define the amount of CO_2 was examined in detailed by Dohle et al. [103]. The author measured methanol flux rates by using the electrochemical method [104], whereas membrane permeability is also identified by changing the concentrations of methanol. Jiang and Chu [105] implemented a gravimetric determination method to calculate the amount of CO_2 content, which indirectly measures the methanol crossover rate. The experimental correction factor for computing methanol crossover through CO_2 permeation in polymeric electrolyte membrane was clearly determined. The work presented by Kin et al. [106] proposed a simple method by measuring current and transient voltage to estimate the methanol crossover rate and DMFC efficiency without the use of fitting analysis. The concentration of CO_2 was measured in different conditions like (i) real time, (ii) different inlet concentrations of methanol, and (iii) various operating conditions, to determine the methanol crossover rate was attempted by Han and Liu [107]. Part et al. [108] adopted mass balance research to measure methanol utilization efficiency, water transport coefficient, and electricity conversion rate. Impact of methanol crossover on the DMFC performance is based on different operating parameters. Temperature influences on the permeability of methanol, crossover rate and performance of fuel cell were studied by Kauranen and Skou [109] and Ravikumar and Shukla [110]. The effects of methanol content and oxygen pressure on the cell performance was studied by Cruickshank and Scott [111], in which the above scenario attributed to fuel crossover phenomena.

Kuver and Vielstich [112] found a new catalyst for obtaining good performance with similar working conditions. The authors also analyzed the effects of fuel cell temperature, catalyst loading, air cathode pressure [113] and concentration of methanol on cell performance of a small-scale DMFC. They also derived that proper selection of methanol content leads to higher power density, which always depends on current density. Gurau and Smotkin [114] implemented gas chromatography based on the function of temperature, fuel flow rate, and methanol content to calculate the crossover rate. Gogel et al. [115] explained that cell temperature and cathode air flow influences the cell performance and methanol permeation rate. The basic analysis shows that dependence of operating conditions on noble metal loading affects the cell performance. Furthermore, the dissimilarity between supported and unsupported catalysts was compared and the near favorable conditions for the DMFC operation were estimated. Du et al. [116] investigated the effect of methanol crossover rate by developing a half-cell, whereas open circuit potentials, cyclic voltammetry profiles, electrochemical impedance spectroscopy, and polarization curves were also measured simultaneously. DMFC with higher concentrations of methanol and lower discharge potential led to the poisoning effect during the oxygen reduction process. A composite electrolyte with a metal hybrid was inserted between proton permeable electronic insulators to study the various electrolyte systems and the influence of Pt particles was investigated by Pu et al. [117]. The obtained results confirmed smaller methanol crossover rate compared to Nafion polymer. Wainright et al. [118] examined the methanol vapor permeability, conductivity, water content of the phosphoric acid-doped polybenzimidazole (PBI), which resulted in low methanol crossover in PEM fuel cell. Wang et al. [119] identified enhanced methanol crossover rate with the reduced water-methanol ratio, temperature, and improved current densities. In this study, acid doped PBI membranes were used to understand the cell performance and methanol crossover at different operating parameters. Cross-linked POP membranes were employed by Kuver and Kamloth [120] to study the methanol crossover by using differential electrochemical mass spectrometry technique and the obtained results were compared with the different commercially available membranes. Lu et al. [121] developed a novel DMFC depends on cathode gas-diffusion layer to construct the hydraulic pressure. The authors utilized commercially available Nafion membranes and MEA materials which exhibit low water flux for analyzing the cell performance under ambient air without pressurization. Sandhu et al. [122] predicted superficial velocity, molar fluxes and concentration profiles using mass flux model. Numerical data has been generated as a function of various parameters using this proposed model in the polymer electrolyte membrane. Liu et al. [123] conducted experiments to attain low water and methanol crossover, and high power density which will explicate the effect of operating parameters, production technique, and material properties. Liu and Wang [124] explained that interfacial liquid coverage affects water diffusion and hydraulic penetration which also leads to an extreme effect on water transport coefficient. Shi et al. [125] developed 1D steady-state model to simulate and analyze the water transport characteristics in PEM, which shows improvement in two-phase countercurrent flow. Xu and Zhao [126] investigated the effects of geometric conditions, design and operating parameters on the water crossover flux through a proposed measurement method. Theoretical study on the effect of structures on saturation level and water transport was attempted by Liu et al. [127] through two-phase water transport model. Reduced liquid saturation level with an increase in MPL contact angle is a key parameter to diminish water crossover was confirmed through modeling studies of anode water transport.

3.5   Flow field design

Arico et al. [128] discussed flow field which improves membrane humidification and mass transport in a DMFC by allowing the highest output when compared with serpentine flow area. The DMFC on the inclusion of serpentine flow area indicated higher fuel utilization, lower methanol crossover, and considerable superior efficiency even at less current density. Yang et al. [129] explain about the serpentine flow area that showed superior performance compared to the parallel flow area. The studies were focused mainly on the special effects of channel length and the open ratio of serpentine flow area on the performance of cell and drop in pressure. These studies indicated that the above said two parameters have a significant effect on the cell performance and pressure drop. Also, care should be taken on designing the serpentine flow field for ensuring optimal flow channel length and open ratio. From the end result, they mentioned that the research focus was shifted to PEM fuel cell, where the higher efficiency of PEM fuel cells up to 60% may be the reason for such a shift. In turn, safety is the major issue in using the PEM fuel cells for Mobiles, Tablets, Laptops, etc. Sulfonated cellulose acetate (SCA) was initially employed for a first time in preparation of PEM for DMFCs, which has higher thermally stability. Remarkably, on studying the important characters essential for PEM in the application of fuel cell depicts that the ion exchange capacity of Nafion 117 is 0.9 meq/g. This indicated that the low-cost SCA membranes can be employed as a polyelectrolyte in the broad application of DMFCs.

4.   Progress and developments on proton exchange membranes

The most challenging objective for DMFC technology is the research and development of novel PEM [130-134]. Generally, the heart of DMFC combines the electrolyte, thermal and barrier properties of the membrane [130-133]. High-cost Nafion, perfluorinated ion-exchange polymers from Dupont combines high hydrophobicity and hydrophilicity from the perfluorinated backbone and functional group of sulfonic acids [134]. The outstanding chemical and thermal stability of Nafion is due to Teflon-like perfluorinated backbone. During swelling in water, Nafion has good proton conductive nature and the sulfonic groups present in it are solvated. In addition, hydrophilic and hydrophobic characters of the same are highly prominent which may be due to nano-separation [134]. The developed DMFC species are also transported across methanol and water [130, 134-136]. This result in the drawback of methanol crossover and the same will be chemically oxidized for diminishing the fuel utilization efficiency and to depolarize the cathode. Consumption of O_2 during methanol oxidation lowers its partial pressure severely affects the cathode performance [137]. As like other authors, a 35% reduction in cell efficiency is found due to methanol crossover [136]. However, high water permeability causes cathode flooding and limitations in mass transport lower the cathode performance [130]. During the methanol crossover, the loss of O_2 from the cathode to anode is eliminated. In contrast, the performance of DMFC does not have any influence due to N_2 and CO_2 mass transfer in the PEM.

4.1   Novel materials

The novelty is to develop an innovative material or alteration of existing ones for DMFC applications which tends to conquer the link of proton conductivity and methanol permeability. Modifying the membranes by using highly hydrophilic oxides or by polymers chemical nature [138-145] and SiO_2 entrapped particles in Nafion polymeric structure [146] works as a physical barrier for methanol crossover. The incorporation of inorganic zirconium phosphate (23 wt.%) lowers methanol crossover compared to modified Nafion [147]. Recently, there are numerous substitute original materials that illustrate talented properties for DMFC applications. Some of the scrutinized membranes so far are (i) sPEEK [148-152], (ii) poly(ether sulfone) [153], (iii) polyvinylidene fluoride [154], (iv) styrene grafted and sulfonated membranes [155], and (v) zeolites gel films and membranes doped with heteropolyanions [156]. Comparatively, new preparation or modification techniques of this developed novel material enable a distinct mass transport mechanism which is also less in cost compared to Nafion. For DMFC applications, sPEEK (non-fluorinated membranes) proved to have a better barrier and electrolyte properties [148, 152]. The sulfonation degree (SD) of plain PEEK optimizes hydrophobic-hydrophilic balance and has better technical properties of the polymer [148-152]. Higher SD increases proton conductivity, cell performance and methanol permeability whereas, it also decreases fuel cell overall efficiency [148] and polymer stability. Recently, Li et al. compared DMFC performances for sPEEK (SD = 39 and 47%) and Nafion. The author found a similar result which is obtained for SD = 42% in the sPEEK membrane. The incorporation of inorganic-ceramic materials enhances the proton conductivity and methanol permeation [158-168] compared to the plain polymer. Researchers found promising results of sPEEK with SD as 68% modified with zirconium phosphate (ZrPh), pre-treated with npropylamine and 11.2 wt.% of polybenzimidazole (PBI) [161, 162] because of superior stability between permeability, proton conductivity, and aqueous methanol swelling.DMFC tests depict similar results with a lower concentration and thus display higher global efficiency. In connection to this, the addition of heteropolyacids in plain polymers [166, 167] has also been experimented. Generally, it is a proton conductor during crystalline form with more number of water molecules [169-171]. Also, this new electrolyte leach put the polymer which results in reducing the fuel cell efficiency [169, 172, 173]. Finally, the addition of zirconium oxide onto the sPEEK polymer decreases the membrane permeability i.e., superior barrier properties and enhances the stability of chemical/thermal properties [163-165]. In some cases, the blending of ZrO_2 has a very elevated impact on the conductivity of proton which literally affects the fuel cell performances [165].

4.2   Methanol oxidation on noble metal catalysts

The methanol oxidation on the addition of Pt catalysts has some limitations in DMFC during fast hydrogen oxidation in PEMFCs. This reaction occurred in an aqueous environment i.e., perfluorinated sulfonic acid of Nafion and furthermore depends on the surface charge. Firstly, before modeling this scenario, methanol oxidation at the interface [174, 175] was thoroughly examined using quantum mechanical density functional theory (DFT). Higher energy is obtained due to CH bond points configure towards the surface. Presence of water molecule stabilizes the energetic dissimilarity and activates the CH bond. Provoked by IRRAS studies based on oxidation of formic acid by Chen et al. [176], DFT studies on the molecules of formic acid [177] and the formate anion [178] on the Pt layer were investigated. The comparable mechanism for oxidation of formic acid as for methanol was observed for the reason that the presence of co-adsorbed molecules of water in it. CH bond points configure towards the metal surface which guide to CO_2, whereas the reaction result will occur at elevated potentials.

Nanomaterials for Alcohol Fuel Cells   Materials Research Forum LLC
Materials Research Foundations **49** (2019) 321--350   doi: https://doi.org/10.21741/9781644900192-11

4.3   Proton transport in polymer electrolyte membranes

Nafion, an aqueous phase and phase-separated polymer with polytetrafluoroethylene backbone afford structural stability and also act as a barrier between the anodic and cathodic chamber of the fuel cell. Based on the number of water molecules in the polymer, enhancement in proton conductance was observed with a rising level of water content. It is evident that proton transport mechanism transfers (bulk Grotthus or structural diffusion theory) the hydronium ion to a nearby molecule of water based on the enormity and temperature of fully humidified membranes of Nafion. The structural diffusion mechanism has been mimicked for molecular dynamics through the empirical valence bond (EVB) model to understand the proton transport in Nafion [178–181]. The motion of polymers on large scale takes larger time, whereas immobile framework contains pores of simple slab on it. Simple pore model with a reasonably realistic description is developed to understand the transport of protons based on structural, dynamical and operational features of enabled fuel cell. Single pore mobility and surface charge density show some experimental tendency on conductance of proton with various weight percentage of water content. Polymeric dynamics at reduced water level present improvements in activation energy for transport of proton in dry membranes, which depends on proton mobility temperature in single pores. Fully atomistic models of Nafion were examined for the above-mentioned pore structure and dynamics [182-184]. In aqueous medium, rigid molecular ions of hydronium and water interact through a simple classical force field [183]. Researchers currently analyzed the topological textures and morphological transitions at reduced and higher water levels. The calculated scattering factor can be evaluated with small angle neutron andX-ray scattering experiments. In recent times, pore morphology of polymer/water mixtures on a diamond lattice has been compared by using Lattice Monte Carlo model [185].

Conclusion

Literature investigation has revealed various approaches involved in the transport of protons and design of PEM for methanol oxidation in DMFC research. These technologies proved to be an alternative for power sources and considered as great challenge in the near future. This finds application in portable electronic devices and has potential to substitute conventional batteries. Nevertheless, still, several drawbacks exist with its basic operation after so much of research on this field. This prevents the global widespread usage of this potentially promising technology. Finally, this chapter also tried to categorize and decisively examine the diverse factors and constraints allied with challenges in the design of PEM for methanol oxidation in DMFC application.

References

[1] C. H. Park, S. Y. Lee, D. S. Hwang, D. W. Shin, D. H. Cho, K. H. Lee, T.W. Kim, T.W. Kim, M. Lee, D.S. Kim, C. M. Doherty, A. W. Thornton, A. J. Hill, M. D. Guiver, Y.M. Lee, Nanocrack-regulated self-humidifying membranes, Nature. 532 (2016) 480–483. https://doi.org/10.1038/nature17634

[2] K.S. Lee, J. S. Spendelow, Y.K. Choe, C. Fujimoto, Y. S. Kim, An operationally flexible fuel cell based on quaternary ammonium-biphosphate ion pairs, Nat. Energy. 1 (2016) 16120. https://doi.org/10.1038/nenergy.2016.120

[3] S. Surampudi, S.R. Narayanan, E. Vamos, H. Frank, G. Halpert, A. LaConti, J. Kosek, G. Prakash, G.A. Olah, Advances in direct methanol fuel cells, J. Power Sources 47 (1994) 377. https://doi.org/10.1016/0378-7753(94)87016-0

[4] M. Winter, R.J. Brodd,What Are Batteries, Fuel Cells, and Supercapacitors?, Chem. Rev. 104 (2004) 4245–4270. https://doi.org/10.1021/cr020730k

[5] F. Lufrano, V. Baglio, P. Staiti, V. Antonucci, A.S. Arico,Performance analysis of polymer electrolyte membranes for direct methanol fuel cells, J. Power Sources 243 (2013) 519-534. https://doi.org/10.1016/j.jpowsour.2013.05.180

[6] N.A. Hampson, M. J. Willars, B. D. McNicol,The methanol-air fuel cell: a selective review of methanol oxidation mechanisms at platinum electrodes in acid electrolytes,J. Power Sources 4 (1979) 191. https://doi.org/10.1016/0378-7753(79)85010-7

[7] X. Ren, M.S. Wilson, S. Gottesfeld, J. Electrochem. Soc. 143 (1996) L12.

[8] Y.F. Huang, L.C. Chuang, A.M. Kannan, C.W. Lin, J. Power Sources 186 (2009) 22.

[9] Honma, H. Nakajima, O. Nishikawa, T. Sugimoto, S. Nomura, Solid State Ionics 162/163 (2003) 237–245.

[10] C. Zhao, H. Lin, H. Na, Int. J. Hydrogen Energy 35 (2010) 2176–2182.

[11] M.A. Hickner, H. Ghassemi, Y.S. Kim, B.R. Einsla, J.E. McGrath, Chem. Rev. 104 (2004) 4587-4611.

[12] Chandan, M. Hattenberger, A. El-Kharouf, S.F. Du, A. Dhir, V. Self, B.G. Pollet, A. Ingram, W. Bujalski, J. Power Sources 231 (2013) 264-278. https://doi.org/10.1016/j.jpowsour.2012.11.126

[13] W.C. Choi, J.D. Kim, S.I. Woo, J. Power Sources 96 (2001) 411.

[14] Küver, W. Vielstich, J. Power Sources 74 (1998) 211.

[15] W.H.J. Hogarth, J.C. Diniz da Costa, G.Q. Lu, J. Power Sources 142 (2005) 223.

[16] P. Krishnan, J.S. Park, C.S. Kim, Eur. Polym. J. 43 (2007) 4019–4027.

[17] P. Dimitrova, Solid State Ionics 150 (2002) 115–22.

[18] L. Wang, Y.Z. Meng, S.J. Wang, X.Y. Shang, L. Li, A.S. Hay, Macromol. 37(2004) 3151–3158.

[19] D. Mecerreyes, H. Grande, O. Miguel, E. Ochoteco, R. Marcilla, I. Cantero, Chem. Mater. 16 (2004) 604–607. https://doi.org/10.1021/cm034398k

[20] L. Wang, Y.Z. Meng, S.J. Wang, X.H. Li, M. Xiao, J. Polym. Sci. Part A: Polym. Chem. 43 (2005) 6411–6418.

[21] P. Xing, G.P. Robertson, M.D. Guiver, S.D. Mikhailenko, S. Kaliaguine, Macromol. 37 (2004) 7960–7967.

[22] P. Xing, G.P. Robertson, M.D. Guiver, S.D. Mikhailenko, S. Kaliaguine, Polym. 46 (2005) 3257–3263.

[23] F. Lufrano, I. Gatto, P. Staiti, V. Antonucci, E. Passalacqua, Solid State Ionics 145 (2001) 47–51. https://doi.org/10.1016/s0167-2738(01)00912-2

[24] K. Okamoto, Y. Yin, O. Yamada, M.N. Islam, T. Honda, T. Mishima, Y. Suto, K.Tanaka, H. Kita, J. Membr. Sci. 258 (2005) 115–122.

[25] Y. Shang, X.F. Xie, H. Jin, J.W. Guo, Y.W. Wang, S.G. Feng, S.B. Wang, J.M. Xu, Eur. Polym. J. 42 (2006) 2987–2993.

[26] T. Watari, J.H. Fang, K. Tanaka, H. Kita, K. Okamoto, T. Hirano, J. Membr. Sci. 230 (2004) 111–120.

[27] Y.S. Choi, T.K. Kim, E.A. Kim, S.H. Joo, C. Pak, Y.H. Lee, Adv. Mater. 20 (2008) 2341-2344.

[28] C. Zhang, X.X. Guo, J.H. Fang, H.J. Xu, M.Q. Yuan, B.W. Chen, J. Power Sources. 170 (2007) 42.

[29] J.A. Kerres, J. Membr. Sci. 185 (2001) 3–27.

[30] D.J. Jones, J.J. Roziere, J. Membr. Sci., 185 (2001) 41–58.

[31] N. Carretta, V. Tricoli, F. Picchioni, J. Membr. Sci. 166 (2000) 189–197.

[32] G.Q. Wang, X.L. Zhu, S.H. Zhang, Y.F. Liang, X.G. Jian, Acta Polym. Sin. 2 (2006) 209–212.

Materials Research Forum LLC
doi: https://doi.org/10.21741/9781644900192-11

[33] M. Gil, X.L. Ji, X.F. Li, H. Na, J.E. Hampsey, Y.F. Lu, J. Membr. Sci. 234 (2004) 75–81.

[34] P.X. Xing, G.P. Robertson, M.D. Guiver, S.D. Mikhailenko, K. Wang, S. Kaliaguine, J. Membr. Sci. 229 (2004) 95–106.

[35] S. Swier, Y.S. Chun, J. Gasa, M.T. Shaw, R.A. Weiss, Polym. Eng. Sci. 45 (2005) 1081–1091. https://doi.org/10.1002/pen.20361

[36] Y.S. Kim, M.A. Hickner, L.M. Dong, B.S. Pivovar, J.E. McGrath, J. Membr. Sci. 243 (2004) 317.

[37] D. Poppe, H. Frey, K.D. Kreuer, A. Heinzel, R. Mulhaupt, Macromol. 35 (2002) 7936–7941. https://doi.org/10.1021/ma012198t

[38] B.J. Liu, G.P. Robertson, M.D. Guiver, Y.M. Sun, Y.L. Liu, J.Y. Lai, S. Mikhailenko, S. Kaliaguine, J. Polym. Sci. Part B: Polym. Phys. 44 (2006) 2299. https://doi.org/10.1002/polb.20867

[39] Basu S (2007) Recent Trends in Fuel Cell Science and Technology, Springer, New York.

[40] Larminie and Dicks A Fuel Cell Systems Explained 2nd Ed., John Wiley, 2003,pp 145.

[41] J.Kerres, A Development of ionomer membranes for fuel cells, J. Membr. Sci. 185 (2001) 3-27.

[42] O.Savadogo, Emerging membranes for electrochemical systems: (I) solid polymer electrolyte membranes for fuel cell systems, J. New. Mater. Electrochem. Syst. 1 (1998) 47-66. https://doi.org/10.1002/chin.199847334

[43] S.Malhotra,R.Datta, membrane-supported nonvolatile acidic electrolytes allow higher temperature operation of proton-exchange membrane fuel cells, J. Electrochem. Soc. 144 (1997) 23-26. https://doi.org/10.1149/1.1837420

[44] Y.Daiko, L.C.Klein, T.Kasuga,M.Nogami, Hygroscopicoxides/Nafion® hybrid electrolyte for direct methanol fuel cells, J. Membr. Sci. 281 (2006) 619-625. https://doi.org/10.1016/j.memsci.2006.04.033

[45] C.Yang, S.Srinivasan, A.S.Arico, P.Creti, V.Baglio,V.Antonucci, composite nafion/zirconium phosphate membranes for direct methanol fuel cell operation at high temperature, Electrochem. Solid State Lett. 4 (2001) 31-34. https://doi.org/10.1149/1.1353157

[46] M.Watanabe, H.Uchida, Y.Seki, M.E.Stonehart, Self-humidifying polymer electrolyte membranes for fuel cells, J. Electrochem. Soc. 143 (1996) 3847-3852. https://doi.org/10.1149/1.1837307

[47] L.Barbora, S.Acharya, R.Singh, K.Scott,A.Verma, A novel composite Nafion membrane for direct alcohol fuel cells, J. Membr. Sci. 326 (2009) 721-726. https://doi.org/10.1016/j.memsci.2008.11.009

[48] K. W.Bo¨ddeker, K.V.Peinemann, S. P. J. Nunes, Membr. Sci. 185 (2001) 1.

[49] M. K. Ravikumar, A. K. J. Shukla, Electrochem. Soc. 143(1996) 2601.

[50] A.Ku¨ ver, W. J. Vielstich, Power Sources.74(1998) 211.

[51] K. D. Kreuer, W. Weppner, A.Rabenau, Angew. Chem. Int. Ed. Engl. 21(1982) 208.

[52] C. Pu, W. Huang, K. L. Ley, E. S. J. Smotkin, Electrochem. Soc. 142(1995) 119.

[53] G. T. Burstein, C. J. Barnett, A. R. Kucernak, K. R. Williams, Catal. Today.38(1998), 425.

[54] V. Tricoli, N. Carretta, M. J. Bartolozzi, Electrochem. Soc. 147(2000) 1286.

[55] J. Kerres, W. Cui, R. Disson, W. J. Neubrand, Membr. Sci. 139(1998) 211.

[56] R. A. Weiss, A. Sen, L. A. Pottick, C. L. Willis, Polym. Commun. 31(1990) 220.

[57] R. A. Weiss, A. Sen, C. L. Willis, L. A. Pottick, Polymer. 32(1991) 1867.

[58] B. M. Sheikh-Ali, G. E. Wnek, U.S. Patent, 6,110,616, 2000.

[59] S. G. Ehrenberg, J. M. Serpico, G. E. Wnek, J. N. Rider, U.S. Patent, 5,679,482, 1997.

[60] S. G. Ehrenberg, , J. M. SerpociWnek, G. E. Rider, J. N. U.S. Patent, 5,468,574, 1995.

[61] Mokrini, J. L. Acosta, Polymer 42 (2001) 9.

[62] S. G. Ehrenberg, J. M. Serpico, B. M. Sheikh-Ali, T. N. Tangredi, E. Zador, G. E. Wnek, O. In Savadogo, P. R. Roberge, Eds.; Proceedings of the second international symposium on new materials for fuel cell and modern battery systems. Montreal Canada, 1997; p 828.

[63] G. E. Wnek, J. N. Rider, J. M. Serpico, A. G. Einset, Proceedings of the First International Symposium on Proton Conducting Membrane Fuel Cells. Electrochem. Soc. Proc. 1995, 247.

[64] W.Grot, Chem. Ing. Technol. 50(1978) 299.

[65] R. A. Weiss, A. Sen, , L. A. Pottick, C. L.Willis, Polymer.32(1991) 2785.

[66] M. Nishide, A.Eisenberg, Macromolecules 29(1996) 1507.

[67] G. E. Green, B. P. Stark, S. A. J. Zahir, Macromol. Sci., Rev. Macromol. Chem. 21(1982), , 187.

[68] J. E. Puskas, G. J. P. J. Kennedy, Macromol. Sci. Chem.1991, A28, 65. Crivello, J. V.; Yang, B. J. Macromol. Sci., Chem. 1994, A31, 517.

[69] H. Le Xuan, C. J. Decker, Polym. Sci. Polym. Chem. Ed. 31(1993) 769.

[70] S. M. Ellenstein, S. A. Lee, T. K. Palit, In radiation curing in polymer science and technology; Fouassier, J. P., Rabek, J. F., Eds.; Elsevier Applied Science: London, 1993; Vol. 4.

[71] C. Decker, T. N. T. Viet, Macromol. Chem. Phys. 1999, 200, 358.

[72] V. Dakin, Radiat. Phys. Chem. 45(1995) 715.

[73] Q. Zhang, O. K. C. Tsui, B. Du, F. Zhang, T. Tang, T.He, Macromolecules 33(2000) 9561.

[74] AarneHalme, JormaSelkainaho, TuulaNoponen, Axel Kohonen, An alternative concept for DMFC Combined electrolyzer and H2 PEMFC, International Journal of Hydrogen Energy 41 (2016) 2154-2164. https://doi.org/10.1016/j.ijhydene.2015.12.007

[75] A.Calabriso, D. Borello, L.Cedola, L. D.Zotto, S.G. Santori , Assessment of CO_2 bubble generation influence on direct methanol fuel cell performance, Energy Procedia. 75 (2015) 1996 – 2002. https://doi.org/10.1016/j.egypro.2015.07.254

[76] B. Kuppan, P. Selvam, Platinum-supported mesoporous carbon (Pt/CMK-3) as anodic catalyst for direct methanol fuel cell applications: The effect of preparation and deposition methods, P. Nat.Science: Mater. I. 22(2012) 616–623. https://doi.org/10.1016/j.pnsc.2012.11.005

[77] C. Suo, X. Liua, J. Duana, G. Dinga,Y. Zhanga, Design of MEMS-based micro direct methanol fuel cell stack, Procedia Chemistry. 1 (2009) 1179–1182. https://doi.org/10.1016/j.proche.2009.07.294

[78] C.C.Yanga, Y.T.Lin, Preparation of a novel composite membrane and PtRu/Hollow carbon sphere (HCS) anode catalyst for alkaline direct methanol fuel cell (ADMFC), Energy Procedia. 61 (2014) 1410– 416. https://doi.org/10.1016/j.egypro.2014.12.137

[79] F.A.Halim, U.A.Hasran, M.S. Masdar, S.K.Kamarudin, W.R.W.Daud, Overview on Vapour Feed Direct Methanol Fuel Cell, APCBEE Procedia. 3(2012) 40 – 45. https://doi.org/10.1016/j.apcbee.2012.06.043

[80] J.Marcel, R. Galloa, G.Gattib, A.Graizzaroc, L.Marcheseb, H.O. Pastorea, Novel mesoporous carbon ceramics composites as electrodes for direct methanol fuel cell, J.Power Sources. 196 (2011) 8188– 8196. https://doi.org/10.1016/j.jpowsour.2011.05.008

[81] J.M. Ogden, M.M.Steinbugler, T.G. Kreutz, A comparison of hydrogen, methanol and gasoline as fuels for fuel cell vehicles: implications for vehicle design and infrastructure development, J.Power Sources. 79 (1999) 143-168. https://doi.org/10.1016/s0378-7753(99)00057-9

[82] M.Umeda, M.Ueda, S.Shironita, Novel O_2-enhancing methanol oxidation at Pt-Ru-C sputtered electrode: Direct methanol fuel cell power generation performance, Energy Procedia. 28 (2012) 102 – 112. https://doi.org/10.1016/j.egypro.2012.08.044

[83] M. Sudarolia, A.K.Kolara, Heat and mass transfer characteristics of direct methanol fuel cell: experiments and model, Energy Procedia. 54 (2014) 359 – 366. https://doi.org/10.1016/j.egypro.2014.07.279

[84] T.Phuttacharta, N. Kreua-ongarjnukoola, R. Yeetsorna, M. Phongaksorna, PMMA/PU/CB composite bipolar plate for direct methanol fuel cell, Energy Procedia. 52 (2014) 516 – 524.

[85] V.S. Silva, A.M. Mendes, L.M. Madeira,S.P. Nunes, Membranes for direct methanol fuel cell applications: Analysis based on characterization, experimentation and modeling, Advances in Fuel Cells, 2005.

[86] A.S.Arico, P. Creti, E. Modica, G. Monforte, V. Baglio and V. Antonucci, Investigation of direct methanol fuel cells based on unsupported Pt–Ru anode catalysts with different chemical properties, ElectrochimicaActa. 45 (2000) 4319-4328. https://doi.org/10.1016/s0013-4686(00)00531-4

[87] X.Ren, P. Zelenay, S. Thomas, J. Davey, S. Gottesfeld, Recent advances in direct methanol fuel cells at Los Alamos National Laboratory, J.Power Sources. 86 (2000) 111-116. https://doi.org/10.1016/s0378-7753(99)00407-3

[88] P.Argyropoulos, K. Scott, W.M. Taama, Gas evolution and power performance in direct methanol fuel cells, J.Appl. Electrochem.29 (1999) 661-669.

Nanomaterials for Alcohol Fuel Cells   Materials Research Forum LLC
Materials Research Foundations **49** (2019) 321--350   doi: https://doi.org/10.21741/9781644900192-11

[89] P. Argyropoulos, K. Scott, W.M. Taama, Carbon dioxide evolution patterns in direct methanol fuel cells, Electrochim. Acta. 29 (1999) 661-669. https://doi.org/10.1016/s0013-4686(99)00102-4

[90] J. Nordlund, C. Picard, E. Birgersson, M. Vynnycky,G. Lindbergh, The design and usage of a visual direct methanol fuel cell, J. Appl. Electrochem.34 (2004) 763-770. https://doi.org/10.1023/b:jach.0000035602.70278.0e

[91] T.Bewer, T. Beckmann, H. Dohle, J. Mergel, D. Stolten, Novel method for investigation of two-phase flow in liquid feed direct methanol fuel cells using an aqueous H_2O_2 solution, J. Power Sources. 125 (2004) 1-9. https://doi.org/10.1016/s0378-7753(03)00824-3

[92] J.P.Meyers, J. Newman, Simulation of the direct methanol fuel cell-ii. modeling and data analysis of transport and kinetic phenomena, J. Electrochem. Soc.149 (2002) A718-A728. https://doi.org/10.1149/1.1473189

[93] G.Q.Lu, C.Y. Wang, Electrochemical and flow characterization of a direct methanol fuel cell, J Power Sources. 134 (2004) 33-40. https://doi.org/10.1016/j.jpowsour.2004.01.055

[94] H. Yang, T.S. Q.Zhao, Ye, In situ visualization study of CO_2 gas bubble behaviour in DMFC anode flow fields, J.Power Sources. 139 (2005) 79- 90. https://doi.org/10.1016/j.jpowsour.2004.05.033

[95] H. Yang, T.S. Zhao, Effect of anode flow field design on the performance of liquid feed direct methanol fuel cells, Electrochim. Acta. 50 (2005) 3243-3252. https://doi.org/10.1016/j.electacta.2004.11.060

[96] Q. Liao, X. Zhu, X. Zheng, Y. Ding, Visualization study on the dynamics of CO_2 bubbles in anode channels and performance of a DMFC, J.Power Sources. 171 (2007) 644-651. https://doi.org/10.1016/j.jpowsour.2007.06.257

[97] M.D.Lundin, M.J. McCready, Reduction of carbon dioxide gas formation at the anode of a direct methanol fuel cell using chemically enhanced solubility, J Power Sources. 172 (2007) 553-559. https://doi.org/10.1016/j.jpowsour.2007.05.074

[98] M.M.Mench, C.Y. Wang, An in situ method for determination of current distribution in PEM fuel cells applied to a direct methanol fuel cell, J.Electrochem. Soc.150 (2003) 79-85. https://doi.org/10.1149/1.1526108

[99] G.Q.Lu, C.Y. Wang, Electrochemical and flow characterization of a direct methanol fuel cell, J.Power Sources. 134 (2004) 33-40. https://doi.org/10.1016/j.jpowsour.2004.01.055

[100] K.S.Chen, M.A. Hickner, D.R. Noble, Simplified models for predicting the onset of liquid water droplet instability at the gas diffusion layer/gas flow channel interface, Int. J. Energ. Res.29 (2005) 1113-1132. https://doi.org/10.1002/er.1143

[101] V. Tricoli, N. Carretta, M. Bartolozzi, A comparative investigation of proton and methanol transport in fluorinated ionomeric membranes, J. Electrochem. Soc.147 (2000) 1286-1290. https://doi.org/10.1149/1.1393351

[102] S. Hikita, K. Yamane, Y. Nakajima, Measurement of methanol crossover in direct methanol fuel cell, JSAE Rev.22 (2001) 151-156. https://doi.org/10.1016/s0389-4304(01)00086-8

[103] J.DohleH, J.Divisek, H.F. Mergel, C.Oetjen, Zingler, D. Stolten, Recent developments of the measurement of the methanol permeation in a direct methanol fuel cell, J. Power Sources. 105 (2002) 274-282. https://doi.org/10.1016/s0378-7753(01)00953-3

[104] K. Ramya, K.S. Dhathathreyan, Direct methanol fuel cells: determination of fuel crossover in a polymer electrolyte membrane, J Electroanalyt. Chem.542 (2003) 109-115. https://doi.org/10.1016/s0022-0728(02)01476-6

[105] R. Jiang D. Chu, Comparative studies of methanol crossover and cell performance for a DMFC, J. Am. Chem. Soc. 151 (2004) 69- 76.

[106] T.H.Kin, W.Y. Shieh, C.C. Yang, G. Yu, Estimating the methanol crossover rate of PEM and the efficiency of DMFC via a current transient analysis, J. Power Sources. 161 (2006) 1183-1186. https://doi.org/10.1016/j.jpowsour.2006.06.009

[107] J. Han, H. Liu, Real time measurements of methanol crossover in a DMFC, J. Power Sources. 164 (2007) 166–173. https://doi.org/10.1016/j.jpowsour.2006.09.105

[108] J.Y.Park, J.H. Lee, S. Kang, J. Sauk, I. Song, Mass balance research for high electrochemical performance direct methanol fuel cells with reduced methanol crossover at various operating conditions, J. Power Sources. 178 (2008) 181-187. https://doi.org/10.1016/j.jpowsour.2007.12.021

[109] P.S.Kauranen, E. Skou, Methanol permeability in perfluorosulfonate proton exchange membranes at elevated temperatures, J. Appl. Electrochem.26 (1996) 909-917. https://doi.org/10.1007/bf00242042

[110] M.K.Ravikumar, A.K. Shukla, Effect of methanol crossover in a liquid-feed polymer-electrolyte direct methanol fuel cell, J. Electrochem. Soc.143 (1996) 2601-2606. https://doi.org/10.1149/1.1837054

[111] K.J.Cruickshan, K. Scott, The degree and effect of methanol crossover in the direct methanol fuel cell, J. Power Sources. 70 (1998) 40-47. https://doi.org/10.1016/s0378-7753(97)02626-8

[112] Kuver, W. Vielstich, Investigation of methanol crossover and single electrode performance during PEMDMFC operation: A study using a solid polymer electrolyte membrane fuel cell system, J Power Sources. 74 (1998) 211- 218. https://doi.org/10.1016/s0378-7753(98)00065-2

[113] K. Scott, W.M. Taama, P. Argyropoulos, K. Sundmacher, The impact of mass transport and methanol crossover on the direct methanol fuel cell, J. Power Sources. 83 (1999) 204-216. https://doi.org/10.1016/s0378-7753(99)00303-1

[114] B. Gurau, E.S. Smotkin, Methanol crossover in direct methanol fuel cells: a link between power and energy density, J. Power Sources. 112 (2002) 339-352. https://doi.org/10.1016/s0378-7753(02)00445-7

[115] V. Gogel, T. Frey, Z. Yongsheng, K.A. Friedrich, L. Jörissen, J. Garche, Performance and methanol permeation of direct methanol fuel cells: dependence on operating conditions and on electrode structure, J.Power Sources. 127(2004) 172-180. https://doi.org/10.1016/j.jpowsour.2003.09.035

[116] C.Y.Du, T.S. Zhao, W.W. Yang, Effect of methanol crossover on the cathode behaviour of a DMFC: A half-cell investigation, Electrochim. Acta. 52 (2007) 5266–5271. https://doi.org/10.1016/j.electacta.2007.01.089

[117] C. Pu, W. Huang, K.L. Ley, E.S. Smotkin, A methanol impermeable proton conducting composite electrolyte system, J. Electrochem. Soc.142 (1995) L119-L120. https://doi.org/10.1149/1.2044333

[118] J.S.Wainright, J.T. Wang, D. Weng, R.F. Savinell, M. Litt, Acid-Doped Polybenzimidazoles: A new polymer electrolyte, J. Electrochem. Soc.142 (1995) L121-L123. https://doi.org/10.1149/1.2044337

[119] J.T.Wang, J.S. Wainright, R.F. Savinell, M. Litt, A direct methanol fuel cell using acid-doped polybenzimidazole as polymer electrolyte, J.Appl. Electrochem.26 (1996) 751-756. https://doi.org/10.1007/bf00241516

[120] Kuver, K. Potje-Kamloth, Comparative study of methanol crossover across electropolymerized and commercial proton exchange membrane electrolytes for the acid direct methanol fuel cell, Electrochim. Acta. 43 (1998) 2527-2535. https://doi.org/10.1016/s0013-4686(97)10114-1

[121] G.Q.Lu, F.Q. Liu, C.Y. Wang, Water transport through Nafion 112 membrane in DMFCs, Electrochem. Solid State. Lett. 8 (2005) A1-A4. https://doi.org/10.1149/1.1825312

[122] S.S.Sandhu, R.O. Crowther, J.P. Fellner, Prediction of methanol and water fluxes through a direct methanol fuel cell polymer electrolyte membrane, Electrochim. Acta. 50 (2005) 3985-3991. https://doi.org/10.1016/j.electacta.2005.02.048

[123] F.Liu, G. Lu, C.Y. Wang, Low crossover of methanol and water through thin membranes in direct methanol fuel cells, J. Electrochem. Soc.153 (2006) A543-A553. https://doi.org/10.1149/1.2161636

[124] W. Liu, C.Y. Wang, Modelling water transport in liquid feed direct methanol fuel cells, J. Power Sources. 164 (2007) 189-195. https://doi.org/10.1016/j.jpowsour.2006.10.047

[125] M.H.Shi, J. Wang, Y.P. Chen, Study on water transport in PEM of a direct methanol fuel cell, J. Power Sources. 166 (2007) 303-309. https://doi.org/10.1016/j.jpowsour.2006.12.036

[126] C.Xu, T.S. Zhao, In situ measurements of water crossover through the membrane for direct methanol fuel cells, J. Power Sources. 168 (2007) 143-153. https://doi.org/10.1016/j.jpowsour.2007.03.023

[127] F. Liu, C.Y. Wang, Water and methanol crossover in direct methanol fuel cells - Effect of anode diffusion media, Electrochim. Acta. 53 (2008) 5517-5522. https://doi.org/10.1016/j.electacta.2008.05.015

[128] A.S.Arico, P. Creti, V. Baglio, E. Modica, V. Antonucci, Influence of flow field design on the performance of a direct methanol fuel cell, J. Power Sources. 91 (2000) 202-209. https://doi.org/10.1016/s0378-7753(00)00471-7

[129] H.Yang, T.S. Zhao, Effect of anode flow field design on the performance of liquid feed direct methanol fuel cells, Electrochim. Acta. 50 (2005) 3243-3252. https://doi.org/10.1016/j.electacta.2004.11.060

[130] S. Aricò, S. Srinivasan, V. Antonucci, Fuel Cell.1 (2001) 133.

[131] T. Schultz, S. Zhou and K. Sundmacher, Chem. Eng. Technol. 24 (2001) 12.

[132] R. Dillon, S. Srinivasan, A.S. Aricò and V. Antonucci, J. Power Sources 127 (2004) 112.

[133] P. Piela and P. Zelenay, The Fuel Cells Review 1(2) (2004) 17.

[134] K. A. Kreuer, J. Membr. Sci. 185 (2001) 3.

Materials Research Forum LLC
doi: https://doi.org/10.21741/9781644900192-11

[135] J. Cruickshank, K. Scott, J. Power Sources 70 (1998) 40.

[136] F. R. Kalhammer, P. R. Prokopius, V. P. Voecks, Status and prospects of fuel cells as automobile engines, State of California Air Resources Board, California, 1998.

[137] D. Chu, S. Gilman, J. Electrochem. Soc. 141 (1994) 1770.

[138] A.S. Aricò, P. Creti, P.L. Antonucci, V. Antonucci, Electrochem. Solid-State Lett. 1 (1998) 66.

[139] T. Schultz, S. Zhou, K. Sundmacher, Chem. Eng. Technol. 24 (2001) 12.

[140] R. Dillon, S. Srinivasan, A.S. Aricò, V. Antonucci, J. Power Sources 127 (2004) 112.

[141] P. Piela, P. Zelenay, Fuel Cells Rev.1(2) (2004) 17.

[142] K. A. Kreuer, J. Membr. Sci. 185 (2001) 3.

[143] J. Cruickshank, K. Scott, J. Power Sources 70 (1998) 40.

[144] F. R. Kalhammer, P. R. Prokopius, V. P. Voecks, Status and prospects of fuel cells as automobile engines, State of California Air Resources Board, California, 1998.

[145] D. Chu, S. Gilman, J. Electrochem. Soc. 141 (1994) 1770.

[146] A.S. Aricò, P. Creti, P.L. Antonucci and V. Antonucci, Electrochem. SolidState Lett. 1 (1998) 66.

[147] C. Yang, S. Srinivasan, A.S. Aricò, P. Cretì, V. Baglio, V. Antonucci, Electrochem. Solid-State Lett. 4 (2001) 31.

[148] V.S. Silva, B. Ruffmann, S. Vetter, M. Boaventura, A. Mendes, M. Madeira, S. Nunes, Chem. Eng. Sci. (submitted, 2005).

[149] X. Jin, M. T. Bishop, T. S. Ellis, F. Karasz, Br. Polym. J. 17 (1985) 4.

[150] T. Kobayashi, M. Rikukawa, K. Sanui, N. Ogata, Solid State Ionic.106 (1998) 219.

[151] S. M. J. Zaidi, S. D. Mikailenko, G. P. Robertson, M. D. Guiver, S. Kaliaguine, J. Membr. Sci. 173 (2000) 17.

[152] S. D. Mikhailenko, S. M. J. Zaidi, S. Kaliaguine, Catal. Today. 67 (2001) 225.

[153] B. Bauer, D. J. Jones, J. Roziere, L. Tchicaya, G. Alberti, M. Casciola, I.Massinelli, A. Peraio, S. Besse, E. Ramunni, J. New Mater. Electrochem. Systems. 3 (2000) 93.

[154] E. Peled, T. Duvdevani, A. Aharon, A. Melman, Electrochem. Solid-State Lett. 3 (2000) 525.

[155] S. Hietala, K. Koel, E. Skou, M. Elomaa, F. Sundholm, J. Mater. Chem. 8 (1998) 1127.

[156] A.S. Aricò, P.L. Antonucci, N. Giordano, V. Antonucci, Mater. Letters 24 (1995) 399.

[157] L. Li, J. Zhang, Y. Wang, J. Membr. Sci. 226 (2003) 159.

[158] B. Kumar, J. P. Fellner, J. Power Sources 123 (2003) 132.

[159] S. P. Nunes, B. Ruffmann, E. Rikowsky, S. Vetter, K. Richau, J. Membr. Sci. 203 (2002) 215.

[160] B. Ruffmann, H. Silva, B. Schulte, S. Nunes, Solid State Ionic. 162-163 (2003) 269.

[161] V. S. Silva, B. Ruffmann, S. Vetter, A. Mendes, L. M. Madeira, S. P. Nunes, Catalysis Today.(accepted, 2005).

[162] V.S. Silva, S. Weisshaar, R. Reissner, B. Ruffman, S. Vetter, A. Mendes, L.M. Madeira, S.P. Nunes, J. Power Sources (accepted, 2005).

[163] V. Silva, B. Ruffmann, H. Silva, A. Mendes, M. Madeira, S. Nunes, Mater. Sci. Forum. 455-456 (2004) 587. https://doi.org/10.4028/www.scientific.net/msf.455-456.587

[164] V.S. Silva, B. Ruffmann, H. Silva, Y. A. Gallego, A. Mendes, L. M. Madeira, S. P. Nunes, J. Power Sources. 140 (2005) 34. https://doi.org/10.1016/j.jpowsour.2004.08.004

[165] V.S. Silva, J. Schirmer, R. Reissner, B. Ruffman, H. Silva, A. Mendes, L.M. Madeira, S.P. Nunes, J. Power Sources 140 (2005) 41.

[166] M. L. Ponce, L.A.S.de A. Prado, B. Ruffmann, K. Richau, R. Mohr, S. P. Nunes, J. Membr. Sci. 217 (2003) 5.

[167] M.L Ponce, L.A.S.de A. Prado, V. Silva, S. P. Nunes, Desalination. 162 (2004) 383.

[168] G. Alberti, M. Casciola, L. Massinelli, B. Bauer, J. Membr. Sci. 185 (2001) 73

[169] P. Saiti, A.S. Aricò, S. Hocevar, V. Antonucci, J. New Mater. Electrochem. Syst. 1 (1998) 1.

[170] O. Nakamura, T. Kodama, I. Ogino, Y. Miyake, Chem. Lett. 1 (1979) 17.

[171] D.E. Katsoulis, Chem. Rev. 98 (1998) 359.

[172] P. Saiti, S. Hocevar, N. Giordano, Int. Hydrogen Energy. 22 (1997) 809.

[173] N. Giordano, P. Saiti, S. Hocevar, A.S. Aricò, Electrochim. Acta 41 (1996) 397.

[174] C.Hartnig, E. Spohr, The role of water in the initial steps of methanol oxidation on Pt(111), Chem. Phys. 319 (2005) 185–191. https://doi.org/10.1016/j.chemphys.2005.05.037

[175] C. Hartnig, J. Grimminger, E. Spohr, The role of water in the initial steps of methanol oxidation on Pt(211), Electrochim. Acta. 52 (2007) 2236–2243. https://doi.org/10.1016/j.electacta.2006.04.065

[176] Y.X.Chen, M. Heinen,Z. Jusys, R.J. Behm, Kinetics andmechanism of the electrooxidation of formic acid—spectroelectrochemical studies in a flow cell, Angew. Chem. Int. Ed. 45 (2006) 981–985. https://doi.org/10.1002/anie.200502172

[177] C. Hartnig, J. Grimminger, E. Spohr, Adsorption of formic acid on Pt(111) in the presence of water, J. Electroanal. Chem. 607 (2007) 133–139. https://doi.org/10.1016/j.jelechem.2007.02.018

[178] P. Commer, C. Hartnig, D. Seeliger, E. Spohr, Modeling of proton transfer in polymer electrolyte membranes on different time and length scales, Mol. Simul. 30 (2004) 775–763. https://doi.org/10.1080/0892702042000270179

[179] E. Spohr, P. Commer, A.A. Kornyshev, Enhancing proton mobility in polymer electrolyte membranes: lessons from molecular dynamics simulations, J. Phys. Chem. B. 106 (2002) 10560–10569. https://doi.org/10.1021/jp020209u

[180] P. Commer, A.G. Cherstvy, E. Spohr, A.A. Kornyshev, The effect of water content on proton transport in polymer electrolyte membranes, Fuel Cells. 2 (2002) 127–136. https://doi.org/10.1002/fuce.200290011

[181] S. Walbran, A.A. Kornyshev, Proton transport in polarizable water, J. Chem. Phys. 114 (2001) 10039–10048. https://doi.org/10.1063/1.1370393

[182] S.J. Jang, V. Molinero, T.Cagin, W.A. Goddard, III Nanophase-segregation and transport in Nafion 117 from molecular dynamics simulations: effect of monomeric sequence, J. Phys. Chem. B. 108 (2004) 3149–3157. https://doi.org/10.1021/jp036842c

[183] D.Seeliger, C. Hartnig, E. Spohr, Aqueous pore struture and proton dynamics in solvated Nafion membranes, Electrochim. Acta. 50 (2005) 4234–4240. https://doi.org/10.1016/j.electacta.2005.03.071

[184] N.P. Blake, M.K. Petersen, G.A. Voth, H. Metiu, Structure of hydrated Na-Nafion polymer membranes, J. Phys. Chem. B 109 (2005) 24244–24253. https://doi.org/10.1021/jp054687r

[185] E. Spohr, Monte Carlo simulations of a simple lattice model of polymer electrolyte membranes, J. Mol. Liquids. 136 (2007) 288–293. https://doi.org/10.1016/j.molliq.2007.08.012

Materials Research Forum LLC
doi: https://doi.org/10.21741/9781644900192-12

Chapter 12

Role of Trimetallic Nanoparticles for Complete Oxidation of Alcohol to CO$_2$

R. Imran Jafri[*], Vasundhara Acharya, S. Akshaya

Department of Physics and Electronics, Christ (Deemed to be University), Hosur Road, Bengaluru – 560029, India

* imran.jafri@christuniversity.in

Abstract

The increasing demand of global energy and extensive use of fossil fuels have created enormous pressure for developing a new renewable energy source which is ultra clean, easily accessible, energy efficient and low cost, hence making way for fuel cells. Among different types of fuel cells, Direct Alcohol Fuel Cells (DAFC) has emerged as a promising technology which has the potential to replace the existing fossil fuel-based machinery. To overcome the various hurdles faced by the commercialization of fuel cells, multimetallic nanoparticles as catalysts have attracted huge attention compared to monometallic as the former can be easily tailored. The present chapter overviews the recent developments in the electrocatalysts (with focus on tri-metallic electrocatalysts) for alcohol (methanol and ethanol) oxidation and review of recent Pt and non-Pt based materials for the DAFC.

Keywords

Alcohol Oxidation, Fuel Cells, Nanomaterials, Electrocatalyst, Trimetallic Electrocatalysts, Carbon Nanomaterials

Abbreviations

Direct Alcohol Fuel Cells - DAFC
Fuel cell - FC
Direct methanol fuel cell - DMFC
Methanol oxidation reaction - MOR
Proton exchange membrane - PEM
Direct ethanol fuel cells - DEFC
Department of energy - DOE
Fuel Cell Commercialization Conference of Japan Protocols - FCCJ

Pt group metals - PGM

Proton exchange membrane fuel cell - PEMFC

Acid electrolytic membranes - AEM

Alcohol Oxidation Reaction - AOR

Oxygen reduction reaction - ORR

Membrane electrode assembly - MEA

Ethanol Oxidation Reaction - EOR

In-situ Fourier transforms infrared - FTIR

Anion conducting electrolyte - ACE

Nano truncated octahedrons - NTOs

Nanotube arrays - NTAs

Nanoassemblies - ANAs

Electrochemical surface area - ECSA

Single membraneless ethanol fuel cell - MLEFC

Mesocellular graphene frame - MGF

cyclic voltammetry - CV

Auxiliary Power Unit - APU

Alkaline-based membrane DEFC - AEM-DEFCs

Contents

Materials Research Forum LLC

doi: https://doi.org/10.21741/9781644900192-12

1.   Introduction

Fuel cell (FC) is an electrochemical device, which directly converts the chemical energy of the fuel (hydrogen, alcohol, natural gas, etc.) into electricity. It is immune to Carnot cycle limitation; therefore high efficiency (about 83%) can be achieved, thus making way for clean and efficient energy compared to existing ones. Due to its high energy density, hydrogen is the most preferred fuel for FC. However, pure hydrogen as a fuel faces certain challenges such as high production cost, transportation, storage as well as handling [1]. Overcoming these hurdles, liquid fuel (methanol, ethanol, etc.) emerges as a promising alternative as it is facile and safe to handle, store and transport, comparatively has higher energy density and is less expensive [2]. Direct alcohol fuel cells (DAFC) are a type of fuel cell which makes use of alcohols such as methanol, ethanol, etc., as a fuel for power generations [3]. Kordesch and Marko (1951) first recognized methanol as a possible fuel for the alkaline-based FC system [4].

Methanol as the simplest alcohol is cheap, has a high theoretical energy density (6.1 kWh/kg), facile to produce, easy to handle, transport, and store. Direct methanol fuel cell (DMFC) uses methanol directly as a fuel which can start-up fast and can operate at low temperature (<100°C). The complete oxidation of methanol can theoretically generate 6 e⁻ per molecule, i.e., three times more than hydrogen (two e- per molecule) [5,6]. DMFC is relevant for applications requiring power under 100 W. The direct methanol-air fuel cell [7] was first developed by Shell Research (England) in the 1960s. They incorporated the acid electrolyte route and utilized sulfuric acid for their experiment. However, methanol's toxicity, low boiling point (65°C), and also sluggish reaction happening at the anode, i.e. methanol oxidation reaction (MOR) and fuel crossover from electrolytic membrane results in short-circuiting the FC, thus decrementing the cell performance [8]. Comparatively, ethanol is less toxic, less permeability for PEM (Proton exchange membrane), has a higher energy density (8.0 kWh/kg) and can be generated from biomass or agricultural waste and the carbon dioxide emitted after the complete oxidation can be absorbed by the plants grown for fuel production [6,9].

Direct ethanol fuel cell (DEFC) utilizes ethanol as a fuel, operates at low temperature, and is an active research area in recent years. However, facile kinetics at the anode and low efficiency is a major drawback faced by DEFC [10]. Platinum is known to be an active metal for alcohol oxidation, however facile anode kinetics, catalyst poisoning (adsorption of CO on the surface) and incomplete oxidation at low temperature, generates a necessity for developing an improved catalyst which can overcome these existing barriers [11]. Various bimetallic and the trimetallic catalyst is being designed to increase the electrocatalytic activity, selectivity, durability, and stability [6]. Some important challenges faced for the commercialization of DAFC are: (1) Slower oxidation kinetics of alcohols as compared to hydrogen. (2) The incomplete oxidation of alcohols with two or more carbon atoms. (3) High cost and scarcity of Pt [12].

To indicate the stature of the technology, the price and performance of FC constituents are benchmarked by various agencies independently and assessed annually. Different FC agencies have set technical targets and test protocols for FC components for FC to be economically viable. The globally used protocols for testing FC are as follows: (1) US DOE (Department of energy) (2) EU Harmonized test protocols and (3) Fuel Cell Commercialization Conference of Japan Protocols (FCCJ) [13]. The US DOE has set certain targets to enhance electrocatalytic performance by 2020 such as Develop electrocatalysts with reduced PGM (Pt group metals) loading (0.125 mg_{Pt}/cm^2 for both electrodes combined), increased-activity, catalyst utilization, tolerance to air, fuel, system-derived impurities during potential cycling and durability/stability. As per 2015 US DOE report, Pt loading of $0.16 mg/cm^2$ is achieved for Proton exchange membrane fuel cell (PEMFC) [14], and DMFC 2013 target is given in ref. [4] proposing that the DMFC system should be able to power up to 250W and operate for 5000 hours.

The electrolyte used in the electro-oxidation of alcohol plays a vital role in selecting the catalyst and the operating temperature of the FC. It should perform certain functions as stated by Cairns and MacDonald in 1964. They are [15]:

1. It should help in complete oxidation of the fuel.

2. Spontaneous reaction with the fuels or oxidation products should not occur.

3. The electrolyte must be immune to CO_2.

4. In order to prevent crossover of the fuel to the cathode, the fuel must have low solubility in the electrolyte.

5. It should possess adequate conductance.

6. It should not react with other components of the FC.

Based on the electrolyte employed, DAFC is mainly of two types:

(i) Acidic Direct Alcohol Fuel Cell

(ii) Alkaline Direct Alcohol Fuel Cell

2.   Acidic direct alcohol fuel cell

The general theoretical reaction governing $C_nH_{2n+1}OH$ mono-alcohol that oxidizes completely to CO_2 are [4]:

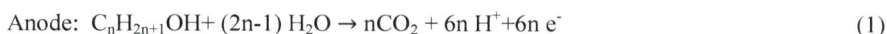

Anode: $C_nH_{2n+1}OH + (2n-1) H_2O \rightarrow nCO_2 + 6n H^+ + 6n e^-$ (1)

Cathode: $3n/2 O_2 + 6n H^+ + 6ne^- \rightarrow 3n H_2O$ (2)

Overall reaction: $C_nH_{2n+1}OH + 3n/2 O_2 \rightarrow n CO_2 + (n+1) H_2O$ (3)

The number of electrons involved in the reaction in case of methanol, ethanol, C_2H_4 $(OH)_2$ (ethylene glycol) and $C_3H_5(OH)_3$ (glycerol) are 6, 12, 10 and 14 respectively.

Practically, the mechanism mentioned above is not obtained. As given by Cameron *et al.*, [16] the complete oxidation process of methanol in an acidic environment comprised of parallel reactions which in principle is as follows:

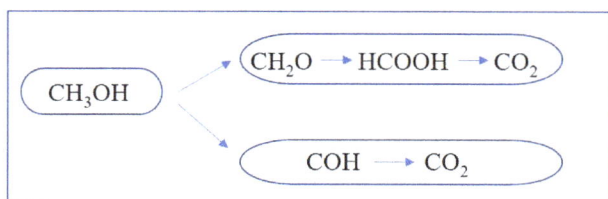

Figure 1. The possible products formed during methanol oxidation in acidic medium [16].

To accomplish both the pathways, a suitable catalyst must be chosen which should be able to (a) disassociate the C-H bond and (b) promote reaction of the formed residue with an oxygen-containing species to form carbon dioxide (or formic acid (HCOOH)). Pure Pt electrode is considered as the ideal catalyst for cleaving the C-H bond. The total oxidation takes place by two processes at discrete potential regions: The initial process involves adsorption of CH_3OH molecules, which requires various neighboring sites at the surface. However, as methanol is unable to dislocate the adsorbed H atoms, the accumulation can begin only at potentials where adequate Pt sites become free from hydrogen. Secondly, the dissociation of water is required, which is the oxygen contributor to the reaction. However, dissociation of water molecules happens around 0.4V vs. RHE in the presence of Pt catalyst. Hence a high potential is required to oxidize methanol to carbon dioxide in the presence of pure Pt, which becomes a hurdle in the commercialization of acidic DMFC [17]. The electro-oxidation of ethanol follows a dual-pathway (1 and 2) mechanism. The reaction involves both pathway 1 (oxidizing CH_3CH_2OH to CO_2) and pathway 2 (oxidizing CH_3CH_2OH to CH_3COOH /CH_3CHO).

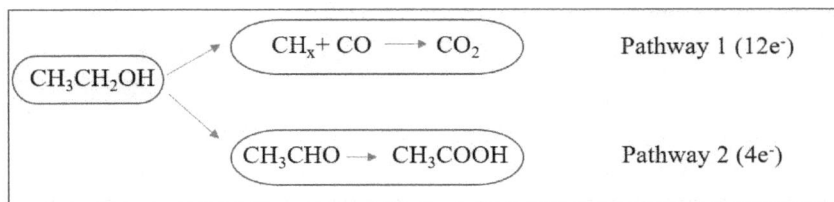

Figure 2.The possible products formed during ethanol oxidation in acidic medium[18].

In contrast to the complete oxidation of ethanol to CO_2 that delivers 12 electrons per molecule, the partial oxidation of ethanol to acetaldehyde and acetic acid yields only 2 and 4 electrons respectively. The products of the ethanol oxidation reaction (EOR) due to incomplete oxidation result in the mixture of acetaldehyde, acetic acid, and CO_2. The amount of CO_2 produced during EOR is <5%. Hence, the main products of the EOR are acetaldehyde (CH_3CHO) and acetic acid (CH_3COOH) with a molar ratio of around 1:1. As CH_3COOH is harmless, highly miscible in water and less corrosive, it is extensively used as a major chemical reagent. E.g., it is used in synthetic fibers, as a food additive for an acidity regulator. Hence, CH_3COOH as the end product is acknowledged. However, CH_3CHO is undesirable as it results in 83.3% loss in the Faradic efficiency. Also, CH_3CHO is a carcinogen and can result in health hazards when applied externally for long periods. Therefore, accomplishing complete oxidation of ethanol-yielding 12 electrons per molecule using state of the art catalyst is a major hurdle to overcome [5,19].

Acidic electrolytes have the ability to reject CO_2 formed in the reaction. However, due to the lack of suitable catalysts and support, most acid radicals get adsorbed on the surface of the catalyst, which detriments the catalytic activity. The commercialization of acidic DAFC faces several issues, mainly due to the sluggish kinetics of alcohol oxidation reaction (AOR) at anode resulting in the loss of activation polarization and hence affecting the net cell performance. Also, acid electrolytic membranes (AEM) are cost inefficient and sufficient amount of Pt loading is essential for the decent performance of FC. Electrocatalyst such as Pt-Sn suffers corrosion when placed in the acid medium, hence shortens the durability of FC[20]. Alkaline DAFC provides: (1) Faster kinetics of AOR (at the anode) and oxygen reduction reaction, i.e., ORR (at the cathode) enables the use of non-noble-metal as an electrocatalyst, hence reducing the cost. (2) Decreases alcohol cross over. (3) Avoids water flooding as the electro-osmotic drag will pull out water produced at the anode via cathode, thus managing water flooding issues. (4) As alkaline electrolyte allows low adsorption of spectator ions onto the membrane electrode assembly (MEA), the electrocatalytic performance may get affected. Therefore, poisoning of Pt is weak in alkaline medium.

Due to the above-mentioned factors, the alkaline DAFC serves better performance and durability when compared to acidic DAFC.

3.   Alkaline DAFC:

DAFC using anion exchange membrane (AEM) as electrolyte, where general reaction is given by [4]:

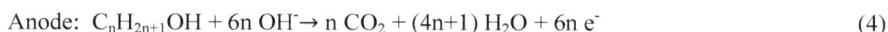

Anode: $C_nH_{2n+1}OH + 6n\ OH^- \rightarrow n\ CO_2 + (4n+1)\ H_2O + 6n\ e^-$ (4)

Cathode: $3n/2\ O_2 + 3n\ H_2O + 6n\ e^- \rightarrow 6n\ OH^-$ (5)

Overall reaction: $C_nH_{2n+1}OH + 3n/2\ O_2 \rightarrow n\ CO_2 + (n+1)\ H_2O$ (6)

Theoretically the number of electrons generated in acidic and alkaline electrolyte should be same, but it is reported different.

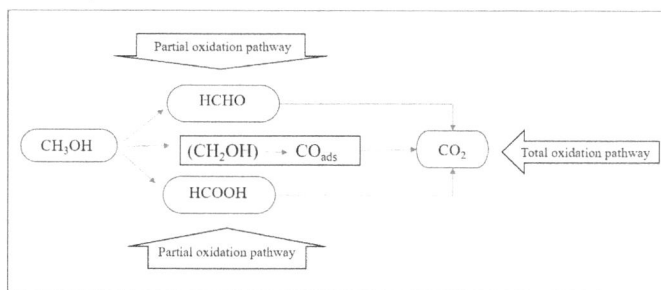

Figure 3.Schematic representation of the parallel pathway for methanol oxidation in alkaline medium [21].

Presently with the existing state of the art catalyst, the product formed during EOR in alkaline medium is acetic acid (acetate) and not CO_2 which is as shown below:

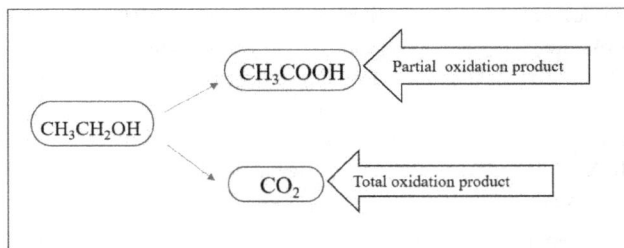

Figure 4.Schematics of the parallel pathway for ethanol oxidation in alkaline medium [19].

In-situ Fourier transforms infrared (FTIR) spectroscopy, cyclic voltammetry, and NMR spectroscopies are used to investigate the product formed during EOR in alkaline media. From the investigation, it was found that along with the electrocatalyst used, the main end product of the EOR is acetic acid (acetate). In contrast to acid media, acetaldehyde is an active intermediate, then the end product. Hence, pathway 2 (oxidizing CH_3CH_2OH to CH_3COOH/CH_3CHO (through 4e⁻ or 2e⁻process respectively)) is predominant in the electro-oxidation of ethanol, which results in the loss of Faradic efficiency (66.7%) of alkaline DEFCs, when compared to pathway 1 (oxidizing CH_3CH_2OH to CO_2) releasing 12 electrons. Therefore, to attain complete electro-oxidation of CH_3CH_2OH to CO_2sustains as a demanding issue in alkaline DEFCs [18,19]. Demerits of alkali DAFC[19]: (1) The carbonate formed during the reaction between the diffused alkali and CO_2 from the air bring down the ionic conductivities of alkaline membranes and its complementary ionomers. (2) The carbonate precipitation lowers the hydrophobicity of the permeable electrode, hence unbalancing the mass transport between oxygen and water. (3) Alkaline membranes illustrate low chemical stability in a strongly alkaline medium.

The problems mentioned above arise due to the addition of a base, implying that the involvement of bases should be avoided, i.e., promoting a base-free alkaline environment that solely depends upon solid ACE (anion conducting electrolyte). The future perspective should be aimed at improving the activity, selectivity of catalysts and increasing the conductivity of the anion-conducting materials.

4.   Thermodynamics of alcohol oxidation

Nernst's equation can be used to calculate the standard potential of the cell. E is given by [22]:

$$E° = (E^+ - E^-) = -\frac{\Delta G°}{nF} \tag{7}$$

Moreover, E^- and E^+ denote the anodic and cathodic reaction potentials respectively.

Where $\Delta G°$ is the Gibbs free energy of the reaction at standard state (298 K and 1bar). ΔG for any chemical reaction is evaluated using the formula:

$$\Delta G = \Delta H - T\Delta S \text{ (at constant temperature)} \tag{8}$$

Where, n is the number of electrons generated per molecule, F is the Faraday constant (96,500 C/mol), ΔH and ΔS is the change in enthalpy and entropy respectively of an overall reaction.

Specific energy is defined as Gibbs free energy per molecular weight of the alcohol given by[5]:

$$W_s = -\frac{\Delta G°}{M} \tag{9}$$

Energy density is defined as energy per unit volume and is given by [18]:

$$W_e = -\frac{\Delta G° \rho}{M} \tag{10}$$

4.1   System efficiency

The various energy-conversion efficiency of FC in a given medium is determined by:

i) Thermodynamic efficiency (η_t) or reversible energy efficiency of FC is a change in Gibbs free energy to the change in enthalpy for an overall reaction.

$$\eta_t = \frac{\Delta G}{\Delta H} \tag{11}$$

Nanomaterials for Alcohol Fuel Cells   Materials Research Forum LLC
Materials Research Foundations **49** (2019) 351-387   doi: https://doi.org/10.21741/9781644900192-12

ii) Faradic efficiency (η_F) is defined as the ratio of a number of electrons released during AOR (n_a) to the theoretical number of electrons produced in complete AOR (n_t):

$$\eta_F = \frac{n_a}{n_t} \tag{12}$$

iii) Voltage efficiency (η_E) is due to the over-potentials developed on the electrode and is defined as:

$$\eta_E = \frac{E_{cell}}{E_o} \tag{13}$$

where E_{cell}, E_o represents the operating voltage and reversible cell voltage respectively [19].

The electro-oxidation reaction of non-aromatic compounds (even the facile one) occurs through a series of successive and parallel pathways liberating many e⁻ during the reactions, thus making oxidation reaction complex. Inefficiency in the cleavage of C-C [23] and the sluggish reaction at low temperature leads to high anodic overvoltage which in turn reduces the operating cell potential. Practically, FC efficiency relies on the current that is obtained by the cell and is less than the thermodynamic efficiency (the electrochemical reaction happening at the electrodes is irreversible).

Therefore, the overall energy-conversion efficiency of FC can be evaluated using the following equation:

$$\eta_{FC} = \eta_F \, \eta_E \, \eta_t \tag{14}$$

The total energy–conversion efficiency (η_{FC}) for an acid DMFCs at 0.5 V is 37%, which is much higher than that of Alkaline DEFC (16% - 23%) and acidic DEFC (11%). The loss in efficiency arises due to the partial oxidation of alcohols in the FC system [19].

Table 1. Reactions involved in Acidic DAFC and the corresponding potential (SHE) [22].

FUEL	ANODE REACTION	CATHODE REACTION	OVERALL REACTION
METHANOL	$CH_3OH + H_2O \rightarrow CO_2 + 6\,H^+ + 6e^-$	$3/2O_2 + 6H^+ + 6e^- \rightarrow 3H_2O$	$CH_3OH + 3/2O_2 \rightarrow CO_2 + 2H_2O + heat$
	$E^+ = 0.016$	$E^- = 1.229$	$E_{emf} = 1.213$
ETHANOL	$CH_3CH_2OH + 3H_2O \rightarrow 2CO_2 + 12H^+ + 12e^-$	$3O_2 + 12H^+ + 12e^- \rightarrow 6H_2O$	$C_2H_5OH + 3O_2 \rightarrow 2CO_2 + 3H_2O + heat$
	$E^+ = 0.084$	$E^- = 1.229$	$E_{emf} = 1.145$
PROPANOL	$CH_3CH_2CH_2OH + 5H_2O \rightarrow 3CO_2 + 18H+ + 18e-$	$9/2O_2 + 18H^+ + 18e^- \rightarrow 9H_2O$	$C_3H_7OH + 9/2O_2 \rightarrow 3CO_2 + 4H_2O + heat$
	$E^+ = 0.098$	$E^- = 1.229$	$E_{emf} = 1.131$
ETHYLENE GLYCOL	$(CH_2OH)_2 + 2H_2O \rightarrow 2CO_2 + 10H^+ + 10e^-$	$5/2O_2 + 10H^+ + 10e^- \rightarrow 5H_2O$	$(CH_2OH)_2 + 5/2O_2 \rightarrow 2CO_2 + 3H_2O$
	$E^+ = 0.026$	$E^- = 1.229$	$E_{emf} = 1.203$

Table 2. Reactions involved in Alkaline DAFC and the corresponding potential (SHE) [13,24,25].

FUEL	ANODE REACTION	CATHODE REACTION	OVERALL REACTION
METHANOL	$CH_3OH + 6OH^- \rightarrow CO_2 + 5H_2O + 6e^-$	$3/2O_2 + 3H_2O + 6e^- \rightarrow 6OH^-$	$CH_3OH + 3/2O_2 \rightarrow CO_2 + 2H_2O + heat$
	$E^+ = -0.81$	$E^- = 0.40$	$E_{emf} = 1.21$
ETHANOL	$CH_3CH_2OH + 12OH^- \rightarrow 2CO_2 + 9H_2O + 12e^-$	$3O_2 + 6H_2O + 12e^- \rightarrow 12OH^-$	$C_2H_5OH + 3O_2 \rightarrow 2CO_2 + 3H_2O + heat$
	$E^+ = -0.77$	$E^- = 0.40$	$E_{emf} = 1.17$
PROPANOL	$CH_3CH_2CH_2OH + 18OH^- \rightarrow 3CO_2 + 13H_2O + 18e^-$	$9/2O_2 + 9H_2O + 18e^- \rightarrow 18OH^-$	$C_3H_7OH + 9/2O_2 \rightarrow 3CO_2 + 4H_2O + heat$
ETHYLENE GLYCOL	$(CH_2OH)_2 + 2H_2O \rightarrow 2CO2 + 10H+ + 10e-$	$5/2\,O_2 + 5\,H_2O + 10\ e^- \rightarrow 10\,OH^-$	$(CH_2OH)_2 + 5/2O_2 \rightarrow 2CO_2 + 3H_2O + heat$
	$E^+ = -0.72$	$E^- = 0.40$	$E_{emf} = 1.12$

5.   Ideal properties for electrocatalyst and catalyst support

Using FC for large scale application, it has to overcome three major criteria, i.e., (1) cost, (2) performance and (3) durability. However, FC suffers from laggard kinetics at the electrodes at low temperature, thus limiting its performance. Hence, numerous research and development are focussed on improving the suitable catalyst for electrodes (anode and cathode) [13]. Catalyst needs to meet the following requirements [26]: (a) At anode: catalyst should possess high- catalytic activity, porosity, electrical conductivity, surface area, stability, durability and low- affinity to carbon monoxide, cost (b) At Cathode: catalyst should be inactive-towards alcohol oxidation and adsorption, high-stability, durability, and activity, less expensive and should perform better ORR.

Platinum (Pt) is identified to have the best electrocatalytic activity when compared to other metals. However, Pt being rare and expensive hinders the commercialization of FC. Reducing the Pt loading by alloying it with other metals has increased the catalytic activity. Also, to increase the electrocatalytic activity and stability, a wide variety of supports have been developed in the past years which exhibits strong resistance to corrosion, possess good electrical conductivity, high surface area and is cost ineffective [6].

Various supports have been developed for FC which are mainly classifiedinto[21]:

1. Carbon-based supports: Major advantages of these supports are- abundant in nature, high- electrical conductivity, surface area, stable in both media (acidic and basic) and inexpensive [27]. Also, precious noble metals can be recovered since carbon can be easily burned off. E.g., Carbon black, CNTs, CNFs, etc. However, carbon suffers from corrosion under normal operating condition because of which catalyst get detached from the support (i.e., Ostwald ripening). This limits the performance of electrocatalyst. Hence, the need for support which overcomes these short comes to have to be incorporated for the effective performance of FC.

2. Non- Carbon supports: Several supports have been examined, which could reduce corrosion and simultaneously retain the desired properties of carbon. They are metal oxides (e.g., MnO_2, ZrO_2, TiO_2, Al_2O_3, etc.), conducting polymers (e.g., PANI(Polyaniline), PPY (polypyrrole)), transition metal carbides (e.g., Tungsten monocarbide (WC)) and nitrides (e.g., CrN, TiN, etc.), Ternary nitrides complexes, etc.

5.1   Role of Platinum as a catalyst

Presently, platinum (Pt) being a noble metal exhibits the highest electrocatalytic activities for AOR on the anode and the ORR on the cathode at low temperature in DAFCs [2].

Based upon the examination of reaction products formed during the methanol oxidation on Pt electrodes using chromatographic and infrared reflectance spectroscopy techniques(in-situ), the below-mentioned reaction mechanism has been developed (This mechanism can be extended to higher alcohol).

The first step involves alcohol adsorption:

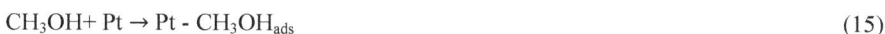

$$CH_3OH + Pt \rightarrow Pt - CH_3OH_{ads} \tag{15}$$

Followed by dehydrogenation of Pt catalyst, given by:

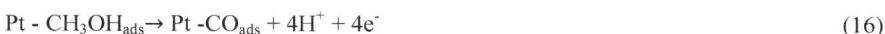

$$Pt - CH_3OH_{ads} \rightarrow Pt - CO_{ads} + 4H^+ + 4e^- \tag{16}$$

Along with the formation of reactive oxygen, water gets dissociatively adsorbed on Pt but at different potential:

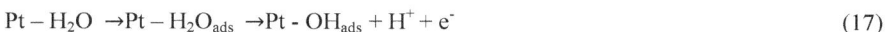

$$Pt - H_2O \rightarrow Pt - H_2O_{ads} \rightarrow Pt - OH_{ads} + H^+ + e^- \tag{17}$$

Followed by a Langmuir-Hinshelwood mechanism, leads to the formation of CO_2:

$$Pt - CO_{ads} + Pt - OH_{ads} \rightarrow 2Pt + CO_2 + H^+ + e^- \tag{18}$$

The trouble with this mechanism is that the presence of $Pt - OH_{ads}$ in the reaction, which requires a higher anodic potential (>0.75 V) this, in turn, will reduce the potential difference between cathode and anode. The partial oxidation of alcohol results in the formation of CO on the Pt surface leads to self-poisoning, thus reducing its catalytic performance. Therefore, Pt being monometallic catalyst has several disadvantages which limit its application in the FC. To overcome this issue, Pt alloyed with other metals to form bi, and trimetallic materials have attracted huge attention for the past few decades [4].

CO gets adsorbed on Pt surface at much lower potential when compared to chemisorption of –OH groups as shown in fig. 4. Therefore, Pt alone cannot be used as a catalyst. Hence other transition metals are alloyed with Pt to oxidize the adsorbed CO by providing oxygen species [13].

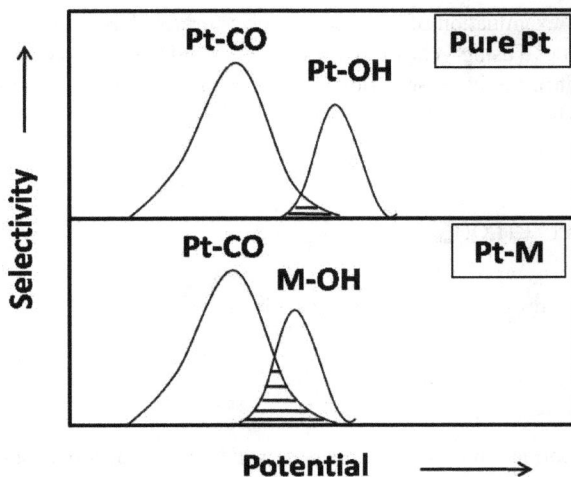

Figure 5. Schematic representation of –CO oxidation on Pt and Pt-M (M – Sn or Ru or Rh) sites by –OH[13].

Metals alloyed with Pt are expected to show:

1. Bifunctional effect - Alloying noble metals (e.g., Pt, Pd, Ru) with more oxophilic metals can reduce electronic binding energy of the electrocatalyst by promoting the adsorption of OH at lower potentials and facilitating the oxidation of the organic species.

2. Intrinsic effect –the incorporated transition metal should reduce the electron density of 5d band of Pt such that it weakens the adsorption strength of formed poisonous species (CO) and hence making it easier to oxidize [2].

To achieve the criteria for attaining electrocatalysts with high performance and low price, many novel structured catalysts have been designed. Among all, the nanostructured materials best fit the requirements of a catalyst with large electrochemical surface area, rising surface-active sites, and low metal loading.

5.2 Role of nanomaterials in bringing down the Pt loading

The breakthrough in the field of commercialization of FC will happen when its cost is reduced, implying that the Pt catalyst used have to be reduced five fold [13]. It has been

Nanomaterials for Alcohol Fuel Cells   Materials Research Forum LLC
Materials Research Foundations **49** (2019) 351-387   doi: https://doi.org/10.21741/9781644900192-12

reported that the 1D platinum-based catalyst such as nanorods, nanotubes, and nanowires may be advantageous than the zero-dimensional Pt nanoparticles as the former suffer less from dissolution and Ostwald ripening/aggregation. Of the various shapes, nanocube which tends to expose (100) facet on all the six faces shows some distinct activity & selectivity toward alcohol oxidation and ORR [13,28].

Metal nanoparticles (NPs) perform electrocatalys is better than their bulk crystals. The NP (clusters of an atom), which have irregular shape are better than their counterpart. There are several reasons for this:

1. The irregularity in the particle surface provides a better environment for the formation of defects and kinks, which in-turn provides the favorable conditions for formation and breaking of bonds.

2. The effective surface area of NPs is more compared to large single crystal, due to the presence of more atoms on the surface of the former than latter.

3. The surface of NPs can be easily restructured compared to single crystals.

The electrode reaction, I is given by:

$$I\ (A\ cm^{-2}) = j\ (A\ cm^{-2})\ s\ (cm^2\ g^{-1})\ w\ (g\ cm^{-2}) \tag{19}$$

Where I - Amperes per surface area of the electrode.

j - specific current density, defined as amperes per real electrocatalyst surface area.

s- specific surface area, defined as real catalyst surface per unit mass.

w-the amount of loaded electrocatalyst defined as mass per electrode surface area.

To develop a catalyst with high performance and low cost, it is more reasonable to increase j and s values keeping PGM loading low [from the equation (19)]. Using multi-metallic catalyst or by addition of adatoms (atoms that are present on the crystal surface), j value can be increased. Dispersing metal catalyst as NPs on a support having high surface area can be used to increase s value [21].

For enhanced electrode reaction, NPs have attracted much attention. The two phenomenon governing the enhanced properties of the nanostructured catalyst is due to [21]:

1. Partially developed band structures in NPs indicate that they possess different electronic properties compared to their bulk.

2. Low coordination number (CN).

Atoms present only on the surface can access the reaction intermediates, while those at the underneath do not participate in the reaction. *Shao etal.,* [29] investigated cubo-octahedralPt particle, they found that as the particle size was reduced, the number of atoms at the edges, surface, and vertices were increased, which implies NPs catalyst can be used on electrodes for FC application. The electrochemical activity and selectivity of Pt depends on its surface structure. Therefore, tuning the surface orientation of Pt single crystal will change the reactivity of the surface toward alcohol oxidation. The change in surface orientation can be achieved by changing the shape of Pt NP [21,29].

Figueiredo et al. [30] showed that by using cubic Pt nanoparticles, where (100) surface sites are predominant, the performance of the fuel cell could be increased from 14 to 24 mW/mg$_{Pt}$ anode catalyst when compared with cuboctahedral Pt. Moreover, the OCP shifts about 50 mV toward more positive potentials. In comparison with commercially available Pt catalysts, the obtained performance by *Figueiredo et al.,* is three times higher for the preferentially oriented (100) nanoparticles. *Figueiredo et al.* also observed the highest activity of (100) surface domains toward C–C bond breaking from ethanol in fuel cell conditions leading to higher fuel cell performances with good stability over 15h of measurements [30].

5.3   Types of NPs as a catalyst

Based on the above observation, various chemical and physical techniques have been proposed and well developed for the preparation of nano-sized catalysts with desired surface properties. The engineered electrocatalyst can be broadly classified into four categories [31]:

(1) Pt-based nano catalysts-By embedding transition metals into Pt, various multi-metallic nanocatalysts (Pt-Ru, Pt-Cu, Pt-Co, etc.) have been designed to enhance the electrocatalytic performance thereby reducing the CO poisoning adsorbed on the surface of Platinum.

(2) Palladium (Pd) based nanocatalysts -Pd can be compared with Pt as it possesses similar intrinsic properties, lower onset potential, and high current density. In-situ Spectro electrochemical investigations reported less poisoning of CO during the electro-oxidation of alcohol. Hence replacing nanostructured Pd will not only enhance the electrocatalytic performance but will also reduce the cost of the FC [32].

(3) Non-noble metal nanoelectro catalysts - many non-noble metals such as coinage metal such as copper, silver, etc. nanoclusters, transition metal oxides,sulfides, macrocyclic complexes, metal incorporating porphyrin compounds, metal nitrides, etc. where reported in the literature which further reduced the cost of FC electrocatalyst.

Nanomaterials for Alcohol Fuel Cells   Materials Research Forum LLC
Materials Research Foundations **49** (2019) 351-387   doi: https://doi.org/10.21741/9781644900192-12

(4) Metal-free and heteroatom-doped carbon materials - due to high- electrochemical stability, catalytic activity, and alcohol tolerance, low cost and lightweight, heteroatom-doped carbon materials have attracted significant attention for FC catalysts. A various heteroatom such as single dopants of N, B and P and multiple dopants of B–N and P–N with a carbon matrix (graphene, carbon nanotubes (CNTs), nanofibers) were synthesized by a plasma process.

Bimetallic and Trimetallic catalysts should be designed in such a way that[18]:

1. Offers suitable surface sites for C–C bond cleavage for complete oxidation of alcohol (except methanol).

2. Enhances the surface composition of electrocatalyst to increase the formed carbon dioxide selectivity.

3. Exhibit bifunctional effect and ligand effect.

6.   Role of Bimetallic catalysts

Pt alloyed with transition metals has been identified for enhanced electrocatalytic activity. The ByBifunctional mechanism, Pt initiates the alcohol oxidation process as mentioned in the above section whereas the second metal/metal oxide should supply the required oxygen to oxidize the adsorbed species and redeem Pt surface for further alcohol oxidation. According to the ligand/electronic effect, the electronic properties of nanostructured Pt is modified through charge transfer processes by the second metal because of which adsorption energy of intermediate species formed during alcohol oxidation is lowered onto the Pt surface [27].

According to the bifunctional mechanism, Pt initiates the catalytic activity in Pt-Ru [33] electrocatalyst for methanol by dehydrogenation. After complete dehydrogenation of methanol, CO gets adsorbed on the surface of Pt thus poisoning the catalyst surface. The second metal Ru may weaken the Pt-CO bond, and supply the oxygen species required to oxidize the CO to CO_2 by activating the water in a neighborhood site to ease the formation of the second C-O bond, thus facilitating the removal of adsorbed CO from Pt surface. Pt-Ru showed enhanced electrocatalytic activity (in agreement with bi-functional effect and the ligand effect between Pt and Ru) when compared with monometallic Pt for MOR but was found less stable than Pt. Also, dissolution of Ru in the FC environment and expensive nature of Ru are the serious drawbacks of this binary system.

Nickel is cost-effective, abundant, and versatile (due to its electrochemical stability) and is resistant to poisoning. So, Pt-Ni [2] is resistant to disintegration (either due to the stability of Ni in the Pt lattice sites or a passivating property of Ni hydroxides) in the potential range used for alcohol oxidation. Disintegration was also observed in some bimetallic catalyst such as Pt-Ru and Pt-Sn [34,35]. The onset potential of Pt-Ni [36] was found to be similar to that of monometallic catalyst Pt. Also, the performance of the catalysts was found to be improved due to the synergetic effect [31]. Increased electrocatalytic activity was shown by Pt-Sn [36]in acid medium and Pt-Ru [37] based nanocatalysts, easing the oxidation of adsorbed CO for EOR, but was recorded with limited improvement due to the restricted breaking of C-C bond and hence resulting in low stability of FC [36].

Various bimetallic catalysts with the oxide phases exhibit high catalytic activity as compared to alloy systems. When Pt-based catalysts along with RuO_x, Co_x, SiO_2, CeO_2, MnO, and NiOare occupied at the anode, it showed enhanced Alcohol oxidation activity for the DAFC.

Recent years have seen a drastic boost in constructing Pd-based catalysts mainly due to easy EOR kinetics in the alkaline medium on Pd. However, Pd alone cannot meet the practical need as Pd suffers from less activity and durability. The activity of Pd based alloy catalyst may be further increased by the inclusion of a second metal or metal oxide. Different Pd-based electrocatalysts have been studied by incorporating it with one or more elements (Ni, Ag, Co, Sn, Ru, Zn, etc.,) or metal oxides (SnO_2, CeO_x, NiO, etc.,) on various supports like graphene, carbon microspheres, rGO, CNT, CNF etc. [31].

Ethanol reflux method was used to form the solid solution of Pt-Sn and Pt-Cu catalyst [3]. The obtained nano-sized particle possessed particle size in the range (2.5nm-3.5nm) and are were relatively spread in equal distribution on the catalyst support. Electrochemical measurements revealed that in terms of electrocatalytically active, Pt-Sn > Pt-Cu and Pt-Sn >Pt in EOR; however-Cu has higher electrocatalytic activity in MOR. Even the electrooxidation of propanol and butanol, show-cased similar traces in terms of electrochemical activities [3].

7.   Role of trimetallic catalyst

Due to the inability of mono and binary metals as catalyst for complete alcohol oxidation which consist of complicated reactions (e.g., dehydrogenation, cleavage of C−C bond (in case of ethanol) and oxidation of adsorbed CO), it is necessary to develop multicomponent catalysts that can multiply catalytic sites in an effective way to improvise the electrocatalytic activity by bifunctional mechanism and ligand effect [38]. The introduction of third metal is expected to yield a series of effects such as reduction in the interatomic distance, the addition of surface sites for the compilation of metal-oxygen bond & adsorption of OH⁻and alteration in the d- band center [39].Fig.6 shows the high

CO_2 selectivity of trimetallic Pt-Pd-Rh indicating the complete oxidation of ethanol into CO_2 [38]. Hence incorporating a third metal/ metal oxide to the bimetallic catalyst can deliver a possible synergistic effect by changing the electronic and structural properties of the alloy, thus overpowering the existing mono and bimetallic system in terms of activity, stability, and durability.

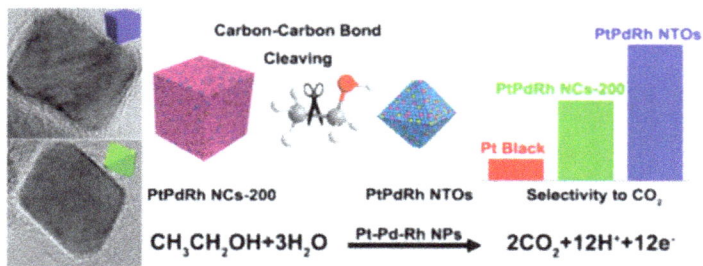

Figure 6. Schematic showing the role of trimetallic Pt−Pd−Rh nanocrystals towards C-C splitting of ethanol. [Reprinted with permission from Ref.[38]. Copyright (2015) American Chemical Society.]

Various Pt-based alloys (Pt-Ru-Sn [40], Pt-Ru-Mo [41,42], Pt-Rh-Sn [43], Pt-Mo-Ir [44], Pt-Sn-Mo [45], etc.)have been reported in the literature.In these catalysts, the existence of the oxophilic element (Sn, Ru) in the alloy boosts the water decomposition, leading to the formation of adsorbed OH^- which acts as oxidants for the actively adsorbed intermediates [46]. Among these ternary mentioned above alloys, *García etal.,* [41]found that the higher ethanol current density could be obtained from Pt-Ru-Mo [41,42]. However, the increased current density was attributed to partially oxidized by-products (higher yields) and not from the total oxidation of ethanol to CO_2. Thus, developing a catalyst for complete oxidation of ethanol to CO_2 at low overpotentials is an open task in the commercialization of the DEFC [46]. The comparison between Pt-Sn-M (M=Co, Ni and, Rh) Wavy nanowires (WNWs) with other monometallic and bimetallic catalyst were studied in [47] for EOR and MOR and is shown in fig.7.

Materials Research Forum LLC
doi: https://doi.org/10.21741/9781644900192-12

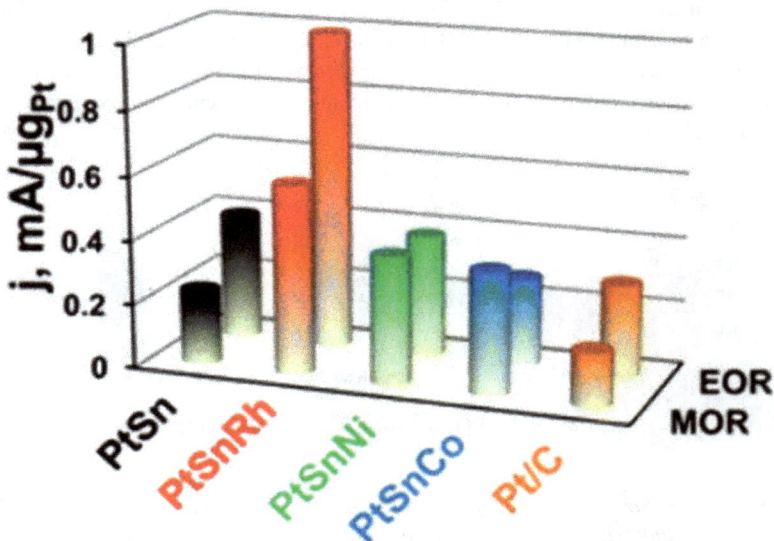

Figure 7. Comparison of electrocatalytic activities of monometallic, binary and ternary catalysts for EOR and MOR. [Reprinted with permission from Ref. in [47]. Copyright (2015) American Chemical Society.]

Fig.7 depicts that Pt-Sn-Rh exhibits much higher electrocatalytic activity than mono and bimetallic catalysts and is found to be more stable than the commercial Pt/C [47]. *Figueiredo et al.,* [46]reported the performance of Pt-Ru/C catalysts altered with varying amounts of irreversibly adsorbed Sb for ethanol oxidation. These trimetallic catalysts display an appreciable activity toward ethanol oxidation, higher power densities, higher open circuit voltages, and good stability when compared to the bimetallic electrocatalysts. *Figueiredo et al.,* have demonstrated ample distribution of the oxophilic metal (at the anode) and the catalyst performance accomplished is 2 times higher than the Pt-Ru catalysts and 6 times higher than pure Pt anodes[46]. The trimetallic catalysts showcase higher (4-5 times) electrocatalytic activity than the mono Pt catalysts, implying that trimetallic catalysts are best suited for alcohol oxidation. further, with the use of non-Pt elements, the cost of the FC can be reduced [39].

Zhu et al., [38] studied composition-varied (100)-terminated Pt−Pd−Rh and (111)-terminated Pt−Pd−Rh nano truncated octahedrons (NTOs) processed in a hydrothermal method by monitoring the reduction ability of the three metal precursors with the aid of halides. A synergistic effect of Pt, Pd, and Rh toward EOR was disclosed, with the help of DFT calculations and experiments. Zhu et al., have demonstrated that the presence of Pd in the Pt−Pd−Rh nanocrystals enhances the activity and durability by providing oxidative hydroxyl groups to promote the oxidation of adsorbed species as well as regulating the electronic structure. The five-fold increment of EOR activity at 0.5 V (vs. NHE) was observed when compared to commercial Pt black. The better performance of Pt−Pd−Rh nanocrystals was due to change in particle size, elemental distribution and overall composition to the alteration of surface composition. Zhu et al., [38] have observed a predominant selectivity to CO_2 on both Pt−Pd−Rh NTOs and Pt−Pd−Rh NCs-200, implying that their increased activity is due to substantial ability in cleavage of C−C bond. Specifically, (111)-terminated Pt-Pd-Rh NTOs holds the capability to completely oxidize ethanol to CO_2 at drastically low potential (0.35 vs. NHE) and the maximum selectivity to CO_2 amidst all the considered catalysts [38].Fig.8 shows the ratios of in situ FTIR adsorption band intensities of CO_2 at 2340 cm^{-1} to that of acetic acid at 1285 cm^{-1} at 0.65 V [38]. The higher ratio implies a higher amount of CO_2 produced thereby indicating the ability of the trimetallic catalyst for complete oxidation of ethanol to CO_2.

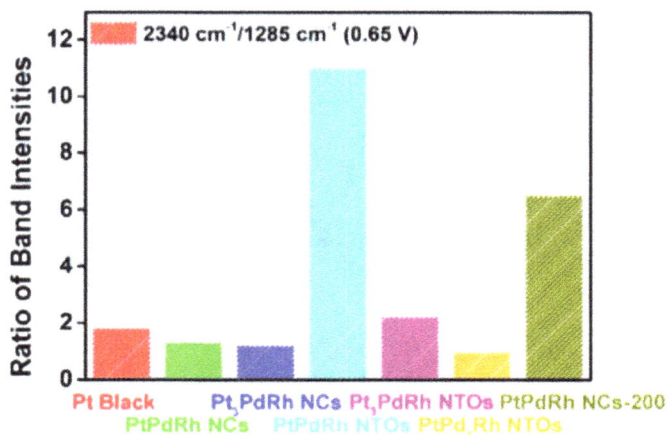

Figure 8.Ratios of in situ FTIR adsorption band intensities of CO_2 at 2340 cm^{-1} to that of acetic acid at 1285 cm^{-1} at 0.65 V. [Reprinted with permission from Ref. [38]. Copyright (2015) American Chemical Society.]

Liu et al., [10] synthesized 3D porous Pt-Rh-Ni alloy nanoassemblies (ANAs) with tunable compositions by facile reduction of an alloyed cyanogel at room temperature and demonstrated that this technique could be utilized to synthesize multimetallic alloy nanostructures with 3D porous architectures [10]. Electrochemical surface area (ECSA) measurements showed that the EOR activity of Pt-Rh-NiANAs was highly reliedon their chemical composition, with the best outcome observed for Pt_3-Rh_1-Ni_2ANAs. *Liu et al.,* reported that compared to commercial Pt black, Pt_3-Rh_1-Ni_2 ANAs showcased appreciable improvement in EOR activity and durability in basic media due to the synergistic effect between Pt, Rh, and Ni atoms, being remarkably favorable anode electrocatalysts for DEFC.

Porous Pt-Ni-P composite Nanotube arrays (NTAs) possess peculiar properties such as porous structures, anisotropic in nature, large surface area, high transport rate of electroactive species, improved utilization of electrocatalysts, low noble metal loading, and synergy among the other elements. Fig.9 shows the SEM image of a ruptured Pt-Ni-P NT, with an inner diameter and wall thickness of ~400 and 70 nm, respectively. The large void volume observed in Pt-Ni-P NTAs will contribute to a 3D space for mass transfer of reactant and product molecules [48].

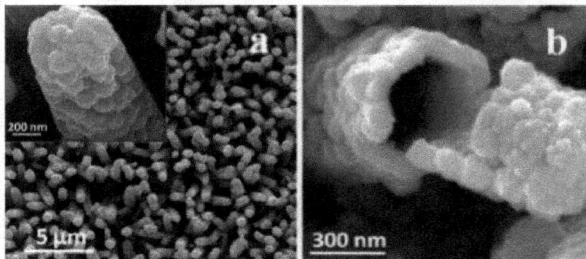

Figure 9. SEM image of (A) Pt -Ni-P NTAs after etching ZnO nanorods, and (B) broken Pt -Ni-P nanorods. [Reprinted with permission from Ref. [48]. Copyright (2012) American Chemical Society].

Ding et al., demonstrated that adding Phosphorous to the Pt-Ni NTAs can result in the distribution of homogeneous nano-size crystal, the immense increase in the ECSA of Pt-Ni-P NTAs, and essentially improves the proportionate content of Pt (0) and the 5d electron density of Pt in Pt-Ni-P NTAs [48].

The peculiar property of Pt-Ni-P NTAs catalyst is that it shows high electrocatalytic activity and stability but a lower I_f/I_b ratio. Generally, a lower I_f/I_b ratio indicates low resistance to the poisoning of carbonaceous (CO) species. Ding *et al.*, attributed this peculiar property to the unique structure of the Pt-Ni-P NTAs catalyst. The porous walls of nanotubes can decrease the diffusion rate of the carbonaceous species and lead to overall oxidation of the carbonaceous (CO) species during the backward scan (Fig.10) [48].

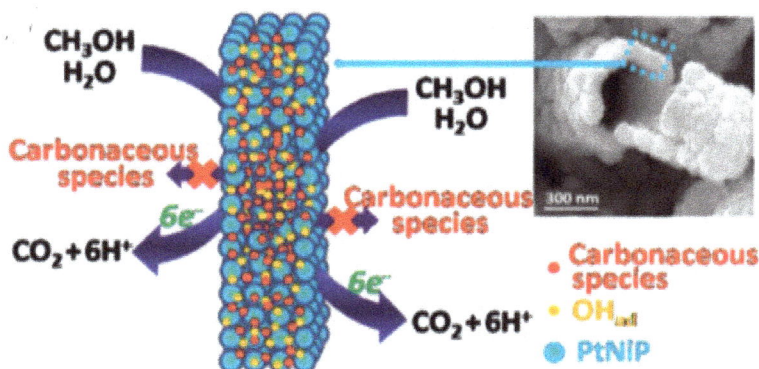

Figure 10. Schematic representation for complete oxidation of carbonaceous species produced during MOR in the porous walls of Pt-Ni-P NTAs.[Reprinted with permission from Ref. [48]. Copyright (2012) American Chemical Society].

The catalyst support is an effective approach to lower the usage of precious metal while parallel improving the electrocatalytic activity by delivering a large surface area, high-conductivity, and porosity. The widely used supports are carbon black (e.g., Vulcan-XC), activated carbon, carbon nanotubes (CNTs) [CNTs have excellent electrical conductivity] [49].

Goel et al studied the effect of various factors such as temperature, electrodes, oxygen flow rate, ethanol concentration on the DEFC polarisation curve., [49] and is reported that the increase in temperature and the oxygen flow rate results in enhanced activity of the cell, in-turn increasing the kinetically controlled region of the polarisation curves, whereas it is reduced by increasing the ethanol concentration using Pt-Re-Sn (20:5:15). Further, the activity of DEFC was not dependent on ethanol flow rate and the anode diffusion layer. Also, the increase in temperature improved the maximum power density linearly. Pt-Re-Sn/C (20:5:15) showed the best cell performance among the three (Pt-Re-Sn/C (20:5:15), Pt-Re-Sn/FuncMWCNT (20:5:15), Pt-Re-Sn/MWCNT(20:5:15)) different compositions examined. The Pt-Re-Sn (20:5:15) catalyst supported with functionalized-MWCNT at anode improves cell performance and the maximum power density obtained is 52.4 mW/cm^2 at 100°C [49].

Hong et al., [50]synthesized ultrathin free-standing ternary-alloy nanosheets (NSs) Pd-Pt-Ag NSs and was measured by using catalytic EOR, and the result was then compared with Pd-Pt NSs, Pd-Pt nanodendrites, Pd-Pt-Ag nanodendrites, and commercial Pt/C and Pd/C catalysts. The ECSA values for Pd-Pt-Ag and Pd-Pt NSs were measured, and it was found approximately two times higher compared to their nanodendrite counterparts due to their peculiar 2D ultrathin sheet-like structure [51]. Though Pd-Pt-Ag and Pd-Pt NSs shows enhanced electrocatalytic properties due to their larger active surface area, Pd-Pt-Ag NSs display greater catalytic activity toward the EOR as the incorporation of Ag improves OH$_{ads}$ formation and CO tolerance. Hong et al examined the electrochemical stability of Pd-Pt-Ag NSs for 400 EOR cycles.,and the results demonstrated that the activity of the Pd-Pt-Ag NSs decreased to 0.92 A mg^{-1} which was higher than Pt/C and Pd/C by a factor of 2.4 and 2.8, respectively.

Wang et al., [18] studied ternary nanocrystals to identify the most favourable combination of greater activity for surface facets and electronic effect. The composition-varied (100)-terminated Pt-Pd-Rh nanocubes (NCs) and (111)-terminated Pt-Pd-Rh nano truncated-octahedrons (NTOs) were synthesized with the aid of halides. Due to the synergistic effects resulting from well-suited surface composition and exposed facets, the Pt-Pd-Rh NTOs exhibited the highest selectivity to carbon dioxide whereas, Pt-Pd-Rh NCs exhibited the best durability.

Dong et al, [52] reported that for both MOR and EOR, Pt-Cr-Co/SWCNT demonstrated the larger ratio of I$_f$/ I$_b$(especially for ethanol oxidation) when compared to Pt/SWCNT and exhibited much better oxidation than Pt–Cr/SWCNT and Pt-Co/SWCNT. Their results demonstrated that Pt-Cr-Co/ SWCNT is a much desirable catalyst for direct alcohols (Methanol and ethanol) FC.

Thilaga et al., [63] prepared Pt, Pt–Sn and Pt–Sn–M (M = Ru, Ni, and Ir) nanocatalysts supported on MWCNTs by ultrasonic assisted chemical reduction. Good dispersion and size distribution were found for the prepared sample. Among the as-prepared catalyst, trimetallic nanocatalysts showed highest ECSA and improved activity than the bimetallic Pt–Sn/MWCNTs for EOR. The effect of the operating conditions on the performance of a single membraneless ethanol fuel cell (MLEFC) with Pt/MWCNTs, Pt–Sn/MWCNTs,

Pt–Sn–Ru/MWCNTs, Pt–Sn–Ni/MWCNTs, and Pt–Sn–Ir/MWCNTs as anode catalysts were studied. Out of all the catalysts, the performance of an MLEFC using Pt–Sn–Ir/MWCNTs was the best, owing to the electronic effect, the bifunctional mechanism, and the hydrogen spillover. These results suggest that MWCNTs could act as a good supporting material in high loading tri metallic catalysts in FC. The durability of the MLEFC was investigated under acid media for a period of about 2 h. MLEFC was able to sustain stable performance with a little decay of cell voltage over the test period. This variation in the cell voltage was either due to the inclusion of the new fuel solution, restarting the experiments, or fluctuation in cell temperature. The durability test showed that the good stability of MLEFC at room temperature makes way for its use in portable power sources.

3D-mesocellular graphene frame (MGF) was synthesised by *Cui et al.,* [61] with pores using zeolite MCM-22 as a template and grew Ultra-small Pd NPs on the zeolite-templated MGF through a stabilizer-free synthesis. The prepared sample (Pd/MGF) possessed a large surface area, layered porous structure and scattered Pd NPs anchoring. Due to these properties, it shows comparably larger electrocatalytic activity and stability than the commercially available Pd/C (10 wt.%) catalyst or the other recorded Pd-based catalysts. Particularly, the electrocatalytic activity of Pd/MGF are found to be more efficient for ethylene glycol and glycerol electrooxidation and is attributed to the 3D-porous graphene network-supported resulting in abundant active sites with rapid mass transfer and super reaction kinetics. Thus, due to the above-mentioned properties, Pd/MGF can be a promising electrocatalyst for DAFC, exclusively for the direct ethylene glycol or glycerol FC with high energy density. *Pan et al.,* [62] suggested a new variety of 3D Ru-Pd-Bi/Nitrogen-doped graphene (Ru-Pd-Bi/NG) catalyst, exhibiting higher catalytic activity (52.4 mA cm^{-2}) for ethylene glycol oxidation, which is about 1.5 times higher than commercial Pd/C catalyst and the Pd loading was only 3.37% in this catalyst which is found to be less than 20% of the loaded metal in the commercially available Pd/C. The loaded metal was Ru in Ru-Pd-Bi/NG catalyst, which is less expensive, reasonably large storage capacity and superior sustainability against toxicity. The synergistic effects of Ru and bismuth activate the production of oxygen species on the catalytic surface of 3D Ru-Pd-Bi catalyst to oxidize the toxic species adsorbed on the surface which are responsible for their high electrocatalytic activity. Thus, making way for palladium-based catalysts in FC for commercialisation.

Mesoporous carbon-supported (MC) Pt–Sn–Ce (75:20:05), Pt–Sn (75:25), Pt–Ce (75:25), and Pt catalysts by co-impregnation reduction method was prepared by *Priya et al.,* [64]. The XRD of Pt–Sn/MC electrocatalyst showed a face-centered cubic structure of the Pt alloys. The Pt metal was the potent material in all the samples, with peaks owing to the fcc crystalline structure. The TEM analysis shows equally dispersed nanoparticles having a size of 2–5 nm. Cyclic voltammetry and Chrono Amperometry results also demonstrate that ternary Pt–Sn–Ce/MC (75:25:05) catalysts give lower onset potential, higher - current, stability compared to all the prepared catalysts and is suitable in MLEFC for ethanol oxidation. It is demonstrated that surface individuality of Pt was altered by the presence of Sn and Ce. The Pt–Sn–Ce/MC electrocatalysts showed higher activity than the prepared Pt-Sn/MC and Pt–Ce/MC. The better performance of MC supported on ternary catalysts is achieved due to the increase of ECSA and the smaller particle size.

Since cyclic voltammetry (CV) is a basic electrochemical characterization technique for the study of any electrocatalyst, Ghosh et al have explained the analysis of electrochemical results., [53].

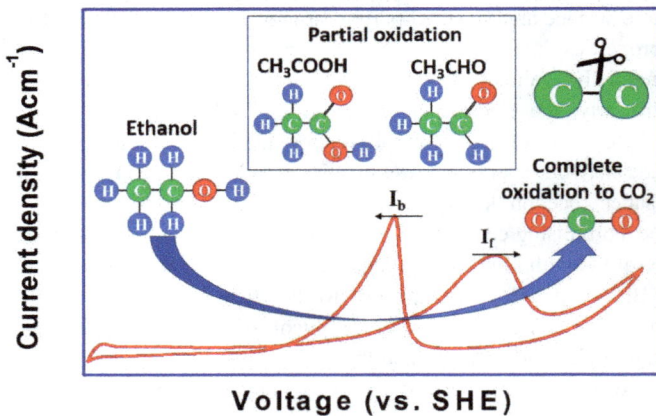

Figure 11. Schematic showing the oxidation of ethanol to CO_2 and CH_3CHO & CH_3CHO (the role of trimetallic catalyst is (i) in the cleavage of C-C bond and oxidize it to CO_2 and (ii) oxidative removal of adsorbed CO species. Also shown is the typical CV plot for ethanol oxidation peak).

The peak current density is denoted by I_f and I_b representing the forward and backward scans as shown in Fig.11, where the I_f peak is attributed to alcohol oxidation representing the presence of various carbonaceous species including CO and CO_2 getting adsorbed on the catalyst surface, and I_b represents the oxidation peak in the backward scan attributing to the oxidation of adsorbed intermediates to CO_2 [27]. The I_f / I_b ratio determines the CO tolerance of any catalyst [54]. A catalyst capable of producing a high value of I_f / I_b is best suited for fuel cell applications.

The following figure gives a summary of some of the studied catalyst with different supports from the literature:

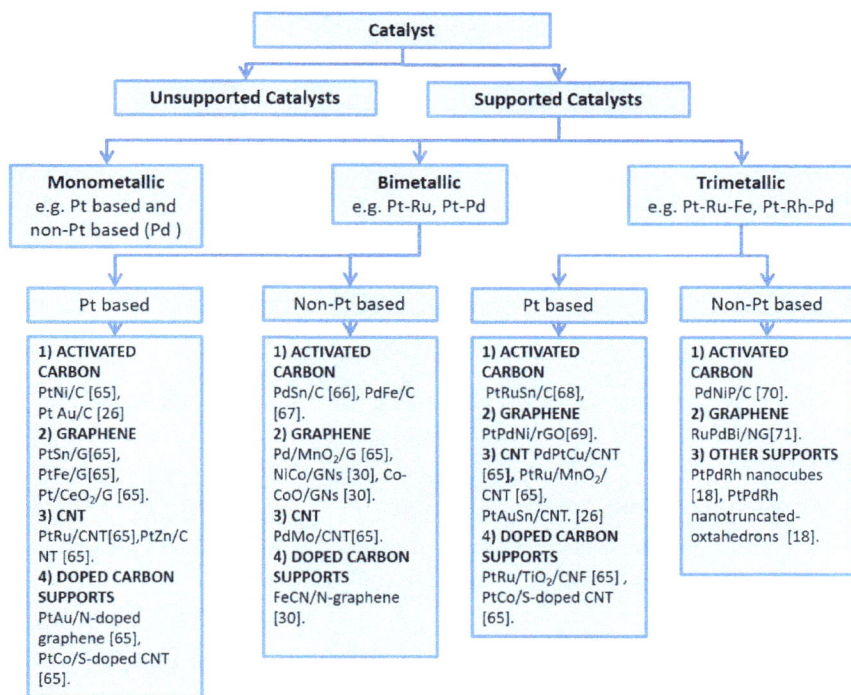

Figure 12. Summary of the electrocatalyst with different support.

8. Application

A summary of different types of a fuel cell with their application is mentioned in ref. [55]. DAFC incorporating PEM in fuel cells can be used for transportation, stationary power appliance, space, military, energy storage systems, and portable power generation-based application. First ever completely commercialized FC systems started to emerge in 2007 as Auxiliary Power Unit (APU) and became more remarkable in 2011, primarily for the stationary application [13]. The PEMFC could perform only for 2000–3000 h but 8000 h and 100000 h are required for transportation and stationary based applications respectively. Mini and portable FC devices are being engineered by quite a few companies. For instance, Toshiba (Tokyo: 6502 JP) has come up with a DMFC for the electronic market, which they tried to incorporate into digital music players (DMP), laptops and computers [13].

DMFC technology has two major applications in the market, i.e., in portable power generation, and alternative to batteries. However, the power density provided by DMFC is low (200-400 mW/cm^2) due to low conversion efficiency, therefore, DMFC is unable to meet the energy production demands for many portable applications. However, commercialization of this technology is still a challenge as it suffers from low performances of the prototypes required for military and consumer electronics [56]. A few prominent players involved in the development of the DMFC device are MTI Micro Fuel Cells, NeahPowerSystems, Ultracell Smart Fuel Cells, Hitachi, Motorola Labs, NEC, Sanyo Electric, Toshiba and Samsung Advanced Institute of Technology, etc. [56].

9. Swot Analysis

To comprehend the opportunities and barriers faced by the DAFC technology for various applications, a SWOT (strengths, weaknesses, opportunities, and threats) analysis have been modeled.

(a) Strengths: They are as follows:

(1)Easily accessible fuels (methanol, ethanol, etc.,) (2). High energy efficiencies (3). Liquid fuels - easy to store, transport, and handle. (4). comparatively low-toxic fuel.

(b) Weaknesses:

1. C-C bond breaking catalyst is required for complete oxidation to CO_2 [30,57].

2. In DAFC reactions is not complete as carbon monoxide gets adsorbed on the catalyst surface

3. CO_2 emissions.

4. When higher alcohols are used energy efficiency goes down.

5. Alcohol crossover- fuel fed from the anode penetrates to the cathode through the existing membrane (Nafion) without contributing to any anodic reaction, resulting in reduced energy conversion efficiency and fuel wastage.

6. Driving range and installation are the main hurdles for the commercialization of DAFC when compared to electric vehicles powered by a battery. The installation becomes even more severe, somewhere in between on-board gas-purification and diesel reformation.

7. No practical vehicle has been demonstrated yet on a large scale, only micro-scale development in case of DEFC.

(c) Opportunities:

Even though DMFC has been commercialized, it suffers from methanol crossover [58] that reduces its durability. So, an electrolytic membrane having lower methanol crossover and higher proton conductivity for PEMFC should be developed to increase the durability and stability for DMFC.

1. To develop Catalyst which has low cost, high ECSA, electrocatalytic activity, stability,and efficiency.

- Increasing performance and durability of FC and simultaneously reducing the PGM loading to less than 0.125 mg Pt cm^{-2} [24].

- Noble metals are considered to be best for catalyst activities. By further enhancing the morphology, composition, and size of the catalyst and studying the degradation mechanism of currently active, the better catalyst can be synthesized for better AOR and ORR [38].

- To Synthesize multimetallic nanosheets which have controlled thickness, size, morphology, composition, and design are attributes of a durable and active catalyst for FC reactions and provide a clear idea on the factors which determine the catalyst stability and activity, and also formulate synthesis method for developing multimetallicnanosheets [59].

- The effect of alloying degree of catalyst for methanol and ethanol oxidation should be specified [18].

2. Alkaline-based membrane DEFC(AEM-DEFCs) has proven to be better than polymer electrolyte membrane FC as cathode does not require any Pt-based catalysts which will reduce the cost and increase the efficiency of FC, So, AEM having high OH$^-$ conductivity and better mechanical, thermal, chemical stability need to be synthesized.

3. Develop stable carbon supports as the existing carbon support block the pores and inhibit the catalytic activity.

4. Domestic fuel production and industry[55].

5. Vehicle implementation for practical application for DEFC[60].

(d) Threats

1. Since battery operated vehicles also come under the category of low emission vehicles, FC based vehicle have to compete with them in terms of cost and performance.

2. ethanol is flammable, toxic and can cause health hazards if not handled with safety, hence limiting its use as a fuel for commercialization purpose.

3. Ethanol large scale production will reduce the lands for farming and hence jeopardize the food chain.

References

[1] A. K. Agrawal, J. Rangarajan, Electro-catalytic activity of nano-sized Pt-Ni bimetallic alloy particles supported on carbon for methanol electro-oxidation, Int. J. Scientific & Engineering Research. 5 (2014) 1544-1551

[2] E. Antolini, Pt-Ni and Pt-M-Ni (M = Ru, Sn) anode catalysts for low-temperature acidic direct alcohol fuel cells: A review, Energies. 10 (2017) 42. https://doi.org/10.3390/en10010042

[3] R.G.C.S. dos Reis, F. Colmati, Electrochemical alcohol oxidation: a comparative study of the behavior of methanol, ethanol, propanol, and butanol on carbon-supported PtSn, PtCu, and Pt nanoparticles, J. Solid State Electrochem. 20 (2016) 2559–2567. https://doi.org/10.1007/s10008-016-3323-3

[4] H.R. Corti, E.R. Gonzalez, Introduction to direct alcohol fuel cells, in: Direct alcohol fuel cells, Springer Netherlands, Dordrecht, 2014, pp. 1–32. https://doi.org/10.1007/978-94-007-7708-8_1

[5] C. Lamy, E.M. Belgsir, J.M. Léger, Electrocatalytic oxidation of aliphatic alcohols: Application to the direct alcohol fuel cell (DAFC), J. Appl Electrochem. 31 (2001) 799–809. https://doi.org/10.1023/A:1017587310150

[6] M.C. Figueiredo, R.M. Arán-Ais, J.M. Feliu, K. Kontturi, T. Kallio, Pt catalysts modified with Bi: Enhancement of the catalytic activity for alcohol oxidation in alkaline media, J. Catal. 312 (2014) 78–86. https://doi.org/10.1016/J.JCAT.2014.01.010

[7] B. McNicol, D.A. Rand, K. Williams, Direct methanol-air fuel cells for road transportation, J. Power Sources. 83 (1999) 15–31. https://doi.org/10.1016/S0378-7753(99)00244-X

[8] S. Zhao, H. Yin, L. Du, G. Yin, Z. Tang, S. Liu, Three dimensional N-doped graphene/PtRu nanoparticle hybrids as high performance anode for direct methanol fuel cells, J. Mater. Chem.A. 2 (2014) 3719–3724. https://doi.org/10.1039/c3ta14809b

[9] J. Tayal, B. Rawat, S. Basu, Bi-metallic and tri-metallic Pt-Sn/C, Pt-Ir/C, Pt-Ir-Sn/C catalysts for electro-oxidation of ethanol in direct ethanol fuel cell, Int. J. Hydrogen Energ. 36 (2011) 14884–14897. https://doi.org/10.1016/j.ijhydene.2011.03.035

[10] H. Liu, J. Li, L. Wang, Y. Tang, B.Y. Xia, Y. Chen, Trimetallic PtRhNi alloy nanoassemblies as highly active electrocatalyst for ethanol electrooxidation, Nano Research. 10 (2017) 3324–3332. https://doi.org/10.1007/s12274-017-1545-z

[11] O. Ugalde-Reyes, R. Hernández-Maya, A.L. Ocampo-Flores, F.A.- Ramírez, E. Sosa-Hernández, C. Angeles-Chavez, P. Roquero, Study of the electrochemical activities of Mo-modified pt catalysts, for application as anodes in direct methanol fuel cells: Effect of the aggregation route, J. Electrochem. Soc. 162 (2015) H132–H141. https://doi.org/10.1149/2.0521503jes

[12] Q. He, Y. Shen, K. Xiao, J. Xi, X. Qiu, Alcohol electro-oxidation on platinum–ceria/graphene nanosheet in alkaline solutions, Int. J. Hydrogen Energ.. 41 (2016) 20709–20719. https://doi.org/10.1016/J.IJHYDENE.2016.07.205

[13] N. Rajalakshmi, R. Imran Jafri, K.S. Dhathathreyan, Research advancements in low-temperature fuel cells, in: Electrocatalysts for low temperature fuel cells, Wiley-VCH Verlag GmbH & Co. KGaA, Weinheim, Germany, 2017, pp. 35–74. https://doi.org/10.1002/9783527803873.ch2

[14] DoE, Fuel cell technologies office multiyear research, development and demonstration plan, Doe. (2016)1–58. https://www.energy.gov/sites/prod/files/2016/06/f32/fcto_myrdd_fuel_cells_0.pdf (accessed 23 December 2018)

[15] S.D. Fritts, R.K. Sen, Assessment of methanol electro-oxidation for direct methanol-air fuel cells, Richland, WA (United States), 1988. https://doi.org/10.2172/7129968

[16] D.S. Cameron, G.A. Hards, B. Harrison, R.J. Potter, Direct methanol fuel cells: Recent developments in the search for improved performance, Platin. Met. Rev. 31 (1987) 173–181

[17] T. Iwasita, Electrocatalysis of methanol oxidation, Electrochim.Acta. 47 (2002) 3663–3674. https://doi.org/10.1016/S0013-4686(02)00336-5

[18] Y. Wang, S. Zou, W.-B. Cai, Y. Wang, S. Zou, W.-B. Cai, Recent advances on electro-oxidation of ethanol on Pt- and Pd-based catalysts: From reaction mechanisms

to catalytic materials, Catalysts. 5 (2015) 1507–1534.
https://doi.org/10.3390/catal5031507

[19] L. An, T.S. Zhao, Y.S. Li, Carbon-neutral sustainable energy technology: Direct ethanol fuel cells, Renew.Sustain.Energ.Rev. 50 (2015) 1462–1468.
https://doi.org/10.1016/J.RSER.2015.05.074

[20] A.R. Khade, Fuel cell technologies and applications, Int. J. Sci. Res. 3 (2014) 978–982

[21] K.I. Ozoemena, Nanostructured platinum-free electrocatalysts in alkaline direct alcohol fuel cells: catalyst design, principles and applications, RSC Adv. 6 (2016) 89523–89550. https://doi.org/10.1039/C6RA15057H

[22] C. Lamy, C. Coutanceau, Electrocatalysis of alcohol oxidation reactions at platinum group metals, in: Catalysts for alcohol-fuelled direct oxidation fuel cells, 2012,pp. 1–70. https://doi.org/10.1039/9781849734783-00001

[23] A. Dutta, S.S. Mahapatra, J. Datta, High performance PtPdAu nano-catalyst for ethanol oxidation in alkaline media for fuel cell applications, Int. J. Hydrogen Energ.. 36 (2011) 14898–14906. https://doi.org/10.1016/J.IJHYDENE.2011.02.101

[24] E.H. Yu, X. Wang, U. Krewer, L. Li, K. Scott, Direct oxidation alkaline fuelcells: from materials to systems, Energ.Environ. Sci. 5 (2012) 5668–5680.
https://doi.org/10.1039/C2EE02552C

[25] E.H. Yu, U. Krewer, K. Scott, Principles and materials aspects of direct alkaline alcohol fuel cells, Energies. 3 (2010) 1499–1528. https://doi.org/10.3390/en3081499

[26] S.S. Munjewar, S.B. Thombre, R.K. Mallick, A comprehensive review on recent material development of passive direct methanol fuel cell, Ionics. 23 (2017) 1–18.
https://doi.org/10.1007/s11581-016-1864-1

[27] B. Singh, L. Murad, F. Laffir, C. Dickinson, E. Dempsey, Pt based nanocomposites (mono/bi/tri-metallic) decorated using different carbon supports for methanol electro-oxidation in acidic and basic media, Nanoscale. 3 (2011) 3334–3349.
https://doi.org/10.1039/c1nr10273g

[28] A. Kongkanand, N.P. Subramanian, Y. Yu, Z. Liu, H. Igarashi, D.A. Muller, Achieving high-power PEM fuel cell performance with an ultralow-pt-content core–shell catalyst, ACS Catal. 6 (2016) 1578–1583.
https://doi.org/10.1021/acscatal.5b02819

Nanomaterials for Alcohol Fuel Cells   Materials Research Forum LLC
Materials Research Foundations **49** (2019) 351-387   doi: https://doi.org/10.21741/9781644900192-12

[29] M. Shao, A. Peles, K. Shoemaker, Electrocatalysis on platinum nanoparticles: particle size effect on oxygen reduction reaction activity, Nano Letters. 11 (2011) 3714–3719. https://doi.org/10.1021/nl2017459

[30] M.C. Figueiredo, J. Solla-Gullón, F.J. Vidal-Iglesias, M. Nisula, J.M. Feliu, T. Kallio, Carbon-supported shape-controlled Pt nanoparticle electrocatalysts for direct alcohol fuel cells, Electrochem.Comm. 55 (2015) 47–50. https://doi.org/10.1016/j.elecom.2015.03.019

[31] M. Liu, R. Zhang, W. Chen, Graphene-supported nanoelectrocatalysts for fuel cells: Synthesis, properties, and applications, Chem.Rev. 114 (2014) 5117–5160. https://doi.org/10.1021/cr400523y

[32] M.K. Debe, Electrocatalyst approaches and challenges for automotive fuel cells, Nature. 486 (2012) 43–51. https://doi.org/10.1038/nature11115

[33] C. Koenigsmann, S.S. Wong, One-dimensional noble metal electrocatalysts: a promising structural paradigm for direct methanol fuel cells, Energ.Environ. Sci. 4 (2011) 1161–1176. https://doi.org/10.1039/C0EE00197J

[34] W. Zhou, B. Zhou, W. Li, Z. Zhou, S. Song, G. Sun, Q. Xin, S. Douvartzides, M. Goula, P. Tsiakaras, Performance comparison of low-temperature direct alcohol fuel cells with different anode catalysts, J. Power Sources. 126 (2004) 16–22. https://doi.org/10.1016/j.jpowsour.2003.08.009

[35] J. Tayal, B. Rawat, S. Basu, Effect of addition of rhenium to Pt-based anode catalysts in electro-oxidation of ethanol in direct ethanol PEM fuel cell, Int. J. Hydrogen Energ. 37 (2012) 4597–4605. https://doi.org/10.1016/J.IJHYDENE.2011.05.188

[36] X. Zhao, M. Yin, L. Ma, L. Liang, C. Liu, J. Liao, T. Lu, W. Xing, Recent advances in catalysts for direct methanol fuel cells, Energ.Environ. Sci. 4 (2011) 2736–2753. https://doi.org/10.1039/c1ee01307f

[37] W. Zhou, Pt based anode catalysts for direct ethanol fuel cells, Appl. Catal.B.Environ. 46 (2003) 273–285. https://doi.org/10.1016/S0926-3373(03)00218-2

[38] W. Zhu, J. Ke, S.B. Wang, J. Ren, H.H. Wang, Z.Y. Zhou, R. Si, Y.W. Zhang, C.-H. Yan, Shaping single-crystalline trimetallic Pt−Pd−Rh nanocrystals toward high-effciency C−C splitting of ethanol in conversion to CO_2, ACS Catalysis. 5 (2015) 1995–2008. https://doi.org/10.1021/cs5018419

[39] C.J. Zhong, J. Luo, P.N. Njoki, D. Mott, B. Wanjala, R. Loukrakpam, S. Lim, B. Fang, Z. Xu, Fuel cell technology: Nano-engineered multimetallic catalysts, Energ.Environ. Sci.. 1 (2008) 454–466. https://doi.org/10.1039/b810734n

[40] Y.H. Chu, Y.G. Shul, Combinatorial investigation of Pt–Ru–Sn alloys as an anode electrocatalysts for direct alcohol fuel cells, Int. J. Hydrogen Energ. 35 (2010) 11261–11270. https://doi.org/10.1016/J.IJHYDENE.2010.07.062

[41] G. García, N. Tsiouvaras, E. Pastor, M.A. Peña, J.L.G. Fierro, M. V. Martínez-Huerta, Ethanol oxidation on PtRuMo/C catalysts: In situ FTIR spectroscopy and DEMS studies, Int. J. Hydrogen Energ. 37 (2012) 7131–7140. https://doi.org/10.1016/j.ijhydene.2011.11.031

[42] Z.B. Wang, G.P. Yin, Y.G. Lin, Synthesis and characterization of PtRuMo/C nanoparticle electrocatalyst for direct ethanol fuel cell, J. Power Sources. 170 (2007) 242–250. https://doi.org/10.1016/j.jpowsour.2007.03.078

[43] M. Li, A. Kowal, K. Sasaki, N. Marinkovic, D. Su, E. Korach, P. Liu, R.R. Adzic, Ethanol oxidation on the ternary Pt–Rh–SnO2/C electrocatalysts with varied Pt:Rh:Sn ratios, Electrochim.Acta. 55 (2010) 4331–4338. https://doi.org/10.1016/J.ELECTACTA.2009.12.071

[44] X.H. Jian, D.S. Tsai, W.H. Chung, Y.S. Huang, F.J. Liu, Pt–Ru and Pt–Mo electrodeposited onto Ir–IrO2 nanorods and their catalytic activities in methanol and ethanol oxidation, J.Mater.Chem. 19 (2009) 1601–1607. https://doi.org/10.1039/b816255g

[45] E. Lee, A. Murthy, A. Manthiram, Effect of Mo addition on the electrocatalytic activity of Pt–Sn–Mo/C for direct ethanol fuel cells, Electrochim.Acta. 56 (2011) 1611–1618. https://doi.org/10.1016/j.electacta.2010.10.086

[46] M.C. Figueiredo, O. Sorsa, R.M. Arán-Ais, N. Doan, J.M. Feliu, T. Kallio, Trimetallic catalyst based on PtRu modified by irreversible adsorption of Sb for direct ethanol fuel cells, J. Catal. 329 (2015) 69–77. https://doi.org/10.1016/j.jcat.2015.04.032

[47] K. Jiang, L. Bu, P. Wang, S. Guo, X. Huang, Trimetallic PtSnRh wavy nanowires as efficient nanoelectrocatalysts for alcohol electrooxidation, ACS Appl. Mater..Interfaces. 7 (2015) 15061–15067. https://doi.org/10.1021/acsami.5b04391

[48] L.X. Ding, A.L. Wang, G.R. Li, Z.Q. Liu, W.X. Zhao, C.Y. Su, Y.X. Tong, Porous Pt-Ni-P composite nanotube arrays: Highly electroactive and durable catalysts for

methanol electrooxidation, J.Am. Chem.Soc. 134 (2012) 5730–5733.
https://doi.org/10.1021/ja212206m

[49] J. Goel, S. Basu, Pt-Re-Sn as Metal Catalysts for Electro-oxidation of ethanol in direct ethanol fuel cell, Energ.Procedia. 28 (2012) 66–77.
https://doi.org/10.1016/J.EGYPRO.2012.08.041

[50] J.W. Hong, Y. Kim, D.H. Wi, S. Lee, S.U. Lee, Y.W. Lee, S.I. Choi, S.W. Han, Ultrathin free-standing ternary-alloy nanosheets, Angew. Chem. Int. Ed. 55 (2016) 2753–2758. https://doi.org/10.1002/anie.201510460

[51] G. Feng, Y. Kuang, P.S. Li, N.N. Han, M. Sun, G.X. Zhang, X.M. Sun, Single crystalline ultrathin nickel-cobalt alloy nanosheets array for direct hydrazine fuel cells, Adv.Sci. 4 (2017) 1600179. https://doi.org/10.1002/advs.201600179

[52] H. Dong, L. Dong, Electrocatalytic activity of carbon nanotube-supported Pt–Cr–Co tri-metallic nanoparticles for methanol and ethanol oxidations, J. Inorg. Organomet. Polymer. Mater. 21 (2011) 754–757. https://doi.org/10.1007/s10904-011-9526-2

[53] S. Ghosh, R. Basu, Electrochemistry of nanostructured materials: implementation in electrocatalysis for energy conversion applications, J. Indian Inst. Sci. 96 (2016) 293–313

[54] S. Sadeghi, H. Gharibi, F. Golmohammadi, Electrooxidation of ethanol and acetaldehyde using PtSn/C and PtSnO2/C catalysts prepared by a modified alcohol-reduction process, Scientia Iranica F. 22 (2015) 2729–2735.
https://doi.org/10.13140/RG.2.1.1021.5288

[55] S. Basu, Proton Exchange membrane fuel cell technology: India's perspective, Proceedings of the Indian National Science Academy. 81 (2015) 865–890.
https://doi.org/10.16943/ptinsa/2015/v81i4/48301

[56] L. Giorgi, F. Leccese, Fuel cells: Technologies and applications, Open Fuel Cell J. 6 (2013) 1–20. https://doi.org/10.1161/01.RES.88.1.117

[57] W. Zhou, B. Zhou, W. Li, Z. Zhou, S. Song, G. Sun, Q. Xin, S. Douvartzides, M. Goula, P. Tsiakaras, Performance comparison of low-temperature direct alcohol fuel cells with different anode catalysts, J. Power Sources. 126 (2004) 16–22.
https://doi.org/10.1016/j.jpowsour.2003.08.009

[58] N.K. Shrivastava, S.B. Thombre, R.B. Chadge, Liquid feed passive direct methanol fuel cell: challenges and recent advances, Ionics. 22 (2016) 1–23.
https://doi.org/10.1007/s11581-015-1589-6

[59] M.A. Zeb Gul Sial, M.A. Ud Din, X. Wang, Multimetallic nanosheets: synthesis and applications in fuel cells, Chem. Soc.Rev. 47 (2018) 6175–6200. https://doi.org/10.1039/C8CS00113H

[60] M.F. Hossain, J.Y. Park, Reduced graphene oxide sheets with added Pt-Pd alloy nanoparticles as a good electro-catalyst for ethanol oxidation, Int.J.Electrochem.Sci. 10 (2015) 6213–6226

[61] X. Cui, Y. Li, M. Zhao, Y. Xu, L. Chen, S. Yang, Y. Wang, Facile growth of ultra-small Pd nanoparticles on zeolite-templated mesocellular graphene foam for enhanced alcohol electrooxidation, Nano Research.12(2019) 351–356. https://doi.org/10.1007/s12274-018-2222-6

[62] Y. Pan, M. Chen, S. Wu, Y. Li, D. Lu, H. Xu, W. Peng , L. Zhou, Development of a Highly Efficient 3D RuPdBi/NG Electrocatalyst for Ethylene Glycol Oxidation in an Alkaline Media, Int. J.Electrochem. Sci. 12 (2017) 11030 – 11041, https://doi.org/10.20964/2017.11.101

[63] S. Thilaga, S. Durga, V. Selvarani1, S. Kiruthika, B. Muthukumaran, Multiwalled carbon nanotube supported Pt–Sn–M (M = Ru, Ni, and Ir) catalysts for ethanol electrooxidation, Ionics. 24 (2018) 1721–1731.

[64] M. Priya, S. Kiruthika, B. Muthukuma, Synthesis and characterization of Pt–Sn–Ce/MC ternary catalysts for ethanol oxidation in membraneless fuel cells, Ionics.23 (2017) 1209–1218. https://doi.org/10.1007/s11581-016-1940-6

[65] H. Huang, X. Wang, Recent progress on carbon-based support materials for electrocatalysts of direct methanol fuel cells, J.Mater. Chem. A.2 (2014) 6266-6291. https://doi.org/ 10.1039/c3ta14754a

[66] L. Ma, H. He, A. Hsu, R. Chen, PdRu/C catalysts for ethanol oxidation in anion-exchange membrane direct ethanol fuel cells, J. Power Sour. 241 (2013) 696-702. https://doi.org/10.1016/j.jpowsour.2013.04.051

[67] W. li, P. Haldar, SupportlessPdFe nanorods as highly active electrocatalyst for proton exchange membrane fuel cell, Electrochem. Comm. 11 (2009) 1195–1198. https://doi.org/10.1016/j.elecom.2009.03.046

[68] A. O. Neto, R. R. Dias, M. M. Tusi, M. Linardi, E. V. Spinace, Electro-oxidation of methanol and ethanol using PtRu/C, PtSn/C and PtSnRu/C electrocatalysts prepared by an alcohol-reduction process, J. Power Sour. 166 (2007) 87–91. https://doi.org/ 10.1007/s11581-009-0396-3

[69] K. Bhunia, Santimoykhilari, D. Pradhan, Monodispersed PtPdNi Trimetallic Nanoparticles-Integrated Reduced Graphene Oxide Hybrid Platform for direct Alcohol Fuel Cell, ACS Sustain. Chem. Eng. 6 (2018) 7769-7778. https://doi.org/10.1021/acssuschemeng.8b00721.

[70] R. Wang, Y. Ma, H. Wang, J. S. Ji, Gas-liquid interface-mediated room-temperature synthesis of "clean" PdNiP alloy nanoparticle networks with high catalytic activity for ethanol oxidation, Chem. Comm.50 (2014) 12877-12879. https://doi.org/ 10.1039/c4cc06026a

[71] T. Li, Y. Huang, K. Ding, P. Wu, S. C. Abbas, M. A. Ghausi, P. Zhang, Y. Wang, Newly designed PdRuBi/N-Graphene catalyst with synergistic effects for enhanced ethylene glycol electro-oxidation, Electrochim.Acta.191 (2016) 940-945. https://doi.org/ 10.3390/catal7070208

Keyword Index

About the Editors

Dr. Inamuddin is currently working as Assistant Professor in the Chemistry Department, Faculty of Science, King Abdulaziz University, Jeddah, Saudi Arabia. He is a permanent faculty member (Assistant Professor) at the Department of Applied Chemistry, Aligarh Muslim University, Aligarh, India. He obtained Master of Science degree in Organic Chemistry from Chaudhary Charan Singh (CCS) University, Meerut, India, in 2002. He received his Master of Philosophy and Doctor of Philosophy degrees in Applied Chemistry from Aligarh Muslim University (AMU), India, in 2004 and 2007, respectively. He has extensive research experience in multidisciplinary fields of Analytical Chemistry, Materials Chemistry, and Electrochemistry and, more specifically, Renewable Energy and Environment. He has worked on different research projects as project fellow and senior research fellow funded by University Grants Commission (UGC), Government of India, and Council of Scientific and Industrial Research (CSIR), Government of India. He has received Fast Track Young Scientist Award from the Department of Science and Technology, India, to work in the area of bending actuators and artificial muscles. He has completed four major research projects sanctioned by University Grant Commission, Department of Science and Technology, Council of Scientific and Industrial Research, and Council of Science and Technology, India. He has published 138 research articles in international journals of repute and eighteen book chapters in knowledge-based book editions published by renowned international publishers. He has published forty-two edited books with Springer, United Kingdom, Elsevier, Nova Science Publishers, Inc. U.S.A., CRC Press Taylor & Francis Asia Pacific, Trans Tech Publications Ltd., Switzerland and Materials Research Forum LLC, U.S.A. He is the member of various editorial boards of the journals and serving as associate editor for journals such as Environmental Chemistry Letter, Applied Water Science, Euro-Mediterranean Journal for Environmental Integration, Springer-Nature, Frontiers Section Editor of Current Analytical Chemistry, published by Bentham Science Publishers, editorial board member for Scientific Reports-Nature and editor for Eurasian Journal of Analytical Chemistry. He has attended as well as chaired sessions in various international and national conferences. He has worked as a Postdoctoral Fellow, leading a research team at the Creative Research Initiative Center for Bio-Artificial Muscle, Hanyang University, South Korea, in the field of renewable energy, especially biofuel cells. He has also worked as a Postdoctoral Fellow at the Center of Research Excellence in Renewable Energy, King Fahd University of Petroleum and Minerals, Saudi Arabia, in the field of polymer electrolyte membrane fuel cells and computational fluid dynamics of polymer electrolyte membrane fuel cells. He is a life member of the Journal of the Indian

Chemical Society. His research interest includes ion exchange materials, a sensor for heavy metal ions, biofuel cells, supercapacitors and bending actuators.

Dr. Tauseef Ahmad Rangreez is working as a postdoctoral fellow at National Institute of Technology, Srinagar, India. He completed his Ph.D in Applied Chemistry, from Aligarh Muslim University, Aligarh, India on the topic "Development of Nanostructure Organic-Inorganic Composite Materials based Sensors for Inorganic Pollutants". He worked as a Project Fellow under the UGC Funded Research Project entitled "Development of Nanostructured Conductive Organic Inorganic Composite Materials based sensors Functionalities for Organic and Inorganic Pollutants". He completed his Masters in Chemistry (Industrial Applications) from Jamia Hamdard, New Delhi. He has published several research articles of international repute. He has edited books with Springer and Materials Research Forum LLC, U.S.A. His research interest includes ion exchange chromatography and biosensor.

Dr. Sen obtained his Ph.D. degree in 2011 from the Middle East Technical University and in the same year joined the Massachusettes Institute of Technology (MIT) in Boston as a postdoctoral research fellow in the group of Prof. Michael S. Strano. He stayed at the MIT for almost two years as post-doc involving different nanomaterials and their applications, such as biosensors, biofuel cells, thermopower energy generation, carbon nanotubes, graphene, etc. At present Dr. Sen is currently an Associate Professor in the Department of Biochemistry (Faculty of Science and Art) at Kütahya Dumlupinar University, Kütahya, Turkey. His main research interests include nanomaterials and their energy, sensor, bioapplications. He has authored and co-authored over 120 publications, 70 book chapters, and 10 patents. Currently serves as the associate editor of International Journal of Environmental Science and Technology, Heliyon Engineering and in editorial boards of Biosensors and Bioelectronics, Heliyon, Current Organocatalysis, Innovations in Corrosion and Material Science and Current Analytical Chemistry etc. Besides, Dr. Sen has been the recipient of multiple prestigious professional award, such as 2017 Science Heroes Association 'The Outstanding Young Scientist Honour Prize', 2015 FABED Eser Tümen Research Incentive Award, 2015 Science Academy Outstanding Young Scientist Award, 2014 METU Mustafa Parlar Foundation Young Investigator Award, Tubitak 3501, Career Grant, 2013 METU Best Thesis Award, 2013-The 11th of Serhat ÖZYAR 'The Outstanding Young Scientist Honour Prize'.

Prof. Abdullah M. Asiri is the Head of the Chemistry Department at King Abdulaziz University since October 2009 and he is the founder and the Director of the Center of Excellence for Advanced Materials Research (CEAMR) since 2010 till date. He is the Professor of Organic Photochemistry. He graduated from King Abdulaziz University (KAU) with B.Sc. in Chemistry in 1990 and a Ph.D. from University of Wales, College of Cardiff, U.K. in 1995. His research interest covers color chemistry, synthesis of novel photochromic and thermochromic systems, synthesis of novel coloring matters and dyeing of textiles, materials chemistry, nanochemistry and nanotechnology, polymers and plastics. Prof. Asiri is the principal supervisors of more than 20 M.Sc. and six Ph.D. theses. He is the main author of ten books of different chemistry disciplines. Prof. Asiri is the Editor-in-Chief of King Abdulaziz University Journal of Science. A major achievement of Prof. Asiri is the discovery of tribochromic compounds, a new class of compounds which change from slightly or colorless to deep colored when subjected to small pressure or when grind. This discovery was introduced to the scientific community as a new terminology published by IUPAC in 2000. This discovery was awarded a patent from European Patent office and from UK patent. Prof. Asiri involved in many committees at the KAU level and on the national level. He took a major role in the advanced materials committee working for KACST to identify the national plan for science and technology in 2007. Prof. Asiri played a major role in advancing the chemistry education and research in KAU. He has been awarded the best researchers from KAU for the past five years. He also awarded the Young Scientist Award from the Saudi Chemical Society in 2009 and also the first prize for the distinction in science from the Saudi Chemical Society in 2012. He also received a recognition certificate from the American Chemical Society (Gulf region Chapter) for the advancement of chemical science in the Kingdome. He received a Scopus certificate for the most publishing scientist in Saudi Arabia in chemistry in 2008. He is also a member of the editorial board of various journals of international repute. He is the Vice- President of Saudi Chemical Society (Western Province Branch). He holds four USA patents, more than one thousand publications in international journals, several book chapters and edited books.

www.ingramcontent.com/pod-product-compliance
Lightning Source LLC
Chambersburg PA
CBHW071317210326
41597CB00015B/1254